Practical Aspects of Industrial Welding Processes

This book covers various materials used by the fabrication industry for the manufacturing of process plant equipment and includes an in-depth review of welding processes widely used in the industry for joining and surfacing. It addresses the practical aspects of welding carbon, Cr–Mo steel, austenitic stainless steel, duplex and super duplex steels, and aluminium. Further, it covers various issues generally encountered during the welding of steels, such as the selection of suitable welding consumables and the good practices adopted by fabrication industries, and so on, along with some case studies.

FEATURES:

- Focuses on the practical aspects of arc welding and on the processes
- Covers recent advances in the field along with addressing the theoretical, practical, and industrial aspects
- Reviews recent trends of implementing DOE techniques and meta-heuristics optimization techniques for setting of accurate parameters
- Contains an in-depth review of the most important advances in the welding process
- Reviews the selection of suitable welding consumables and the good practices adopted by the fabrication industry by including case studies.

This book is aimed at professionals and researchers in materials science and industrial manufacturing.

Practical Aspects of Industrial Welding Processes

Edited by
Krishnan Sivaraman, J. Krishnan
and Vishvesh J. Badheka

CRC Press
Taylor & Francis Group
Boca Raton London New York

CRC Press is an imprint of the
Taylor & Francis Group, an **informa** business

First edition published 2026
by CRC Press
2385 NW Executive Center Drive, Suite 320, Boca Raton FL 33431

and by CRC Press
4 Park Square, Milton Park, Abingdon, Oxon, OX14 4RN

CRC Press is an imprint of Taylor & Francis Group, LLC

ISBN: 978-1-032-40306-9 (hbk)
ISBN: 978-1-032-40307-6 (pbk)
ISBN: 978-1-003-35241-9 (ebk)

DOI: 10.1201/9781003352419

Typeset in Times
by Apex CoVantage, LLC

Contents

Preface

Welding is a dynamic and essential discipline within the engineering landscape, continually adapting to meet the evolving demands of modern industries. From aerospace to power generation, the role of welding is pivotal in ensuring structural integrity and performance. As materials and technologies advance, so too must our understanding and application of welding processes.

This book is conceived as a comprehensive guide to the practical aspects of welding, aiming to bridge the gap between theoretical knowledge and real-world application. It is tailored for a diverse audience—including welding engineers, designers, inspectors, managers, and researchers—who seek to deepen their expertise and adapt to the challenges of contemporary welding practices.

Structured into nine detailed chapters, the book delves into:

- the intricacies of welding various steel types, emphasizing factors affecting weldability and presenting case studies on fabrication failures;
- the impact of aluminide coatings on TIG welding, supported by experimental analyses and characterization techniques;
- the properties and welding considerations of duplex and super duplex stainless steels, highlighting their industrial applications;
- hard facing techniques to enhance wear resistance, evaluating various arc welding processes;
- the complexities of welding Inconel alloys, focusing on industrial practices and challenges;
- joining techniques for aluminum alloys in aerospace applications, analysing the relationship between process parameters and mechanical properties;
- methodologies for applying corrosion-resistant overlays to substrates, enhancing durability and protection;
- welding considerations for nickel and nickel-based alloys, discussing alloy types, welding characteristics, and quality assurance.

Each chapter is enriched with practical insights, real-world examples, and case studies to provide a holistic understanding of the subject matter.

In compiling this work, the intention is to equip professionals and students alike with the knowledge and tools necessary to navigate the complexities of modern welding applications. It is hoped that this book will serve as both a reference and a catalyst for innovation in the field.

Editor Biographies

Krishnan Sivaraman is an experienced academician and industry expert based in Mumbai. He is a professor in Mechanical Engineering department and associate dean of research and development at Fr. C. Rodrigues Institute of Technology (FCRIT), Navi Mumbai. He earned his PhD in mechanical engineering from the Indian Institute of Technology (IIT) Bombay and holds the esteemed certification of International Welding Engineer (IWE).

With over 20 years of rich industry experience at Larsen & Toubro Heavy Engineering, Dr. Krishnan possesses profound expertise in welding technology, metallurgy, and quality assurance. His scholarly research spans advanced welding processes, materials engineering, and manufacturing optimization, with numerous publications and presentations at national and international journals and conferences. He is widely recognized for his mentorship and leadership in cultivating future engineering innovators.

A Fellow of the Institution of Engineers (India), Dr. Sivaraman maintains strong industry–academia linkages by providing high-impact technical consultancy and spearheading collaborative projects. He is an active and influential member of various national and international welding societies and forums, driving advancements in welding science and fostering professional excellence within the engineering community.

J. Krishnan is a distinguished metallurgist and former research scientist/technologist at the Bhabha Atomic Research Centre (BARC), where he served for nearly four decades. His career has been dedicated to the development of advanced materials and joining technologies for nuclear and allied applications in the Department of Atomic Energy (DAE), Government of India. Among his many contributions, Dr. J. Krishnan is credited with the development of numerous dissimilar metal joints, essential for high-performance engineering systems. He holds a patent for an aluminium–stainless steel diffusion-bonded joint, and he developed a floor-type electric heat treatment furnace that earned him the prestigious Prime Minister's Award for Excellence in Public Administration.

In addition to his research work, Dr. J. Krishnan has played a key role in training and mentoring future engineers, delivering lectures as part of BARC's one-year specialized training programme in nuclear engineering. Following his retirement from BARC, Dr. Krishnan continued his contributions to industry and academia. He served as a senior general manager at Bharat Fritz Werner (BFTL), Indore, and later as the L&T Chair Professor at the Maharaja Sayajirao University, Baroda, for two terms.

Currently, Dr. J. Krishnan is active as a consultant, academic examiner, and research guide. He holds a PhD in welding engineering from the Indian Institute of Technology Bombay (IIT Bombay).

Vishvesh J. Badheka studied metallurgical engineering at the Maharaja Sayajirao University of Baroda, India, and received his bachelor's, master's, and PhD degrees in metallurgical engineering. He obtained the Diploma of International Welding Engineer in December 2011 from the International Institute of Welding.

Dr. Badheka is Professor of Mechanical Engineering in the School of Technology and Dean of Academic Affairs at Pandit Deendayal Energy University (PDEU), Gandhinagar, India.

He has received funding from various funding agencies, including DST-Young Scientist (twice), ISRO, DRDO, and DAE, for his research in the areas of advanced welding processes, narrow gap welding using metal-cored and flux-cored wires, friction stir welding of similar and dissimilar metals, friction stir processing for surface composites and superplasticity, friction welding of pipes for dissimilar metal combinations, and ATIG of reduced activation martensitic steels and 9Cr–1Mo steels. So far, he has completed ten sponsored R&D projects and guided 45 MTech and 14 PhD research projects in the area of advanced welding processes.

He has published in 150+ international journals and 60 international and national conferences. Moreover, he holds two Indian process patents on narrow gap torch and oscillation devices. He is the co-editor of *Advances in Welding Technologies for Process Development* and *Additive Manufacturing for Advance Applications: Technologies, Challenges and Case Studies*, both published by CRC Press.

He is a member of various professional bodies, including IIW, ISNT, ISTE, IIM, IIF, MRSI, and AWS. He is the chairman of the IIW-PDEU students' chapter and a young professional volunteer of the International Institute of Welding on behalf of IIW-INDIA.

Dr. Badheka has received several academic awards. He received the Dr. Vikram Sarabhai Award for the year 2009–2010 for "Friction Stir Welding of Aluminum and Its Alloys" from the Gujarat Council on Science and Technology (GUJCOST), Government of Gujarat. He was also awarded the Bronze medal by the Center for Innovation, Incubation and Entrepreneurship (CIIE) of the Indian Institute of Management Ahmedabad (IIMA). He has received several Best Oral Presentation and Best Research Papers awards from the Indian Institute of Welding.

He received the National Technical Teacher's Award from AICTE in 2022 in recognition of teaching excellence in technical education. He received Teachers Associateship for Research Excellence (TARE)—DST, 2022.

Contributors

Vishvesh J. Badheka
IWE, Professor, Department of Mechanical Engineering, School of Technology
Pandit Deendayal Energy University (PDEU)
Gandhinagar, Gujarat, India

S. Bhattacharya
Principal Consultant, Dr. Bhattacharya & Associates
Bengaluru, Karnataka, India

Chakravarthy P.
Aerospace Building
Indian Institute of Space Science and Technology
Thiruvananthapuram, Kerala, India

Raghavendra Darji
Lead Engineer – Welding
PD&T, Materials Engineering Department
Matter Motor Works Pvt. Ltd
Ahmedabad, Gujarat, India

Pradip Goswami
Consulting Welding, Materials and Corrosion Engineering Specialist
Canada

Nirav I. Jamnapara
SO-G and Head - Outreach Division, IPR
Head—Projects & Technology Transfer Section, IPR
Associate Professor, Homi Bhabha National Institute (HBNI)
Head, AIC—IPR Plasmatech Innovation Foundation
Gandhinagar, Gujarat, India

Gaurang Joshi
Department of Mechanical Engineering
Marwadi University
Rajkot, Gujarat, India

Agilan M.
Welding Development Section (WDS)
Vikram Sarabhai Space Centre
Thiruvananthapuram, Kerala, India

Aishwary Mishra
Aerospace Building
Indian Institute of Space Science and Technology
Thiruvananthapuram, Kerala, India

Purvesh K. Nanavati
Assistant Professor, (Metallurgy)
Government Engineering College
Gandhinagar, Gujarat, India

R. Vaideeswaran
Senior Manager (Rtd)
Bharat Heavy Electricals Limited
Tiruchirappalli, Tamil Nadu, India

Divij Vaishnav
Senior Manager
Matter Motor Works Private Limited
Ahmedabad, Gujarat, India

Arunsinh B. Zala
Senior Research Fellow
National Institute of Education (NIE)
Nanyang Technological University (NTU)
Singapore

1 Introduction to Practical Aspects of Industrial Welding Processes

*Krishnan Sivaraman, J. Krishnan,
and Vishvesh J. Badheka*

CONTENTS

1.1 INTRODUCTION

Welding is an ever-growing discipline that presents challenges and work opportunities for new generations of engineers. There is a rapid development in the welding industry demanding the development of new and innovative methods for the welding of many critical materials and thereby becoming more and more competitive to handle the fierce market competition effectively.

The primary purpose of this book is to bring out the practical aspects of the welding processes and welding techniques which have evolved over periods in the industry. This practical reference guide would help immensely the young welding engineers and the academic and research community to understand the various practical aspects of welding. This book is designed for the individual who needs to expand his/her core competence in welding. Designers, inspectors, managers, or welders with a need to understand the fundamentals of welding will benefit from the practical aspects of welding technology covered here.

The book content has been broadly classified into nine different chapters. While the first chapter gives a brief introduction regarding the book content, the second chapter addresses the practical aspects in welding of steels. This chapter, moreover, presents an in-depth review of welding processes widely used in the industry for joining while fabricating many types of critical equipment. For welding of various kinds of steels, the chapter describes in detail the factors affecting weldability, namely composition, heat affected zone, stress, and welding procedure have been described in detail [1]. Also, the chapter depicts a few case studies to highlight fabrication and site failures.

Aluminide coatings are extensively used to improve the lifespan of equipment in a wide range of industries [2, 3]. In Chapter 3, autogenous TIG-welded samples are produced to evaluate the impact of coatings using bead-on-plate experiments.

DOI: 10.1201/9781003352419-1

Different characterization techniques such as X-ray diffraction for phase identification, scanning electron microscopy for microstructure analysis, and microhardness tests are analyzed to examine the weld metallurgy.

Duplex and super duplex stainless steels are extremely high-grade engineering alloys combining good corrosion resistance with high strength and ease of fabrication [4]. Chapter 4 narrates various technical aspects of duplex and super duplex stainless steels, namely materials, welding, corrosion, and end user experiences.

Hard facing is the deposition of a special alloy material on a metallic part by various arc welding processes predominantly to obtain wear-resistant properties [5, 6]. The property usually sought is greater resistance to wear from abrasion, impact, adhesion (metal-to-metal), heat, corrosion, or any combination of these factors. Chapter 5 describes the various advantages and disadvantages of each of these methods in detail.

Inconel retains strength over a wide temperature range and is attractive for high-temperature applications, whereas Inconel alloy is used in applications where resistance to corrosion and oxidation is required, such as chemical processing, aerospace, and power generation [7–9]. Chapter 6 investigates Inconel welding techniques and related challenges with the most industrially practised arc welding processes, such as gas tungsten arc welding (GTAW), gas metal arc welding (GMAW), shielded metal arc welding (SMAW), and submerged arc welding (SAW) processes for joining of Inconel/Inconel alloys.

Aluminium has many advantages that make it an attractive material in various industrial applications [10]. Chapter 7 discusses joining techniques employed in the various aluminium alloys in the industrial (aerospace) sectors, focusing specifically on the correlation between process parameters, microstructure evolution, and achieved mechanical properties.

Corrosion resistant overlay involves depositing a layer of corrosion-resistant material onto the surface of a substrate to enhance its durability and protect against corrosive agents [11, 12]. Chapter 8 delves into the realm of corrosion resistance overlay, exploring its principles, methodologies, applications, and advantages.

Nickel and nickel-based alloys are an integral part of the modern industry because of their ability to withstand a wide variety of severe operating conditions [13]. Chapter 9 covers all aspects which influence the welding of nickel alloys to achieve an excellent quality weld. This chapter also covers the types of alloys, the alloying elements, their impact on weldability, and welding characteristics.

REFERENCES

1. G.E. Dieter, *Mechanical metallurgy*, McGraw Hill Book Company, 2017.
2. A. Rafiee, K.R. Khalilpour. Renewable hybridization of oil and gas supply chains, in: *Polygeneration with polystorage for chemical and energy hubs*, Academic Press, 2019, pp. 331–372.
3. A.B. Zala, N.I. Jamnapara, C.S. Sasmal, P. Chaudhuri, M. Ranjan. Investigation of alumina film formed over aluminized RAFM steel by plasma assisted heat treatment, *Fusion Engineering and Design*, 2019, 146, pp. 2002–2006.
4. J. Charles, S. Bernhardsson, *Duplex stainless steels conference (DSS)*, J. Charles, ed., Éd. de Physique, Beaune, Bourgogne, France, 1991, pp. 151–168.

5. American Welding Society, *Welding handbook*, vol. 4, 9th edition, chapter 7, Surfacing Material, American Welding Society.
6. S.S. Babu, S.A. David, J.W. Park, J.M. Vitek, *American society of metals handbook, vol. 6: Welding-brazing-soldering-hardfacing, weld cladding, and dissimilar metal joining*, Science and Technology of Welding and Joining, American Society of Metals, 2004, pp. 9–14.
7. J.N. DuPont, J.C. Lippold, S.D. Kiser, *Welding metallurgy and weldability of nickel-base alloys*, John Wiley & Sons, Hoboken, NJ, 2009.
8. R.C. Reed, *The superalloys: Fundamentals and applications*, Cambridge University Press, Cambridge, 2006.
9. D.G. Eskin, *Physical metallurgy of direct chill casting*, CRC Press, Taylor & Francis Group, 2008.
10. M. Cabibbo, C. Paoletti, High-temperature equal-channel angular pressing of a T6-Al-Cu-Li-Mg-Ag-Zr-sc Alloy, *Journal of Manufacturing and Materials Processing, MDPI*, 2021, 5, p. 6. https://doi.org/10.3390/jmmp5010006
11. C.P. Alvarães, et al., Performance of the INCONEL 625 alloy weld overlay obtained by FCAW process, *Materia (Rio J.)*, 2019, 24 (1), https://doi.org/10.1590/S1517-707620190001.0627
12. M. Amani, et al., A comprehensive review of corrosion and its inhibition in the oil and gas, in: *SPE Kuwait oil and gas show and conference*, Kuwait, 2015, pp. 1–14, https://doi.org/10.2118/175337-MS
13. G.M. Mudd, S.M. Jowitt, A detailed assessment of global nickel resource trends and endowments, *Economic Geology*, 2014, pp. 1813–1841.

2 Practical Aspects of Welding of Carbon Steels

R. Vaideeswaran

CONTENTS

 DOI: 10.1201/9781003352419-2

2.1 INTRODUCTION

2.1.1 Weldability

Weldability is the ease with which a material may be joined by as many as possible of the common welding processes to produce a joint having properties suitable to the conditions of service to be imposed upon it.

2.1.2 Type of Steels

Steels are classified into plain carbon mild steels, carbon–manganese steels, low-alloy steels, high-alloy steels, high-strength steels, and corrosion-resistant steels. Each group is defined by a specification which is usually confined to its chemical composition and mechanical properties.

2.1.3 Factors Influencing Weldability

Broadly, the factors that influence the weldability are classified into the following four groups:

a) Heat-affected zone
b) Composition
c) Residual stress
d) Welding procedure

2.1.3.1 Heat-Affected Zone

The most commonly used welding process is fusion welding, wherein local application of heat energy is required to bring the materials to fuse. This energy is not dissipated into the surrounding atmosphere but into the parent plate material on either side of the weld. During welding, temperature gradient is set up in the material ranging from the fusion temperature in the weld pool to the initial plate temperature some distance away from the weld. The material adjacent to the weld, even though it does not melt, undergoes a severe thermal cycle as a result of welding. Because of this, microstructure changes may occur in those regions or heat-affected zone (HAZ).

The extent of these changes in microstructure will depend on the composition of the material and the rate of heating and cooling during the thermal cycle. The rate of cooling in the HAZ is largely dependent on the section thickness of the material and the heat input. If a material is of a composition which is hardenable, then a hardened HAZ will be produced by the cooling rates normally associated with welding. This hardening of HAZ microstructure is one of the potential problemsthat arise when steel is welded. The original properties of the material may be drastically altered by the application of welding.

2.1.3.2 Composition

The chemical composition of a material is usually specified within certain limits. The composition of the weld pool is another variable which is difficult to predict, as it is made partly from filler metal and partly from the parent metal. The proportion of the parent metal in the molten pool, or dilution, depends on the types of joints and the characteristics of the welding processes, but the problems which may arise and the mechanical properties of the weld metal will depend, at least in part, on the chemical composition of the parent metal.

It is clear that difficulties arising during welding will become more severe as the thickness of the materials is increased. This is true not only of the metallurgical factors but also of other contributory factors such as welding procedure and residual stress.

2.1.3.3 Residual Stress

A residual stress system is a system of internal stress in the body which can exist in the absence of any external loading and which is balanced within the body itself. Residual stresses arise as a result of permanent plastic strain in the material and the magnitude of the stress at any point and depends on the amount of strain at that point. In welded structures, these strains are the result of the local heating and cooling cycles associated with welding and especially the shrinkage of the weld metal. This shrinkage has to be largely accommodated by the plastic straining of the weld metal. As a result of the plastic strain, it follows that the level of residual stress in the weld metal must be at the yield point of the material unless it is modified by any subsequent heat treatment such as heating or stretching. The plastic deformations which give rise to residual stress may also produce metallurgical effects that alter the physical properties of the material.

2.1.3.4 Welding Procedure

The most practical factor having considerable influence on the weldability of steel is the welding procedure and the welding process employed in that procedure. The various fusion welding processes normally used for welding are shielded metal arc welding (SMAW) or manual metal arc welding (MMAW), gas tungsten arc welding (TIG), gas metal arc welding (GMAW), submerged arc welding (SAW), and electroslag welding (ESW). Obviously, each of these processes covers a wide range of heat input. For a given plate thickness, the greater the energy input per unit length of weld, the slower the subsequent cooling rate will be in the weld metal and HAZ.

Since weldability can be very much a function of cooling rate, the choice of welding process is important. In processes where a sufficient cooling rate is not possible, it may be required to resort to preheating, maintaining the interpass temperature or post heating the joint being made. Thus, a degree of control over the weldability of a material ultimately depends on continuous attention to the welding procedure.

2.2 WELDING METALLURGY OF CARBON STEELS

Carbon steel is a term applied to a broad range of steels containing upto

Carbon 1.70% max
Manganese 1.65% max
Silicon 0.60% max

Alloying elements such as Cr, Ni, and Mo are not present in these steels in anysignificant amount. Based on the carbon content, carbon steels can be classified intofour groups:

- Low-carbon steels (upto 0.15% C)
- Mild-carbon steels (0.15–0.29% C)
- Medium-carbon steels (0.30–0.59% C)
- High-carbon steels (0.60–1.70% C)

Low-carbon steels can be supplied as rimmed, semi-killed, or killed steels (killed with Si or Al additions) with Mn normally present from 0.30% to 0.60%. These steels are produced in both cold- and hot-rolled conditions and are often joined by

TABLE 2.1
Preheat Details

Low-carbon steel	20 mm thick	100 mm thick
Preheat	Nil	80°C
Low-carbon steel	12 mm	Low-alloy steel 12 mm
Preheat	Nil	120°C

resistance welding. These grades have low yield strengths and can be easily formed. These materials can be easily joined by arc welding processes.

Mild steels, whose carbon content ranges from 0.15% to 0.29% with Mn upto 1.60%, are extensively used for structural applications. These steels have a yield strength of 36–65 ksi. In the heat-treated condition, these steels have excellent notch toughness, and their yield strength can be as high as 100 ksi. Little difficulty is encountered in welding these steels if proper welding procedures are adopted. Table 2.1 gives the preheat details for various thicknesses of low-carbon and low-alloy steels.

The medium-carbon steels (0.30% to 0.59% C) have yield strength of 50–150 ksi, depending on their heat treatment. Welding of these steels may require special procedures like pre heat, post heat, and stress relieving.

The fourth type of steels, namely high-carbon steels with 0.60–1.70% C, is generally classified as tool steels. These steels have high hardness and when heat treated, their hardness can be raised even higher. They are used as dies, cutting tools, mill rolls etc. These are difficult to weld because of their high carbon content. Elaborate welding procedures are required for welding these steels.

2.2.1 Characteristic Features of Welds

2.2.1.1 Single Pass Weldments

In the case of a single-pass bead, the weldment is generally divided into two main regions, namely (i) the fusion zone, or weld metal, and (ii) the heat-affected zone (HAZ) as shown in Figure 2.1. Within the fusion zone, the peak temperature exceeds the melting point of the base steel, and the chemical composition of the weld metal depends on the choice of welding consumables, the base steel dilution ratio, and the operating conditions. Under conditions of rapid cooling and solidification of the weld metal, alloying and impurity elements segregate extensively to the centre of the interdendritic or intercellular regions and to the centre parts of the weld, resulting in significant local chemical in-homogeneities. Therefore, the transformation behaviour of the weld metal can be quite different from that of the base steel, even when the bulk chemical composition is not significantly changed by the welding process [1]. The typical anisotropic nature of the solidified weld structure is also shown in Figure 2.1.

The chemical composition remains largely unchanged in the HAZ region since the highest temperature remains below the melting point of the base steel. However, considerable microstructural change takes place within the HAZ region during

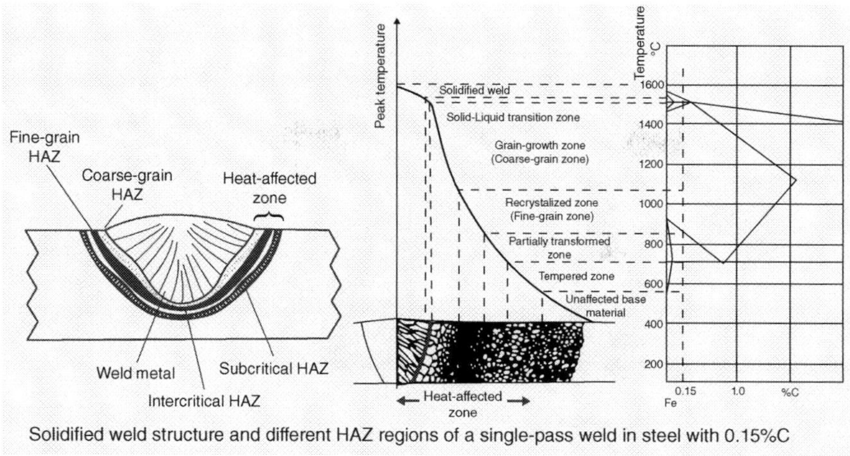

FIGURE 2.1 Weld structure and HAZ regions in a single-pass low-carbon steel butt weld.

welding as a result of the extremely severe thermal cycles. The material immediately adjacent to the fusion zone is heated high into the austenitic temperature range. The micro-alloy precipitates which have developed in the previous stages of processing get generally dissolved, and unpinning of austenite grain boundaries occurs with substantial growth of the grains, forming the coarse grain in the HAZ region.

The average size of the austenite grains, which is a function of the highest temperature attained, decreases with increasing distance from the fusion zone. The cooling rate also varies from point to point in the HAZ region. It increases with increasing highest temperature at constant heat input and decreases with increasing heat input at constant highest temperature. Because of varying thermal conditions as a function of distance from the fusion line, the HAZ region is essentially composed of coarse grain zone (CGHAZ), fine grain zone (FGHAZ), inter-critical zone (ICHAZ), and subcritical zone (SCHAZ). The various HAZ regions of a single-pass low-carbon steel butt weld are shown in Figure 2.1 along with the iron–carbon diagram of temperature versuscarbon percentage.

2.2.1.2 Multi-Pass Weldments

The situation in the multi-pass weldments is much more complex because of the presence of reheated zones within the fusion zone (Figure 2.2). The partial refinement of the microstructure by subsequent weld passes increases the inhomogeneity of the various regions with respect to the microstructure and the mechanical properties. Re-austenitization and subcritical heating can have a profound effect on the subsequent structures and properties of the HAZ region. Toughness property deterioration is related to small regions of limited ductility and low cleavage resistance within the CGHAZ which are identified as the localized brittle zone (LBZ).

LBZs consist of unaltered CGHAZ, inter-critically reheated coarse grain (IRCG) HAZ, and sub-critically reheated coarse grain (SRCG) HAZ. LBZs can be aligned at an adjacent fusion line. Figure 2.2 shows overlapping of HAZ to form localized

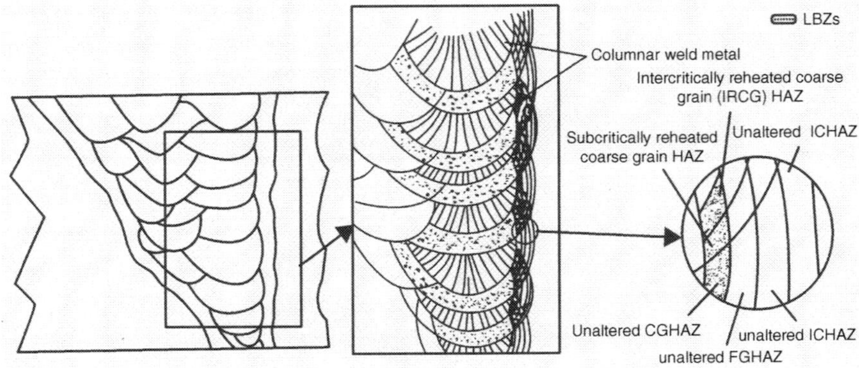

FIGURE 2.2 Overlapping of HAZ to form localized brittle zones aligned along the fusion line.

brittle zones aligned along the fusion line. The aligned LBZs offer short and easy paths for crack propagation. Fracture occurs along the fusion line.

2.2.2 Metallurgical Factors Affecting Weldability

The metallurgical factors which affect the weldability are described here.

2.2.2.1 Hardenability

Hardenability in steels is generally used to indicate austenite stability with alloy additions. The transformation of austenite to martensite by a diffusionless shear type transformation in quenching of steel is one of the most common strengthening processes used in engineeering materails [2]. It is also being used as an indicator of weldability and as a guide for selecting a material and welding process to avoid excessive hardness and cracking in the HAZ region. Steels with high hardness often contain a high-volume fraction of martensite, which is exceedingly susceptible to cracking during the processing. Hardenability is also used to indicate the susceptibility of steel to hydrogen (H_2)-induced cracking. Traditionally, empirical equations have been developed experimentally to express weldability of steels. Carbon equivalent (CE) is one such terminology. It has been developed to estimate the cracking susceptibility of a steel during welding and to determine whether the steel needs pre-weld or/and post-weld heat treatments to avoid cracking.

CE equations include the hardenability effect of the alloying elements by expressing the chemical composition of the steel as a sum of weighted alloy contents. Many CE terminologies with different coefficients for the alloying elements are being used. The equation for the CE as given by the International Institute of Welding (IIW) is given here.

$$\mathbf{CE = C + Mn/6 + (Ni + Cu)/15 + (Cr + Mo + V)/5}$$

In this equation the concentration of the alloying elements is given in weight percent. It can be seen in the equation that C is the element which has maximum effects on the weldability. Along with other chemical elements, C can affect the solidification temperature range, hot tear susceptibility, hardenability, and cold cracking behaviour of the steel weldment.

The application of CE terminologies is also empirical. As an example, the CE equation of IIW has been used effectively with usual medium-carbon low-alloy steels. Steels with lower CE values generally show good weldability. When the CE of the steel is less than 0.45%, weld cracking is unlikely, and no heat treatment is needed.

When the CE of the steel is between 0.45% and 0.60%, weld cracking is likely to take place, and preheating in the range of around 100 Deg C to 400 Deg C is generally suggested. When the CE of the steel is more than 0.60%, there is a high probability for the weld to crack, and in such a case both preheating and the post-weld heat treatments are needed to get a sound weld.

However, the CE equation of IIW does not accurately correlate with the microstructures and properties of low-carbon micro-alloyed steels over extended alloy ranges. Thus, new terminologies based on solution thermodynamics and kinetic considerations have been developed to obtain better predictions of the alloy behaviour and weldability of low- C low-alloy steels. Complex interactive terms, rather than simple additive forms, are included in these equations. Equations with these non-linear terms are more useful in predicting electric arc welding behaviour.

Several terminologies are also available for other steel groups with a wider range of alloying elements and with different prior heat treatments, H_2 contents, and weld hardnesses. One such equation (given here) includes fabrication conditions such as heat input, cooling rate, and joint design, and restraint conditions have also been proposed.

$$Ph = Pcm + 0.075 \log10 H + Rf/40{,}000$$

In this equation, Ph is the cracking susceptibility parameter, H is the concentration of H_2 (in parts per million, ppm), R_f is the restraint stress (in N/sq mm), and Pcm is as per the following equation.

$$Pcm = C + (Mn + Cu + Cr)/20 + Si/30 + Ni/60 + Mo/15 + V/10 + 5B$$

The thickness of the steel being welded can also be related to CE as a compensated carbon equivalent (CCE), which is CCE = CE + 0.00254t, where t is the thickness of the part in millimeters.

These equations are valid only for specific ranges of chemical composition and welding conditions. However, despite the different forms and terms included in the predictive equations, the main objective remains that of estimating the weldability and cracking susceptibility of the steel.

2.2.2.2 Weld Metal Microstructure

Inherent in the welding process is the formation of a pool of liquid steel directly below a moving heat source. The shape of this liquid pool is determined by the flow of both heat and metal, with melting occurring ahead of the heat source and solidification occurring behind it. Heat input determines the volume of liquid steel and therefore the dilution and weld metal composition, as well as the thermal conditions under which solidification takes place. Also important to solidification is the crystalline growth rate, which is geometrically related to weld travel speed and weld pool shape.

Thus, weld pool shape, weld metal composition, cooling rate, and growth rate are all the factors which are interrelated with heat input, which in turn affects the solidification microstructure and the tolerance of the weldment to hot cracking.

Incipient melting at base steel grain boundaries immediately adjacent to the fusion zone allows these grains to serve as seed crystals for epitaxial (an oriented overgrowth of crystalline material upon the surface of another crystal of different chemical composition but similar structure) grain growth during weld metal solidification. The continuous growth of the epitaxial grains results in large columnar grains whose boundaries provide easy paths for crack propagation. An elongated weld pool yields straight and broad columnar grains, which promote the formation ofcentreline cracking because of impurity segregation, mechanical entrapment of inclusions, and the shrinkage stresses that develop during solidification. Epitaxial columnar growth is particularly harmful in multi-pass welds, where grains can extend continuously from one weld bead to another.

Hot tears originate near the liquid–solid interface when strains from solidification shrinkage and thermal contraction cause rupture of the liquid films of low melting point located at grain boundaries. The susceptibility of steel to hot tearing is related to its inability to accommodate strain through dendrite interlocking as well as the tendency of tears to backfill with the remaining liquid. The time interval during which liquid films can exist in relation to the rate of strain generation can also play a role in hot tear susceptibility. Steels can be hot tear sensitive depending on the amount of phosphorus (P) and sulphur (S) impurities they contain. Carbon and nickel (Ni) are also known to influence hot cracking in steel welding.

When the solidified steel weld metal cools down, solid-state transformation reactions can occur. As in solidification, the two main factors which determine the final microstructure are the chemical composition and thermal cycle of the weld metal. In maximum structural steels, weld metal solidifies as deltaferrite. At the peritectic temperature, austenite forms from the reaction between liquid weld metal and deltaferrite, and subsequent cooling leads to the formation of alpha-ferrite. During the austenite to ferrite transformation, proeutectoid ferrite forms first along the austenite grain boundaries. This is known as grain boundary ferrite.

Subsequent to grain boundary ferrite formation, ferrite side plates develop in the form of long needle-like ferrite laths that protrude from the allotriomorphs. A coarse austenite grain size and low carbon content, in combination with a relatively high degree of super cooling, are found to promote ferrite side plate formation. These laths can be properly characterized by their length-to- width aspect ratios (usual values are more than 10:1).

As the temperature continues to drop, intra-granular acicular ferrite nucleates and grows in the form of short laths separated by high angle boundaries. The inclination between orientations of adjacent acicular ferrite laths is usually larger than 20°.

The random orientation of these laths provides good resistance to crack propagation. Acicular ferrite laths have aspect ratios ranging from 3:1 to 10:1.

During proeutectoid ferrite formation, C is rejected continuously from the ferrite phase, enriching the remaining austenite, which later transforms into a variety of constituents, such as martensite (both lath and twinned), bainite, pearlite, and retained austenite. Because of the acicular nature of the bainite laths, they can alsobe described by their aspect ratio, with values similar to those of Widmanstätten sideplates. Very frequently, however, bainite laths occur in the form of packets associated with grain boundaries.

2.2.2.3 HAZ Microstructure

In terms of microstructure, long bainite laths with alternate layers of connected martensite islands are generally found in the CGHAZ of HSLA steel weldments. Martensite islands (martensite-retained austenite constituents) are formed because of the enrichment of C in austenite in the inter-critical zone. Coarse austenite grain size in the near-fusion region of the HAZ can suppress high-temperature transformation products in favour of martensite and bainite on cooling. Upper bainite has a relatively high transformation temperature and is stable relative to the thermal cycles subsequent to those of the first pass. Fluctuation of the chemical composition of the micro-alloying elements can also contribute to CE change and to the amount of hard martensite present in the CGHAZ.

In the case of FGHAZ, even though the peak temperature attained is above thermal cycle Ac3, it is still well below the temperature for grain coarsening. The smaller prior austenite grain size and subsequent ferrite transformation produce a refined microstructure having grains smaller than those of the base steel. The microstructure is similar to that of normalized steel, with considerable toughness.

In case of ICHAZ, only partial transformation takes place resulting in a mixture of austenite and ferrite at the peak temperature of the thermal cycle. Upon cooling, the austenite in a matrix of soft ferrite decomposes, and the final microstructure depends on the bulk and local composition of the alloying elements. The cooling rate is also an important factor in determining the amount of martensite and bainite in the ferrite matrix.

In the case of SCHAZ, no observable microstructure changes are observed. Some spheroidization of carbides can occur. Upon reheating by subsequent weld passes, precipitates or pre-precipitate clusters can form, reducing the toughness. Irregularly shaped particles can also coalesce and strain the surrounding matrix, further lowering the toughness.

During HAZ thermal cycles between Acl and Ac3, the austenite gets enriched with carbon, which, upon cooling, transforms to martensite islands. In the as-welded condition, this transformation affects the IRCG region more than the other reheated zone.

2.2.2.4 Effect of Chemical Composition

The presence of a certain phase in the final microstructure of a weldment is generally explained by means of a continuous cooling transformation (CCT) diagram, which is formed by two sets of curves namely (i) the percent transformation curves, and (ii) the

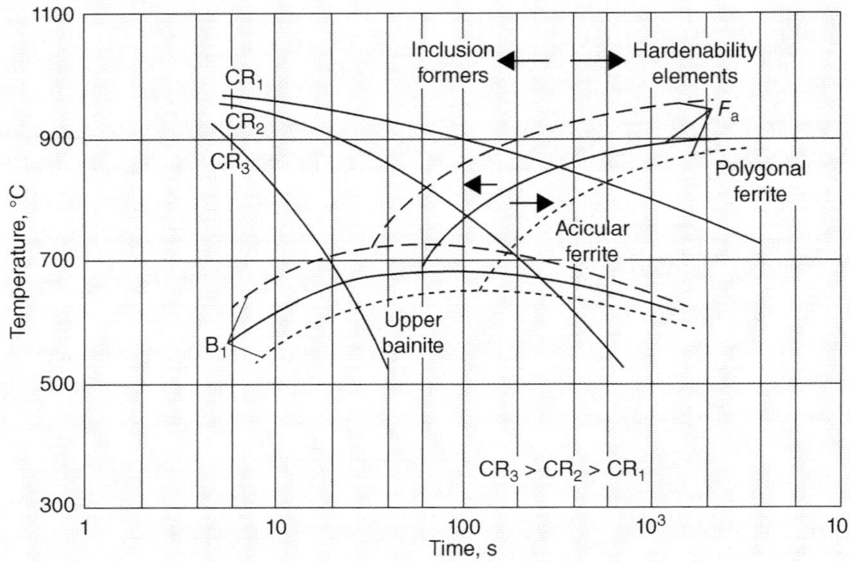

FIGURE 2.3 CCT diagram for an HSLA steel weld metal.

cooling curves. The percent transformation curves define the regions of stability of the different phases. The cooling curves represent the actual thermal conditions that the weld experienced. The intersection of these two sets of curves determines the final microstructure of the different weld zones [3]. A typical CCT diagram for an HSLA steel weld metal is given in Figure 2.3, which describes the temperature versustime plot showing the continuous cooling transformation for an HSLA steel weld metal.

Hardenability elements, such as C, manganese (Mn), chromium (Cr), and molybdenum (Mo), suppress the start of austenite decomposition to lower temperatures. This is equivalent to pushing the transformation curves to the right side of the CCT diagram, resulting in a refined microstructure. Inclusion formers, such as oxygen (O_2) and S, accelerate the austenite to ferrite transformation byproviding more nucleation sites for the reaction to initiate at higher temperatures. Faster cooling has the same effect as an increase in hardenability elements, while a slower cooling rate acts in the same direction as a decrease in hardenability agents or an increase in nucleation site providers. Since the cooling rate varies from point to point in the HAZ, the microstructure also changes accordingly, with martensite and bainite in regions close to the fusion line.

2.2.3 Pre-Weld and Post-Weld Heat Treatments

In the welding of C and low-alloy steels, the final microstructure of the weldment is primarily determined by the cooling rate from the peak temperature. Since the alloy level in C and low-alloy steels is low, the main physical properties of the steel are not affected. Thus, temperature gradient and heat input become the important parameters in weld metal microstructural evolution. A slower cooling rate decreases shrinkage stress, prevents excessive hardening, and allows time for H_2 diffusion.

Cooling rate is mainly important and is a function of the difference in temperature as well as the thermal conductivity of the steel.

During preheating, the initial temperature of the steel plate increases, decreasing the cooling rate and the amount of the hard phases, such as martensite and bainite, in the weld microstructure. For the welding of hardenable steels, it is important to determine the critical cooling rate (CCR) that the base steel can tolerate without cracking. The higher is the CE of the steel, the lower is the critical or allowable cooling rate. The use of a low-H_2 welding electrode also becomes more important. Preheating is to be applied to adjust the cooling rate accordingly.

2.3 POTENTIAL WELDING PROBLEMS

During welding of steels, problems may arise which may be classified into two categories:

1. Problem of weld metal or HAZ cracking directly attributable to the welding process. This category covers:
   i) Weld metal hot cracking ii) HAZ burning or hot tearing iii) Lamellar tearing iv) HAZ hydrogen-induced cracking
2. Problems which occur at welded joints during service. This category covers:
   i) Heat treatment cracking ii) Corrosion iii) Factors affecting joint efficiency

Of these problems, hydrogen cracking, solidification cracking, and lamellar tearing are discussed in detail.

2.3.1 Hydrogen Cracking

Hydrogen dissociation (H_2 $_{--->}$H + H) takes place in the material and the atomic hydrogen dissolves in the weld metal. Hydrogen solubility is high in liquid state, and it gets super saturated upto the solubility limits present in the interstitial solution. Beyondthe solubility limit, it is retained in traps. Then localization of hydrogen takes place at the defect sites. Adsorption of hydrogen reduces the surface energy required for crack initiation. Recombination of atomic hydrogen leads to rupture. Figure 2.4

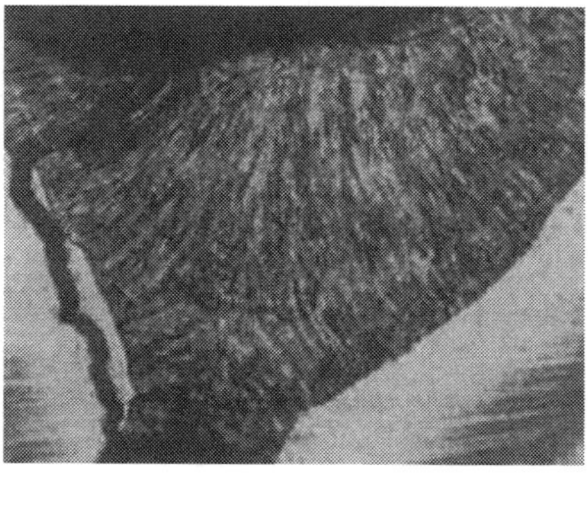

FIGURE 2.4 Cold crack delayed crack.

FIGURE 2.5 Underbead crack.

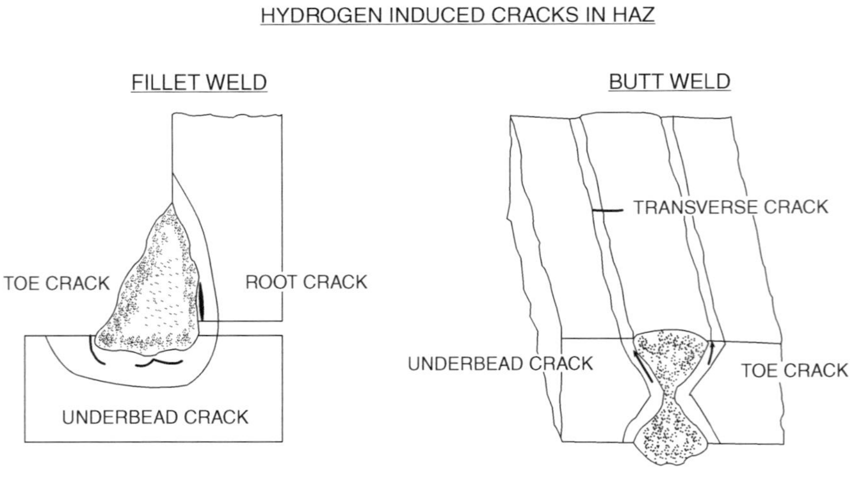

FIGURE 2.6 Crack occurring in HAZ and weld.

shows the cold cracker delayed crack; Figure 2.5 shows the underbead crack; and Figure 2.6 shows crack occurring in HAZ and weld portion.

2.3.1.1 Sources of Hydrogen

Hydrogen may come from

i) the flux coating material in electrodes and SAW welding flux
ii) consumables exposed to atmosphere

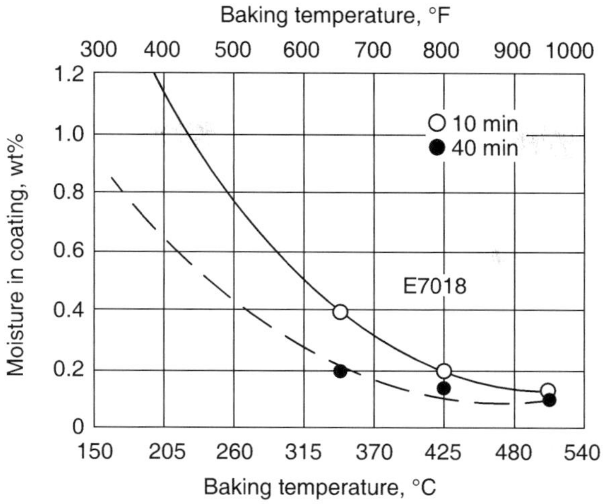

FIGURE 2.7 Moisture versus baking temperature.

iii) moisture in shielding gas
iv) improper baking of weld consumables
v) hydrocarbons in base metal and
vi) rusted consumable/base metal

The diffusible hydrogen content can be controlled by using low-hydrogen welding process, by using basic coated low-hydrogen welding electrodes and by preheating and post heating.

Baking of electrodes prior to welding should be carried out (Rutile electrode 250 C Basic electrode 350 C). Figure 2.7 shows the graph of baking temperature versus moisture in the coating of electrodes. The electrodes should not be kept in open atmosphere and be kept in hand quivers.

2.3.1.2 Preheat

The preheat depends upon

i) Carbon equivalent
ii) Thickness of the base metal
iii) Diffusible hydrogen content
iv) Restraint
v) Welding process, consumables and conditions

The preheating reduces the cooling rate of weld metal, reduces hydrogen concentration, and reduces residual stresses.

Precautions to be Taken for Preheating

i) No welding on the base metal covered with ice.
ii) No welding should be done at or below –18 Deg C.

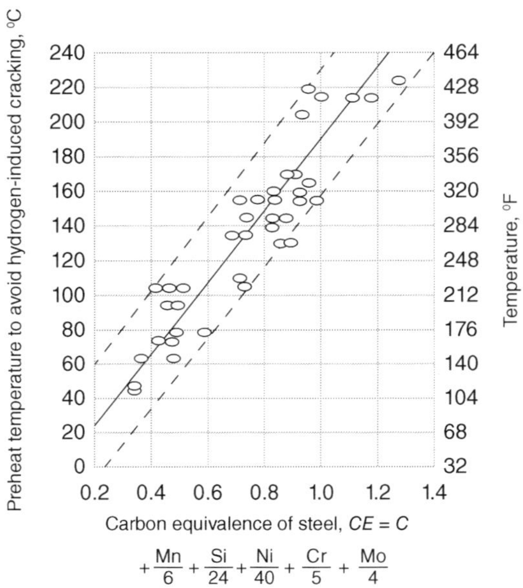

FIGURE 2.8 Preheat versus carbon equivalent.

(Raise the base metal temperature to minimum 16 C when the ambient temp is 0 to –18 Deg C)

iii) Preheat temperature should be measured at a distance of 2*t* or 4″ whichever is higher.

iv) Preheat temperature shall be maintained during welding. The correlation between preheat temperature and carbon equivalent is given in Figure 2.8.

2.3.2 Solidification Cracking

Hot cracking is also known as solidification cracking. It occurs at higher temperatures and is generally located in the weld metal. Hot cracking also can be found in the HAZ, where it is known as liquation cracking. Hot cracking in weld deposits during cooling occurs predominately at the weld centreline or between columnar grains.

The fracture path of a hot crack is intergranular. The causes of hot cracking are well understood. The partition and rejection of alloying elements at the columnar grain boundaries and ahead of the advancing solid/liquid interface produce significant segregation. The elements of segregation form low melting phases or eutectic structures to produce highly wetting films at the grain boundaries. They weaken the structure to the extent that cracks form at the boundaries under the influence of the tensile residual stresses during cooling. Liquation cracking is also associated with grain boundary segregation and is aggravated by the melting of these boundaries near the fusion line. These impurity-weakened boundaries tend to rupture as the weld cools because of the high residual stresses.

Solidification cracking occurs at high temperatures close to liquidus temperature. Figure 2.9 shows the weld depth and weld width in an incorrect weld and correct weld.

i) Fully Austenitic weld metal are more susceptible.
ii) Hot cracks may be macro cracks or micro cracks.

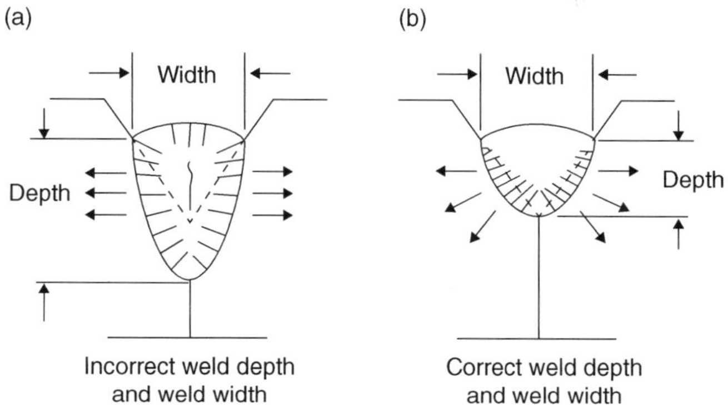

FIGURE 2.9 Weld depth and weld width.

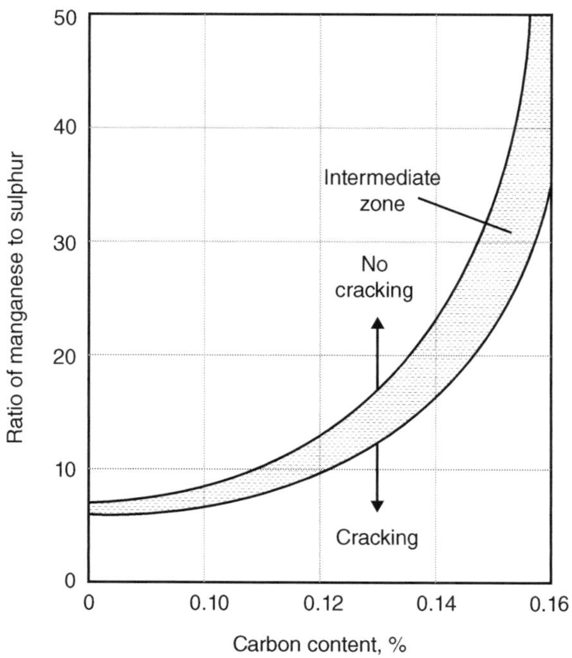

FIGURE 2.10 Ratio of Mn:S vs carbon content.

The causes for solidification cracking or hot cracking are

i) Segregation of impurities to the interdentritic regions
ii) Formation of low-melting eutectic along the grain boundary
iii) Shrinkage stresses

S, P, Si, and O form liquid films along the grain boundaries. Figure 2.10 shows the graph pertaining to the ratio of manganese to sulphur versus carbon content. Below the solidus temperature of the alloy, the metal has not developed strength. The contraction stresses cause tearing along the grain boundaries.

2.3.3 Lamellar Tearing

Lamellar cracking is also known as lamellar tearing. It is characterized by a step-like crack parallel to the rolling plane. Figure 2.11 shows the appearance of fracture face of a lamellar tear and Figure 2.12 shows the lamellar tearing in a T butt weld. The problem occurs particularly when making tee and corner joints in thick steel plates such that the fusion boundary of the weld runs parallel to the plate surface. High tensile stresses can develop perpendicular to the mid plane of the steel plate, as well as parallel to it. This tearing is usually associated with inclusions in the steel and progresses from one inclusion to another [4].

There is some indication that sensitivity to lamellar tearing is increased by the presence of H_2 in the steel. Inclusions which contain low melting compounds, such as those of P and S, also increase the sensitivity of steel to lamellar tearing by wetting the prior austenite grain boundaries, since this makes them too weak andfragile to withstand the thermal stresses during cooling. Some approaches which can minimize lamellar tearing are (i) changing the location and design of the welded joint to minimize through thickness strains, (ii) using a lower-strength weld metal,

FIGURE 2.11 Appearance of fracture face of a lamellar tear.

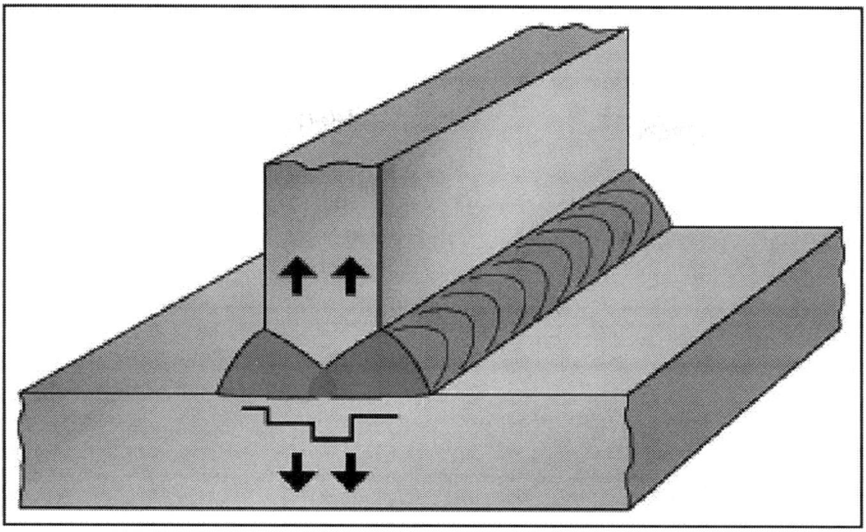

FIGURE 2.12 Lamellar tearing in a T butt weld.

(iii) reducing available H_2, (iv) buttering the surface of the steel plate with weld metal prior to making the weld, (v) using preheat and interpass temperatures of minimum 100 Deg C, and (vi) using base plates with inclusion shape control.

The main causes for lamellar tearing are:

i) High tensile stresses parallel to the weld
ii) Stringer-type inclusions in the base plate
ii) Presence of diffusible hydrogen

2.3.3.1 Causes

Three conditions which must be satisfied for lamellar tearing to occur are the following:

i) Transverse strain—the shrinkage strains on welding must act in the short-direction of the plate, i.e., through the plate thickness.
ii) Weld orientation—the fusion boundary will be roughly parallel to the plane ofthe inclusions.
iii) Material susceptibility—the plate must have poor ductility in the through-thickness direction.
iv) The risk of lamellar tearing will be greater if the stresses generated on welding actin the through-thickness direction.
v) The risk will also increase the higher the level of weld metal hydrogen.

2.3.3.2 Factors to Be Considered to Reduce the Risk of Tearing

The choice of material, joint design, welding process, consumables, preheating, and buttering can all help reduce the risk of tearing.

Lamellar tearing is due to restraint stress which leads to poor transverse ductility. Joint thickness, joint design, welding procedure, and inclusion content in the material contribute to lamellar tearing.

2.3.3.3 Inclusions

Large amounts of P and S are added to some steels to provide in them the free machining characteristics. These steels have relatively poor weldability because of hot tearing in the weld metal caused by low-melting compounds P and S at the grain boundaries.

Iron oxide and iron sulphide inclusions, if present, are also harmful because of their solubility change with temperature and their tendency to precipitate at grain boundaries, contributing to low ductility, cracking, and porosity.

Laminations, which are flat separations or weaknesses which sometimes occur beneath and parallel to the surface of rolled steels, have a slight tendency to open up if they extend to the weld joint.

The inclusions in steel depend on the type of steel, whether it is obtained by deoxidation practice, its composition, and the position of ingot.

Type of inclusions (whether silicates, oxides, sulphides), inclusion size and shape (globular or lamellar), and distribution (volume fraction or mean spacing between inclusion) also account for lamellar tearing.

2.4 WELDABILITY OF LOW-ALLOY STEELS

Low-alloy steels, commonly used in fabrication industries, contain small amounts of varying alloying elements to impart some special properties to those steels. The composition of the weld metal may broadly match the composition of the parent metal or sometimes may differ widely depending upon the properties required for theservice conditions and post-weld heat treatment that may follow the welding. Thecomposition of the filler materials normally used may contain less than 6% alloyingelements except a few of the newly developed steels that may contain up to 22% of alloying elements. Most of the filler materials used in the manual arc welding process are covered under the specification AWS A 5.5 [3].

2.4.1 As Rolled or Normalized Low-Alloy Steels

These are normally referred to as high-strength structural steels since these were the developments from structural carbon steels where it was noticed that relative amounts of certain alloying elements improved mechanical properties. The addition of elements like Cr, Cu, Mn, Mo, Ni, P, V, and Zr produces a fine microstructure in these steels during air cooling, thus increasing the strength and toughness of these steels. The addition of 0.01% to 0.05% Columbium or Vanadium further increases the yield strength.

In the higher-tensile strength steels which are stress relieved, the precipitation of copper or alloy carbides adds to the strength. For improved crack resistance during welding, these steels are produced with the carbon content less than 0.25%.

The steels are available in structural shapes, plates, pipes, and in the form of castings.

Example: i) ASTM specifications A 242, A 441, A 572, A 588 (structural fabrications)

ii) ASTM specifications A 203, A 204, A 225, A 553 (for pressure vessel fabrications)

These steels with carbon levels between 0.08% and 0.25% with alloys such as V, Ni, Cr are produced to get fine microstructure, strength, and other desired properties and are normalized to improve notch toughness. These steels in as rolled condition have good notch toughness at room temperature, but the transverse notch values are poorer. Some grades of these categories of steels are used for low-temperature service also. Nickel additions improve notch toughness at low temperature, and these steels are produced with strict control on chemistry and with specified de-oxidizing practices (ASTM A 203, A353, and A553 Q&T) [3].

ASTM A353 and ASTM A553 are used upto a service temperature of –198 Deg C. The mechanical properties of these steels vary slightly depending on thicknesses.

2.4.2 Welding

Compared to structural carbon steels which have C and Mn, these alloys contain varying amounts of Cr, V, Ni, and P, which strengthen the ferrite, increase hardenability, and refine the grain size.

The main consideration during welding of these steels is to avoid "Under Bead" or "cold Cracking," which is caused by the entrapment of hydrogen in steel during welding, and such cracks appear in the HAZ region.

At the strength level prescribed in these group of steels, the hydrogen may not produce crack in the weld metal.

Cold crack usually occurs below 93 Deg C and is caused by the stresses produced due to

i) presence of hydrogen,
ii) austenite martensite transformation, and
iii) the shrinkage of weld and HAZ.

Hydrogen having dissolved in steel at high temperature is released as the temperature falls down. The hydrogen atoms that fail to escape upon cooling form molecular hydrogen, occupying fissures and micro voids in the hardened HAZ and developing high aerostatic pressures that tend to initiate and propagate further cracks in the highly stressed HAZ. By keeping the steel above 100 Deg C or by eliminating martensite and hydrogen, one can avoid cold cracking. One can eliminate or reduce hydrogen content by a suitable welding process or by a proper selection of consumable (electrode). The formation of martensite depends on the chemistry of steel, the thickness of the part welded, the heat input, and preheat levels, i.e., factors controlling the cooling rates.

The chemical composition in this group of steels makes them susceptible to cold cracking, and increased section thickness adds to the problem. The underbead cracks are difficult to inspect and therefore proper procedures like preheat, interpass temperature control, and correct selection of low-hydrogen electrodes will reduce chances for the occurrence of the defect.

Use of low-hydrogen electrodes ClassEXX15, EXX16, and EXX18 (low-hydrogen type) are preferred over EXX10, EXX12, and EXx13 (cellulose, Rutile types). The preheat levels can be as used for structural steels, but care is to be taken to avoid welding when ambient temperature falls very much below normal. A slight preheat, upto 70 Deg C may be necessary in such cases.

Steels can be welded by all welding processes. Root welding on critical pressure vessel applications are done by TIG welding processes. Due to higher hardenability of some of the steels, special preheat and post heat cycles may have to be incorporated in the resistance welding operation for retaining the mechanical properties.

When welding 2.25%Cr and 3% Ni steels used for low-temperature applications, a preheat of 100 Deg C is prescribed, and a matching consumable is used in the manual arc or TIG process.

Pressure vessels manufactured to ASME boiler and pressure vessels code may require stress-relieving heat treatment at 620 Deg C (2.25% Ni), 590 Deg C (3.5%Ni). The 9% Ni steels are welded with consumables equivalent to 25%Cr 20%Ni type or with consumables equivalent to 70% Nickel type (if strength levels are to be matched). When making weldments to pressure code practice, a stress relief at 565 Deg C is recommended.

2.4.3 Quenched and Tempered Low-Alloy Steels

Steels in this group have strength levels from 50,000 psi yield strength to 1,80,000 psi yield strength depending upon the chemical composition, thickness, and heat treatment. These steels are used in fabrication without an additional post-weld heat treatment.

Some noted examples of this group are ASTM A353, A533, A537, A543, A514, A517, A553 (Q&T version of A353 Ni steel) and high-strength steels like Hy130, Hp 9/4/20, and 10Ni/Cr/Mo/Co steels (Alloy content ~22%).

A careful study of the isothermal transformation diagram of one grade of 415/417 will illustrate that very long time elapses begin in the pearlitic transformation temperature range of 700/590 Deg C. Hence this high-temperature pearlitic transformation, which is undesirable, may not occur unless the cooling rate is extremely slow. Also, it will be found that moderately long time elapses before transformation beginsin the ferrite–upper bainite transformation temperature range of 590-150 Deg C. This factor helps when quenching the steel without the occurrence of this undesirable transformation, even in thicker sections. Ferrite–upper bainite transformation is to be avoided, since the surrounding austenite, rich in carbon will either remain or form a high-carbon austenite or a high-carbon bainite upon further cooling, resulting in a heterogeneous microstructure of soft ferrite with hard upper bainite and high-carbon martensite, together with retained austenite. This microstructure even after tempering does not produce a high level of toughness.

At low temperatures (480 Deg C), only a very small period is required to produce a homogeneous bainitic structure having excellent toughness, particularly when tempered.

Martensite forms at 390 Deg C. A lower temperature is not required in these steels for the martensite to form—this is a significant factor for quenching to prevent cracking.

2.4.4 Welding

i) High strength of these steels is to be taken to advantage when designing themembers and the weldments to obtain optimum economy.
ii) But extreme care must be taken for such designs since abrupt changes incontour and notches will not be tolerated as much as could be permitted inlow-strength carbon steels.
iii) Grooved fillets are preferred.
iv) Butt joints could be welded from both sides.
v) Thermal cutting methods can be employed without much preheating andoverheating and the preheat levels can be the same as used for high-carbonsteels.
vi) Arc cutting is preferred over oxy-acetylene cutting for high-strengthsteels like HP9/4/20 and 10 Ni/Cr/Mo/Co steels.
vii) Utmost care must be taken to remove oxidized layers and all reground edges to remove hardened surface. Edges must be free from moisture and oil. All-welding should be done with full penetration.
viii) Normal edge preparation with U and V shapes is applicable.
ix) The welds must be of smooth contour, and care is to be taken to avoidcracking due to poor contouring and improper penetration, since only a smallvolume of metal at the stress raisers participates in the plastic deformation, and the ductility of the metal in that particular area may get exhaustedresulting in cracks.
x) Manual arc welding, TIG welding, and MIG welding are done with greatperfection by proper choice of filler material.
xi) Submerged arc welding is practised at lower strength levels in this group, butsuch high productive processes and multiwire methods may lead toexcessive heating thus reducing ductility of high-strength level steels.
xii) The rapid cooling occurring in manual arc welding gives adequatemechanical properties.
xiii) If welds are made with higher heat input and subsequently a slow coolingrate, the weldments should be given a post-weld quenching and temperingtreatment to restore mechanical properties.
xiv) Filler materials are available to match the various grades of steel for manualarc welding.
xv) In submerged arc welding, it is necessary to ensure dryness of the agglomerated flux to avoid defects.

Moreover, maintaining proper arc voltage is very essential to get repeatedly the desired mechanical properties when using agglomerated fluxes since the alloying elements are provided through the flux, and the melting point of the flux is related to the arc voltage.

As these steels are susceptible to hydrogen cracking, a moisture level of 0.4% is tolerated for steels at lower strength levels, and stipulations to control moisture upto 0.2% and 0.1% are necessary when welding is done with steels at high strength levels.

i) Also, total hydrogen level is to be controlled below 5 ppm for steels upto 1,00,000 psi strength levels and 3 ppm for steels at higher strength levels.
ii) Low-hydrogen electrodes are to be used with extreme care since the moisture-proof packing and the cardboard carton do not prevent the absorption of moisture in the coating, unless they are vacuum packed.
iii) Normally all hydrogen electrodes should be baked at 420 Deg C for one hour and then kept in holding ovens at 125 Deg C before use.
iv) Further exposure after taking it out of the oven should not exceed more than 20 minutes when welding with steels of 1,30,000 psi strength levels and above.
v) Precaution must be taken to avoid welding when the ambient temperature is very low. Preheating at 70 Deg C or slightly high is recommended, depending upon the thickness.
vi) For steels in the range of 130,000 psi strength level, maximum preheatof upto 150 Deg C is recommended for 3 mm thickness. For certain higher-strength steels, no preheat is required.
vii) Heat input is to be very much controlled since excessive heat and subsequent slow cooling rates will reduce notch toughness.
viii) Often, small multi-pass welds are preferred, and stringer bead method is employed. Weaving techniques, which result in the heat source lying for longer period in the same area, are to be avoided.
ix) If arc gouging is carried out, it is necessary to move the arc as rapidly as possible and grind the surface before welding.
x) Post-weld stress relief heat treatment is to be avoided since notch toughness suffers as the steel is held at high temperature for a long time.

2.4.5 Post-Weld Heat Treatment

Exposure to temperature in the range of 500–700 Deg C may impair the toughness of weld and the parent material, and the damage is greater due to slow cooling which is associated with stress relief treatment. Also, intergranular cracking may occur in the coarse grained HAZ (stress rupture) in the early stages of such heat treatment before the welding stresses are fully reduced.

By proper contouring at the toes and thus reducing stress concentration and by peening at the toe of the weld, and by deposition of filler materials, whose elevated temperature strength is considerably less than that of the parent material, one can eliminate cracking. Still, if stresses are to be relieved through the thermal treatment on components which go for further machining and where dimensional stability is required, it is done at the cost of loss of ductility at a temperature 30 Deg C lower than the recommended temperature of the particular grade.

Thermal cutting is employed when preparing material with suitable preheat as applied to carbon steels (150–200 Deg C). If cutting edges are hard, a local tempering treatment may be given.

2.4.6 Steels Quenched and Tempered after Welding

The carbon content in these steels is higher (0.25%–0.45%) compared to the previously discussed group, and these steels can be heat treated to have extremely high strength and hardness. Also with the addition of various alloying elements, these steels can be hardened through heavy sections.

To produce sound welds in these steels, which contain higher carbon content and alloy content, preheating is necessary to prevent cracking in the weld and HAZ. Susceptibility to cracking is more influenced by the presence of sulphur (hot cracking in weld metal) and phosphorus (cold cracking in parent metal and weld metal due to reduction in ductility and toughness). Hence, the base metal chemistry is to be controlled to less than 0.15% each of sulphur and phosphorus. Hydrogen level is to be kept very low to prevent cracking.

Materials in this group are heat treated to the required hardness and strength, and the mechanical properties depend on each melt and the subsequent quenching and tempering temperatures. A lower tempering temperature will rise the strength and hardness levels at the loss of ductility and toughness.

The process of quenching and tempering consists of heating the steel above austenizing temperature and holding the metal at that temperature for a length of time and subsequent quenching in water or oil followed by tempering in the range of 250–700 Deg C depending on the final product requirements. As the metal is heated above the autenitizing temperature and subsequently rapidly cooled to interpass temperature levels during welding, formation of hard and brittle high-carbon martensite in the weld and HAZ is promoted.

The stresses caused by the contraction of the solidifying weld metal and the dimensional changes that take place during the formation of martensite may cause the brittle materials to crack. Also, the hydrogen that is present lowers the stress level at which cracks occur.

Therefore, the procedure for welding should be one that introduces the minimum amount of hydrogen in the weld. Still it would require an additional preheat and interpass temperature to be maintained to allow the hydrogen to escape. Preheat levels to prevent cracking will depend on the chemistry and condition of heat treatment of steel and the amount of hydrogen. This means a lower preheat will be sufficient when carbon and alloy contents are lower. And a low hydrogen welding process is followed.

The recommended preheat prevents cracking by decreasing the cooling rate of weld and by allowing the hydrogen to escape. It raises the steel above the brittle/ductile transition temperature and even by promoting transformation of the structure in the fusion and HAZ to microstructures having resistance to cracking than the normally formed quenched martensite.

The post-weld heat treatment that is done before the final quenching and tempering treatment consists of holding the steel above interpass temperature or above for a considerable time before allowing it to cool to room temperature. Some steels that are preheated to 150 Deg C are raised on completion of welding to 350 Deg C and held for sometime before placing them on a stress-relieving furnace and treating them at 650 Deg C. In some other different applications, preheating is done at 350

Deg C, and the interpass temperature for a considerable time after welding is maintained and is then cooled to room temperature. Such treatments allow hydrogen to escape and also promote transformation to finer bainite or martensite. These steels are welded by MIG, TIG, MMAW, SAW, FBW, spot welding, seam welding, and EBW, and minor repairs are done by gas welding.

2.4.7 Stress Relief Cracking

Stress relief cracking is also known as stress rupture cracking and reheat cracking. When quenched and tempered steels and heat resistant steels containing substantial amounts of carbide formers, such as Cr, MO and V, are heat treated for stress relieving, this type of cracking occurs. When weldments of these steels are heated to more than around 500 Deg C, intergranular cracking along the prior austenite grain boundariescan take place in the CGHAZ. Figure 2.13 shows the reheat cracking associated with the CGHAZ. Stress relief cracking is thought to be closely related to the phenomenon of creep rupture. Figure 2.14 also indicates reheat cracking associated with the CGHAZ, and Figure 2.15 shows the intergranular morphology of reheat cracking. Furthermore, during reheating, the re-precipitation of carbides is likely to occur, further increasing the hardness. The precipitation of carbides during stress relaxation alters the delicate balance between resistance to grain boundary sliding and resistance to deformation within the coarse grains of the HAZ.

Certain procedures which can be used individually or in combination to decrease stress relief cracking in steels include the selection of a more appropriate weld joint design, weld location, and sequence of assembly to minimize restraint and stress concentrations. Selecting a filler metal which provides a weld metal that has significantly

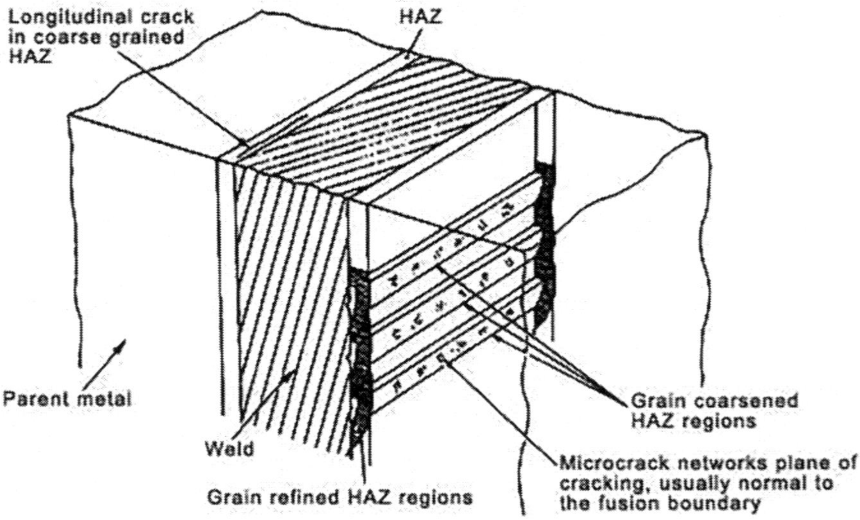

FIGURE 2.13 Reheat cracking associated with CGHAZ.

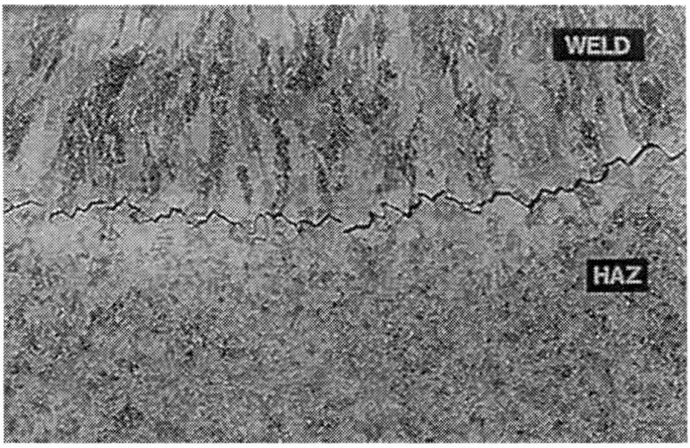

FIGURE 2.14 Reheat cracking associated with CGHAZ.

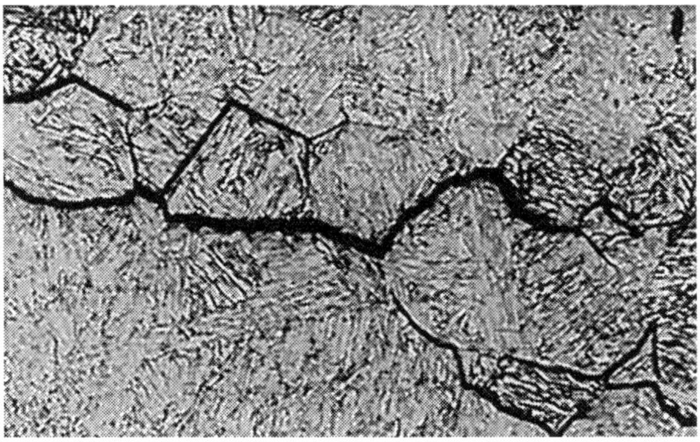

FIGURE 2.15 Intergranular morphology of reheat cracking.

lower strength than that of the HAZ at the heat treating temperature is another way to minimize stress relief cracking. Peening each layer of weld metal to generate a surface compressive stress state that counteracts shrinkage stresses is also very effective.

2.5 CHROMIUM AND MOLYBDENUM STEELS

Chromium and molybdenum steels are used in steam-generating boiler industries due to their high-temperature strength provided by the alloying contents (chromium for preventing oxidation and molybdenum for providing high-temperature strength). They are also used in petroleum industries due to their corrosion resistance against sulphur and sulphur compounds and in heaters due to their resistance to oxidation.

The chromium content ranges from 0.9% to 1.1%, and the molybdenum content ranges from 0.45% to 0.65%. Oxidation resistance, high-temperature strength, and resistance against sulphur increase with increased Cr and Mo contents. These Cr–Mo steels are air hardenable and undergo both high- and low-temperature transformation and affect the mechanical properties, and therefore proper weld procedures are important to restore properties. 2.25%Cr 1%Mo steels (ASTM A542) have very high tensile strength in the range of 1,05,000 psi to 1,15,000 psi. When these steels are heated and cooled rapidly as in welding, the hardnewithss and strength show an increase, and the toughness and ductility suffer. But the absence of large quantities of carbides gives increased ductility, compared to that of plain carbon steels at these strength levels. These steels after being heated above the transformation temperature during welding or forming require further heat treatmentto restore toughness, ductility, and the required mechanical properties.

These steels are givenannealing, normalizing and tempering, andquenching and tempering treatments to restore proper grain size and desired mechanical properties.

Tempering is applied after welding to relieve stress and also to soften the already-hardened material. This treatment is done before the lower critical temperature (760 Deg C) and held for sometime before cooling in air to remove the residual stresses caused during welding and also to temper the hardened HAZ. No further hardening due to rapid cooling happens during this tempering treatment, since tempering is done below lower critical temperature.

However, care has to be taken to avoid introducing new stresses and causing cracks at sharp corners.

The ability of these steels to be hardened to high strength levels is used advantageously in hydro treating plants where high-pressure hydrogen is handled. After the heat treatment process, be it normalizing and tempering or quenching and tempering, the resultant microstructure is low-carbon martensite or bainite having excellent ductility. The time and temperature of tempering are important to restore good properties.

2.5.1 Welding Metallurgy

MMAW, SAW, GMAW, TIG, and ESW are employed for welding Cr–Mo steels. The air hardening tendency necessitates proper welding procedures such as preheat, post-weld heat treatment, and choice of filler metal. The failure to adhere to precautions will lead to cracking in the weld and HAZ. When preheating is low and cracking happens, this can be avoided by raising the preheat and holding the weld at interpass temperature for sometime after welding is completed before allowing the metal to cool. The preheat, which depends on the material thickness and chemical composition, enables hydrogen to escape and thus prevents cracking.

- A low-hydrogen process like TIG welding will reduce the level of preheat.
- Use of low-carbon filler metal may avoid cracking, but when mechanical properties are critical, lower-carbon content filler metals cannot be employed. Normally, fillers with higher Cr–Mo content are used in welding these Cr–Mo steels.

Austenitic stainless steels 25% Cr 20% Ni or 25% Cr 12% Ni types are used when components do not require post-weld heat treatment, as these welded structures would have adequate ductility (but less strength). But this welding does not hold good for welded components subjected to thermal cycles due to carbon migration in the base metal or sigma phase formation in the weld. Also, the difference in theco-efficient of thermal expansion between base metal and weld metal may cause fusion line cracking. This happens in dissimilar weldments especially in 2.25 Cr 1 Mo steel with 18% Cr 8% Ni austenitic stainless steel. The carbon migration from lower Cr side to higher Cr side is more pronounced at elevated temperatures (550 Deg C) and results in weakening the weldment.

2.5.2 Need for Stress Relieving Heat Treatment

1. When section thickness is small, and proper preheating is done, and when lower-carbon filler metals are used for welding, some steels in CrMo group do not require any post-weld heat treatment.
2. This must be ascertained by comparing the as-welded properties with code requirements on mechanical properties.
3. The subcritical tempering treatment removes residual stresses, reduces hardness, and increases ductility. Some weldments are annealed at 850–900 Deg C and cooled to 550 Deg C at 30 Deg C per hourand then air-cooled to room temperature to produce a soft ferrite microstructure in the HAZ and base metal.
4. Post-weld heat treatment or stress relieving can be given either immediately after welding as in the case of materials with higher alloy content, higher thickness, and high restraint or sometime after the weldment is cooled to room temperature after welding. The procedure must be so chosen to avoid cracking.
5. If stress relieving cannot be done immediately after welding, then raise the metal temperature to 550 Deg C and slowly cool and later take up for stress relieving. The weldment is very susceptible to hydrogen-induced cracking.
6. Low-hydrogen electrodes with proper rebaking and drying are preferred to rutile types.
7. Further, since the electrodes are synthetic, removing the flux and using them as filler materials for TIG welding should be avoided.
8. Edge preparation may be wide enough to permit proper manipulation, since the basic electrodes have thick coating and are difficult to clean. Surface cleanliness is a must in these steels.

2.5.3 Pre-Coated Steels

Thin steel plates and sheets are frequently pre-coated to protect them from oxidation and corrosion. The coatings usually used are aluminium, zinc, and zinc-rich primers. As anticipated, the coating originally at the weld region is destroyed during fusion welding, and the effectiveness of the coating adjacent to the weld is significantly reduced by the welding heat. In the case of aluminized steels, the formation of aluminium oxide can adversely affect the wetting and weld pool shape. The welding electrode and filler metals are to be selected carefully. A basic coating shielded metal arc welding (SMAW) electrode is suggested [1].

For galvanized steels, weld cracking is generally attributed to intergranular penetration by Zn. Zn dissolves considerably in iron to form an intermetallic compound at temperatures close to the melting temperature of Zn. Thus, molten Zn penetrates along the grain boundaries, leaving behind a brittle champagne fracture during cooling with the onset of a tensile stress state. Cracking occurs primarily at the throat region of a fillet weld, where shrinkage strain is more significant. The use of hot-dipped coatings results in more severe cracking, while thin electro galvanized coatings are the least susceptible to cracking. Both low-silicon (Si) electrodes and rutile-based SMAW rods are good for galvanized steel welding. Specific welding and set-up procedures need to be followed, such as removing the Zn coating by an oxy-fuel process or by grinding, ensuring a large root opening, and using a slower welding speed to allow Zn vapourization and to prevent Zn entrapment in the weld metal.

2.6 WELDABILITY TESTS

Weldability tests are conducted to provide information on the service and performance of the welds. However, the data obtained in these tests can also be applied to the design of useful structures. Often, these data are obtained from the same type of test specimens used in determining the base steel properties. Predicting the performance of structures from a laboratory-type test is very complex because of the nature of the joint, which is far from homogeneous, metallurgically or chemically. Along with the base steel, the weld joint consists of the weld metal and the HAZ. Thus, a variety of properties are to be expected throughout the welded joint. Careful interpretation and application of the test results are needed [1].

There are presently many tests that evaluate not only the strength requirements of steel structures but also the fracture characteristics and the effect of atmospheric conditions on early failure of the weldments. Certain major tests are described in the following.

2.6.1 Weld Tension Test

Several tension test specimens can be used for obtaining an accurate assessment of the strength and ductility of welds. Both transverse weld specimens and longitudinal weld specimens are used. In the all weld metal tests, the base steel dilution needs to be minimized if the test is to be representative of the weld metal. However, the resulting properties may not be easy to translate into those properties achievable from welds made in an actual weld joint.

Interpreting test results for the transverse butt weld test is complicated by the different strengths and ductilities normally found in the various regions of the joint. The primary information gained from the test is the ultimate tensile strength (UTS). YS and elongation requirements are usually not specified.

Tests of HAZ properties which are unaffected by the presence of either base steel or weld metal are not easy to conduct since it is practically impossible to obtain specimens made up entirely of the HAZ. In addition, the HAZ is composed of various regions, each with its own distinct properties. Simulated HAZ specimens that are generated and tested using a Gleeble thermo-mechanical testing system can be used to provide a more accurate assessment of the tensile properties of this region.

2.6.2 Bend Test

Different types of bend tests are used to evaluate the ductility and soundness of welded joints [4]. Bend test results are expressed in various terms, such as percent elongation, minimum bend radius prior to failure, go/no-go for specific test conditions, and angle of bend prior to failure. Various specimen designs, both notched and un-notched, and testing techniques have been used. Nowadays, un-notched specimens can be used in quality control tests, while notched specimens may be used to predict in-service behaviour. However, most notched bend tests are used for research purposes and are not in common industrial use. Transverse bend tests are useful since they quite frequently reveal the presence of defects which are not detected in tension tests. However, the transverse specimen suffers from the same weakness as the transverse weld tension test specimen in that non-uniform properties along the length of the specimen can cause non-uniform bending, although this is often compensated for by the use of a wrap-around bend fixture.

2.6.3 Hardness Testing

Hardness testing can be used to complement information gained through tension or bend tests by providing information about the metallurgical changes caused by welding. Routine methods for the hardness testing of metals are well established. In C and low-alloy steels, the hardness near the fusion line in the HAZ may be much higher than in the base steel because of the formation of martensite. In the HAZ areas, where the temperature is low, the hardness may be lower than in the base steel because of tempering effects.

2.6.4 Pellini Drop Weight Test

The drop weight test design is based on service failures resulting from brittle fracture initiation at a small flaw located in a region of high stress. The drop weight test can be considered a limited deflection bend test that uses a crack starter to introduce a running crack in the specimen. The specimen is a steel bar on which a brittle crack starter weld is deposited. This overlay cracks when the bar is deflected by the drop weight. A series of test is performed at different temperatures to determine the testing temperature below which the crack will propagate to the edges of the specimen. This critical temperature is also called the nil ductility temperature (NDT), defined as the highest temperature at which the propagating crack reaches the edge of the specimen. Therefore, the drop weight test is also known as the NDT test. This test is carried out as per ASTM E 208 specification.

2.6.5 Charpy V-Notch (CVN) Impact Test

The Charpy V-notch (CVN) impact test is the most popular technique for evaluating the impact properties of welds. The energy absorbed by a sample at fracture determines the toughness of the specimen. In this test, specimens at different temperatures are broken using a pendulum hammer. A typical plot of the CVN results for a

C and low-alloy steel shows that there is a transition from low-energy to high-energy fracture over a narrow temperature range. This is associated with a change from trans-crystalline to ductile fracture. Therefore, material quality can be defined in terms of this transition temperature.

In the CVN impact testing of welds, the notch is typically located at the weld centerline. For CVN impact testing of the HAZ, the notch is more typically introduced at the CGHAZ. However, because precise location of a notch is never simple in the HAZ, simulated weld samples are used instead.

2.6.6 Crack Tip Opening Displacement Test

The crack tip opening displacement (CTOD) test measures toughness, primarily for elastic–plastic conditions. In CTOD tests, the clip gauge opening at the onset of fracture is measured and used to calculate the crack opening displacement at the crack tip. The critical value of CTOD at fracture is a critical strain parameter that is analogous to the critical stress-intensity parameter. The CTOD test provides a useful method of determining the critical flaw size.

The CTOD test is very sensitive to changes in sample thickness, hardness, and strength, and it is difficult to obtain valid results in practical specimen thicknesses. The application of fracture mechanics to the prevention of catastrophic failure in weldments is, however, complicated by the nature of the weldment. In addition to their metallurgical heterogeneity, weldments often contain high residual stresses. Therefore, it is inadequate to fracture test the base steel and assume that the critical crack length thus determined is valid when the base steel is made into a weldment. The fracture toughness criterion needs to be determined for the base steel, the HAZ, and the weld metal. By first determining the zone with the lowest toughness value, it is then possible to evaluate a more realistic critical flaw size. However, the plane-strain fracture toughness tests are preproduction or pilot plant type tests that provide a rational means for designs and for estimating the effects of new designs, materials, or fabrication practices on the fracture-safe performance of structures.

Other popular tests include compact tension (CT) and wedge opening load (WOL) tests, which are normally used in the evaluation of structural weldments.

2.6.7 Stress Corrosion Cracking Test

The presence of corrosive atmospheres in a steel weldment can accelerate the initiation of a crack. Generally, the higher the strength of the steel, the more susceptible it becomes to stress corrosion cracking. The C and low-alloy steels are not generally exposed to severely corrosive atmospheres but rather to the normal atmosphere, moisture, hydrocarbons, fertilizers, and soils. However, welding can lower corrosive resistance by the introduction of (i) compositional differences which promote galvanic attack between weld metal, HAZ, and base steel when the joint is immersed in a conducting liquid, (ii) residual stresses that can cause stress corrosion cracking, and (iii) surface flaws that can act as sites for stress corrosion cracking.

Stress corrosion cracking is normally delayed cracking, with longer time to failure at lower stresses. Most stress corrosion tests are fairly long in terms of time because of the slow crack initiation that occurs in the un-notched test specimen. However, it has been found that the long initiation period can be eliminated by testing pre-cracked samples.

2.6.8 Fabrication Weldability Tests

There are various types of tests for determining the susceptibility of the weld joint to different types of cracking during fabrication. These tests are (i) restraint tests, (ii) externally loaded tests, (iii) underbead cracking tests, and (iv) lamellar tearing tests.

2.6.9 The Lehigh Restraint Test

The Lehigh restraint test is particularly useful for quantitatively rating the crack susceptibility of a weld metal as affected by electrode variables. This test provides ameans of imposing a controllable severity of restraint on the root bead that is deposited in a butt weld groove with dimensions suitable to the application. Slots are cut in the sides and ends of a steel plate prior to welding. By changing the length of the slots, the degree of plate restraint on the weld is varied without significantly changing the cooling rate of the weld. Therefore, a critical restraint for cracking can be determined for given welding conditions. This sample is also useful for H2 cracking.

2.6.10 The Varestraint Test

The Varestraint test determines the susceptibility of the welded joint to hot cracking. The test utilizes external loading to impose controlled plastic deformation in a steel plate while a weld bead is being deposited on the long axis of the steel plate. The specimen is mounted as a cantilever beam, and a pneumatically driven yoke is positioned to force the test piece downward when the welding arc reaches a predetermined position. By the choice of the radius to which the steel plate is bent, the severity of deformation causing cracking can be determined. Strain from 0% to 4% can be chosen according to the susceptibility of the joint to hot cracking. When the bending moment is applied transverse to the weld axis, the test is termed trans-varestraint test. A spot varestraint test can also be conducted by keeping the arc stationary with bending applied at the moment the arc is extinguished.

2.6.11 The Controlled Thermal Severity Test

The controlled thermal severity test is designed to measure the cracking sensitivity of steels under cooling rates controlled by the thickness of the steel plates and the number of paths available for dissipating the welding heat. It is conducted with a steel plate bolted and anchor welded to a second steel plate in a position to provide two fillet (lap) welds. The fillet located at the plate edges has two paths of heat flow. The lap weld located near the middle of the bottom plate has three paths of heat flow, thus inducing faster cooling. The fillet welds are made first and allowed to cool, followed by the lap welds. After a holding time of 72 hours at room temperature, the degree of cracking is determined by measuring the crack length on metallographic specimens.

A number of other tests have been developed that contain welds in a circular configuration. The circular patch test has probably the most severe testing conditions.

The two varieties are the navy circular patch restraint test and the segmentedcircular patch restraint level. Cracking is detected by visual, radiographic, and liquid penetrant inspection. The cracking susceptibility of a material is measured as the total crack length and expressed as a percentage of the weld length. These tests can be used to determine both hot and cold cracking in the weld metal and the HAZ. Depending on the results, a go/no-go criterion is established for weld qualification.

2.7 FAILURE CASE STUDIES

2.7.1 Case Study I: Failure of Straight Tube Butt Welded Joint

In Case I, an STBW joint of a platen super heater coil assembly leaked in the shop floor during hydro testing. The material grade of the superheater coil was SA 213 T91 with Ø51 × 7.1 mm dimensions. The weld joint leaked during hydro test.

Primary observation: The tube leaked at the toe of the STBW joint (Refer to Figure 2.16 and Figure 2.17).

Root cause analysiswas carried out by macro examination and micro examination of the failed portion.

Macro Examination: The failed portion was sectioned and etched with 10% Nital and macro examined. The section at the leaked region reveals absence of root fusion, lack of side wall fusion, and porosity as indicated by Figure 2.18 and Figure 2.19.

Figure 2.20 and Figure 2.21 indicate 180° to leaked region showing insufficient root penetration and lack of fusion of side wall.

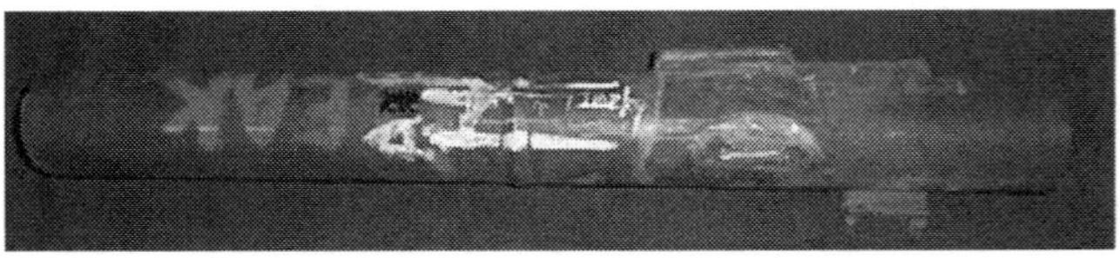

FIGURE 2.16 Tube showing the leaked region (arrowhead).

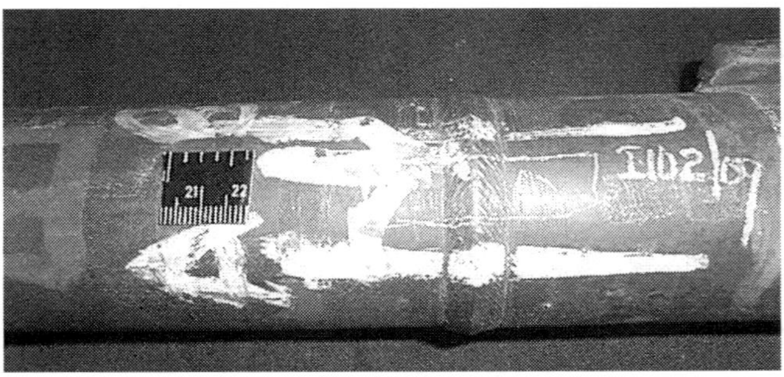

FIGURE 2.17 Close-up view.

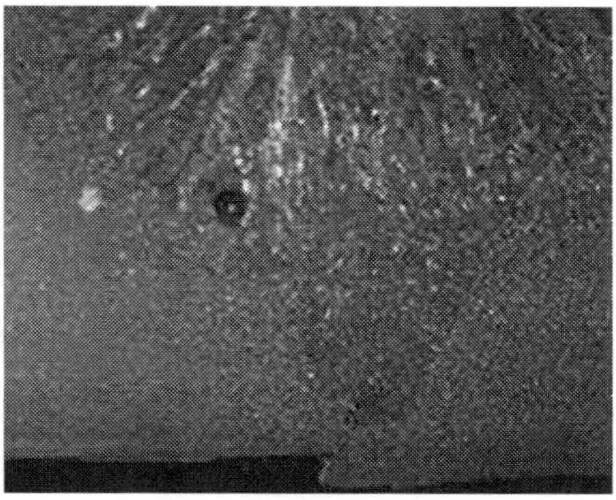

FIGURE 2.18 Leaked region.

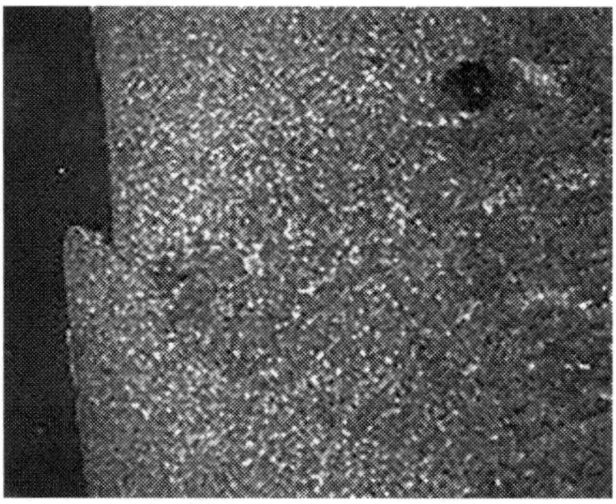

FIGURE 2.19 Leaked region.

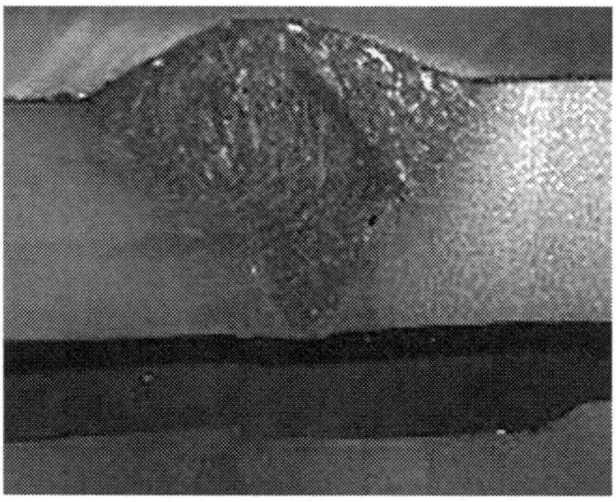

FIGURE 2.20 180° to leaked region.

Micro Examination: The failed portion wassectioned and polished with various grades of emery papers and polishing cloth and etched with 3% Nital and micro examined in a metallurgical microscope.

Transverse section of the STBW joint at the leaked region revealed a slight mismatch of the members, absence of root penetration, and root fusion. Figure 2.22

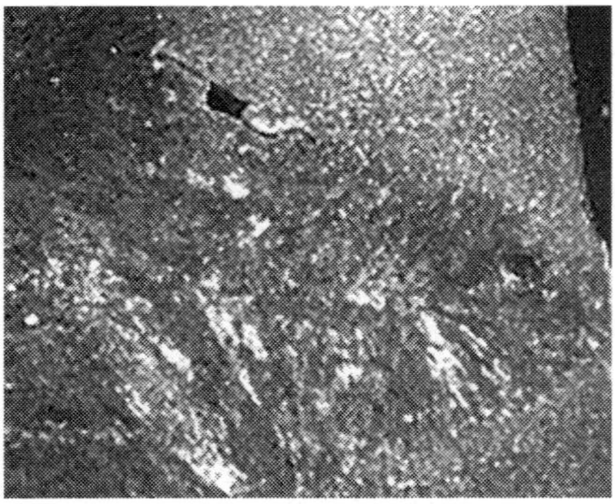

FIGURE 2.21 180° to leaked region.

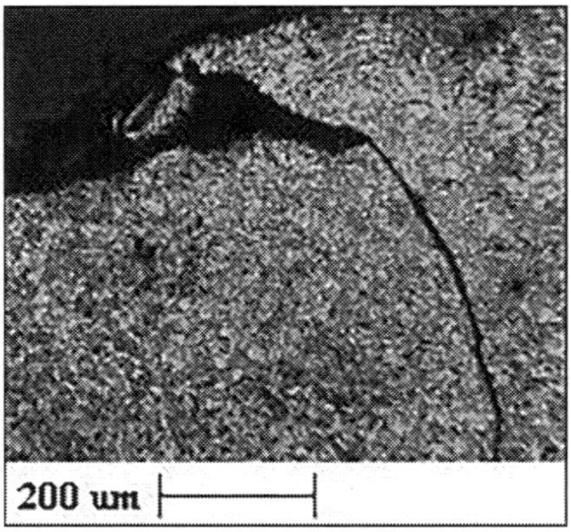

FIGURE 2.22 Absence of root penetration.

shows the absence of root penetration, and Figure 2.23 shows the absence of root fusion. Also, side wall lack of fusion and porosity in weld metal were observed.

One more section at about 180° to the leaked region was also micro examined. It also revealed insufficient root penetration and side wall lack of fusion as shown in Figure 2.24.

A small oxidized crack in the weld metal near this lack of fusion was observed. The weld metal/HAZ showed coarse martensite. Tube metal structure showed fine-tempered martensite as shown in Figure 2.25.

Reason for Failure: The leak observed during shop hydro test near the STBW joint is attributed to STBW defects, i.e., absence of root fusion and side wall lack of fusion.

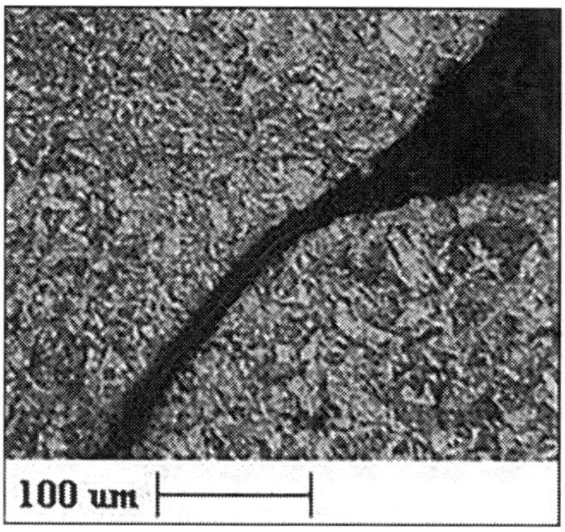

FIGURE 2.23 Absence of root fusion.

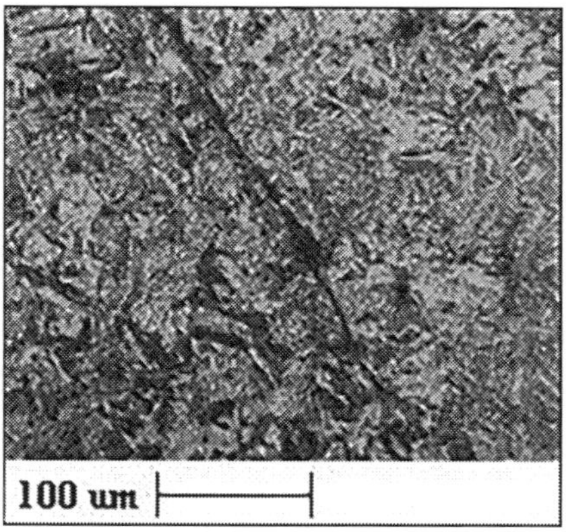

FIGURE 2.24 Side wall lack of fusion.

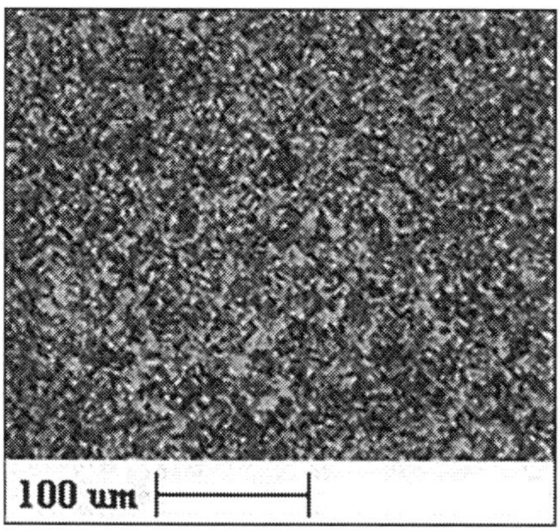

FIGURE 2.25 Tube shows tempered martensite.

2.7.2 Case Study II: Stress Corrosion Cracking in SS Weld Joint

In Case II, an insert SS347H tube with dimensions of Ø 51 × 6.0 mm and Ø 63.5 × 12.5 mm between an SS 347H tube of 63.5 × 8.0 mmand a bifurcate assembly with a straight tube butt joint (STBW) was brought from the site for failure investigation (Figure 2.26).

Lpi Examination: Liquid penetrant inspection was carried out on the STBW joint. It revealed fine cracks on the insert near the toe of the weld joint (Figure 2.27).

Swab Analysis: Swab analysis was carried out on the STBW joint qualitatively. It showed the presence of chloride on the inside of the insert tube.

Macro Examination: The cracked portion was sectioned, and it was macro examined after etching with 1:5 HNO_3:HCl. Fine-branched cracks were seen on the ID of the insert tube (Figure 2.28).

Micro Examination: A section of the failed portion of the insert was polished with various emery sheets and velvet cloth and etched with 1:5 HNO_3:HCl. It showed several fine, branched cracks originating from the inside (steam-side) surface in theinsert at a region close to the butt joint, the main crack being about 6 mm from the fusion line (Figure 2.29).

Cracks were seen on the weld and HAZ portions of the insert tube (Figure 2.30 and Figure 2.31). Similar cracks were not seen on the tube side HAZ or on either side of the bifurcate weld.

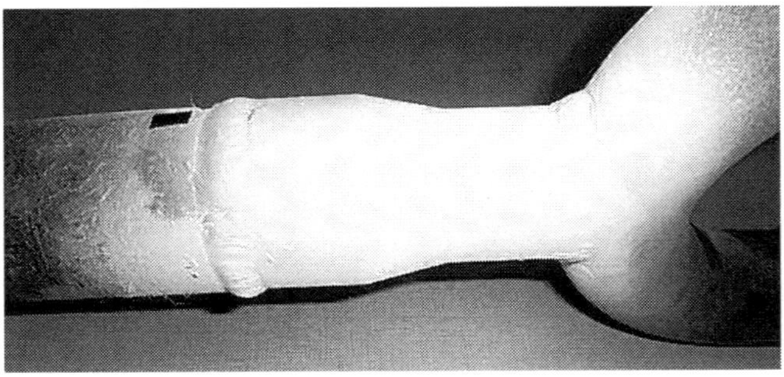

FIGURE 2.26 Tube-insert-bifurcate tube assembly.

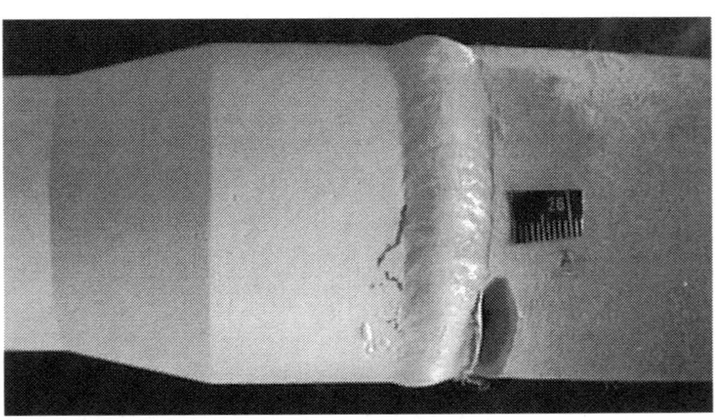

FIGURE 2.27 Fine crack (LPI) on the insert near the toe of theweld joint, as seen on the outside surface.

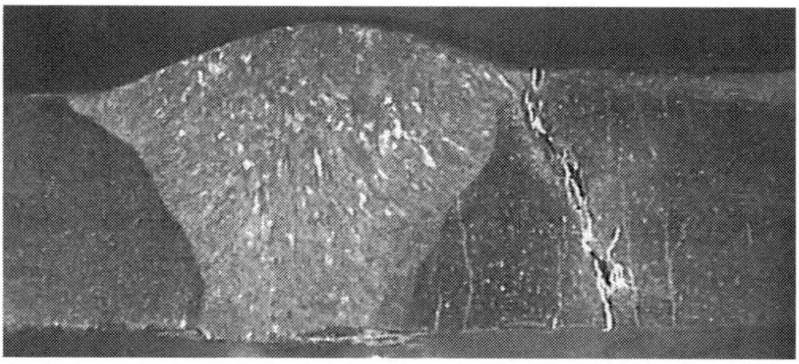

FIGURE 2.28 Fine-branched cracks on insert ID.

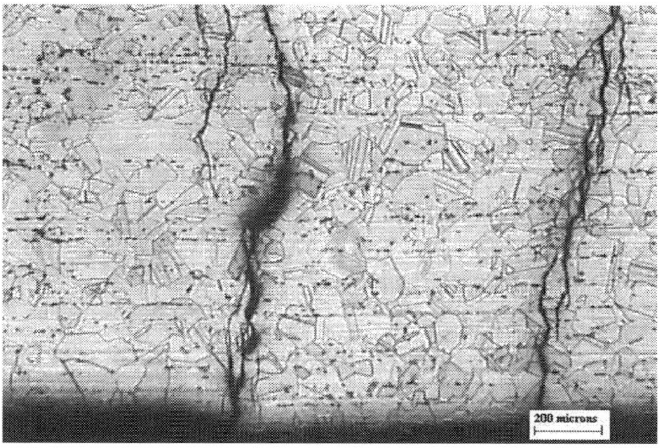

FIGURE 2.29 Insert ID-Near STBW.

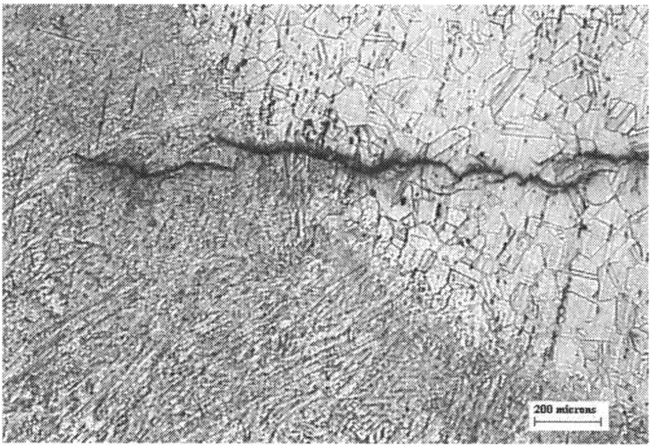

FIGURE 2.30 Crack in insert HAZ and weld.

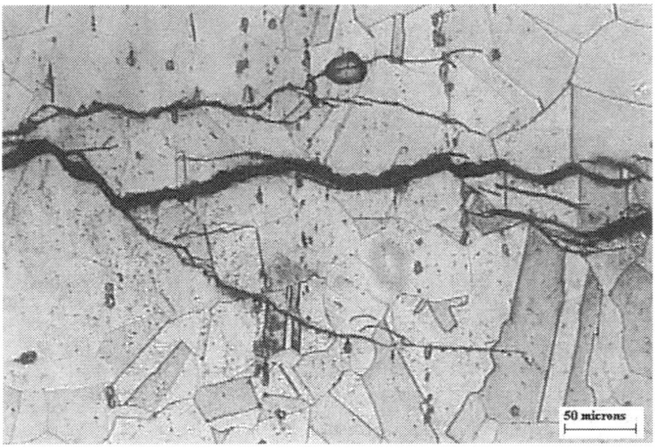

FIGURE 2.31 Cracks in insert-HAZ.

2.7.1.1 Chemical Composition

The chemical composition of the tube and insert were found out, and they met the requirements of SA213 TP347H specification. The weld metal chemistry was typical of Nb-stabilized austenitic stainless steel filler material.

Reason for Failure: The indication of chloride in swab analysis and the nature of the cracks which are transgranular and branched attribute the insert failure to stress corrosion cracking caused in the material, which is typical of stainless steel grades.

Annexure 2.1

TABLE 2.1
Types of Steels and Welding Processes Used

Carbon steels		SMAW, TIG, MIG,
Low-alloy steels	----->	SAW
Stainless steels		
Aluminium	----->	MIG
Dissimilar metals	----->	Solid phase welding
Differing melting temperature		

Annexure 2.2

TABLE 2.2
Conventional Boiler Materials

Area of Application	Material Type	Typical Spec. for Plates, Tubes, Pipes	Upper Limit Temp. (Heat- Absorbing Surface)	Guiding Reason for Upper Limit
Drum	Carbon/low-alloy steel	SA 299	425°C	
Waterwalls, economizer	Carbon steel	SA 192, SA 210, SA 106	425°C	Graphitization
Superheater, reheater	C-1/2 Mo	A 209T1	465 °C	Graphitization
	1 Cr ½ Mo	SA213 T11 SA213 P11	565 °C	Oxidation/corrosion, flue gas
	2 ¼ Cr 1 Mo	SA 213 T22 SA 213 P22	580°C	Oxidation/corrosion, flue gas
	18Cr 8 Ni	SA213 TP304 H	704°C	
	18Cr 10NI Cb	SA213TP 347H	704 °C	
	Modified 9 Cr	SA 213 T91/P91	650°C	ASME code
	12% Cr	X20CrMoV12 1	700°C	German code

TABLE 2.3
Nominal Chemical Composition of Cr–Mo Steels

Type	C	Cr	Mo	Mn
½Cr–1/2Mo	0.10–0.20	0.50–0.80	0.45–0.65	0.30–0.60
1Cr–1/2Mo	0.15	0.80–1.25	0.45–0.65	0.30–0.60
1¼ Cr–1/2Mo	0.15	1.00–1.50	0.45–0.65	0.30–0.60
2Cr–1/2Mo	0.15	1.65–2.35	0.45–0.65	0.30–0.60
2¼ Cr–1Mo	0.15	1.90–2.60	0.87–1.13	0.30–0.60
3Cr–1Mo	0.15	2.65–3.65	0.80–1.06	
5Cr–1/2Mo	0.15	4.00–6.00	0.45–0.65	
9Cr–1Mo	0.15	8.00–10.00	0.90–1.10	
9Cr–1Mo–Nb– V	0.08–1.12	8.00– 9.50	0.85–1.05	

TABLE 2.4
Common Specifications based on Application

Type	Forging	Pipe	Plate
3Cr–1Mo	A182 F21	A335 P21	A387 Gr 21
	A336 F21	A369 FP21	
5Cr–1/2Mo	A182 F5	A335 P5	A387 Gr 5
	A336 F5	A369 FP5	
5 Cr –½oSi		A335 P5b	
5 Cr – ½ Mo Ti		A335 P5c	
7 Cr–1/2 Mo	A182 F7	A335 P7 A369 FP7	A387 Gr 7
9 Cr–1Mo	A 182 F9	A335P9 A369 FP9	A387 Gr 9
9 Cr–1 MoVNbN	A182 F91	A335 P91 A369 FP91	A387 Gr 91

TABLE 2.5
Typical Gas Contents of Fusion Welds in Steels (ppm)

	H	N	O
SMAW basic	3–9	120	150
SMAW rutile	30–45		
Cellulose	50–65		
SAW	3–13	60	200–700
Self-shielded	3–9	50/170*	130

*Combined hydrogen

TABLE 2.6
Code Relevant for Process Plant

Pressure vessels	ASME Sec VIII
Heat exchangers	TEMA
Structures	AWS D1.1
Tankage	API 650, API 620

TABLE 2.7
Oxidation Limit of Steel

Material	ASME spec.	Temp limit (C)
• Carbon steel	SA 178, SA 210, SA 192 454	
• Carbon + 1/2 Mo	SA 209 T1	480
• 1 ¼ Cr–½ Mo	SA 213 T 11	550
• 2 ¼ Cr–1 Mo steel	SA 213 T22	580
• 25 Cr–20 Ni	1050	

TABLE 2.8
Minimum Preheat Temperatures with Low Hydrogen-Covered Electrodes

Type	Forging	Pipe	Plate
	13 mm	13–25 mm	25 mm
Steel		Temperature in Deg C	
½ Cr–1/2Mo	38	93	149
1 Cr–1/2Mo	121	149	149
1 ¼ Cr–½ Mo	121	149	149
2 Cr–½ Mo	121	149	149
2 ¼ Cr–1 Mo	121	149	149
3 Cr–1 Mo	149	177	177
5Cr–½ Mo	149	177	177
7Cr– Mo	149	177	177
9Cr–1Mo	149	177	177
9Cr–1Mo–V–Nb–N	177	204	204

Annexure 2.3

Details about Spec.

DeSteel HP 9–4–20 is a high-strength steel alloy with a high concentration of cobalt and nickel. It is known to have exceptional fracture toughness which makes it an excellent choice as filler metal for joints of exceptional quality in the aircraft industry, as well as for aircraft structure components such as rocket motor casings, armour plate, pressure vessels, and gears. More precisely, Steel HP 9–4–20 was designed explicitly for the B1-B2 Bomber programme. This alloy is also known to have excellent weldability and hardenability.

HP 9–4–20 has a yield strength in the region of 200–205 ksi but can be brought to 235 ksi with some treatments. That said, if yield strength in the region of 235 ksi is achieved, the fracture toughness falls off quickly. Other desirable properties of HP 9–4–20 are through hardening, Charpy Impact Toughness of 84 ft-lbs, and elongation values of 14–18%. Tech Steel and Materials offers Steel HP 9–4–20 in three sub-type specifications and multiple shapes/forms:

- Thermal treatment AMS 6468 (Wire) AMS 6523 (Plate, Sheet, and Strip)

HY 130 steel: This quenched and tempered low-carbon alloy steel finds uses as pressure vessels, heavy construction equipment, and in large steel structures. It has high tensile strength (yield and ultimate), good ductility, notch toughness, atmospheric corrosion resistance, and weldability.

Tensile strength, yield	>= 896 MPa	>= 130,000 psi
Elongation at break	14–15%	14–15%
Modulus of elasticity	205 GPa	29,700 ksi
Poissons ratio	0.28	0.28
Shear modulus	80.0 GPa	11,600 ksi

Element	Composition	Element	Composition
Carbon, C	<= 0.12%	Nickel, Ni	4.75–5.25%
Chromium Cr	0.40–0.70%	Phosphorus, P	<= 0.010%
Iron, Fe	91.9–93.8%	Silicon, Si	0.15–0.35%
Manganese Mn	0.60–0.90%	Sulphur, S	<= 0.015%
Molybdenum Mo	0.30–0.65%	Vanadium, V	0.050–0.10%

AMS 6525 (Bar, Forging, Ring, and Tubing)

REFERENCES

1. Satyendra, *Weldability of Steels*, Handbook, ISPAT Guru, 2016.
2. George E. Dieter, *Mechanical Metallurgy*, McGraw Hill Book Company, 2017 (p. 227).
3. IIW, *Basics of Welding Technology*, Welding edn., Subcommittee, Handbook, IIW, Tiruchi, 2000.
4. Robert W. Messler, *Principles of Welding*, Wiley-VCH, Verlag GmbH & Co., KGaA, 2008 (pp. 603–609).

3 Weldability Studies of Aluminide-Coated 9Cr Steels

Arunsinh B. Zala, Vishvesh J. Badheka, and Nirav I. Jamnapara

CONTENTS

3.1 INTRODUCTION

The energy demand is growing exponentially with an annual consumption of near to 778 Etta Joule by end of 2035 [1]. To cater to this demand, collective efforts from different sectors such as coal, oil, gas, and nuclear industries are essential. The refiners of conventional energy sectors are putting efforts to increase the efficiency and, consequently, the profit at same time. Gasification process which produces significantly lower quantities of air pollutants and uses less water than traditional coal-based power plants has gained more attention in recent years. This procedure has economic advantages in addition to environmental benefits. It can successfully compete to supply both electricity and goods in high-priced energy situations. In a similar vein, nuclear fusion will be essential for the creation of clean energy.

For instance, the development of hot-gas filtration in conventional energy production is attracting more attention in a wide variety of processes over the last few years. This filtration process produces significantly lower quantities of air pollutants

DOI: 10.1201/9781003352419-3

and also uses less water than the traditional coal-based power plants. However, the ceramic filters used for hot gas filtration process inhibit the production due to low ductility at high temperature. Low ductility of ceramic filters promotes cracking during the process, and therefore, porous iron aluminide (FeAl) has shown promising results for the hot gas filtration process compared to the existing ceramic filters. The ongoing research for fabrication sequence (welding techniques) for iron aluminide will pave the way to establish aluminide filters for gasification process.

Additionally, the research and development related to blanket modules is widely going on to establish the fusion power plants commercially. Engineering advancements are required to validate the existing blanket modules. One such fine example of advancement is aluminide coating process which has been optimized microstructurally for Pb–Li cooling blanket structures for nuclear fusion [2–5]. Although, the fabrication sequence of aluminide-coated 9Cr–1Mo steels is scarcely reported. Conventional fabrication processes would be a very important aspect in the development of new coating processes which can be implemented for the structural components. In the fusion blanket module, there is an uncertainty like whether the module will be fabricated first and then coated or it will be coated first and then fabricated. It seems that manufacturing followed by coating may not be a smart alternative when it comes to aluminide coatings for blanket applications since diffusion heat treatment after the coating may cause distortions.

The weldability of coated steels is questionable in the second method, which involves coating followed by manufacturing. Aluminide-coated 9Cr–1Mo steel's weldability is also seldom ever recorded. Iron aluminides (FeAl) and the top coating are likely to have an impact on the weld metallurgy, which will then deteriorate the mechanical properties. The effects of coating on the weld microstructure will be also an important aspect as far as weldability is concerned. Some of the critical issues are listed in Figure 3.1, and all are elucidated further with respect to the weldability of coated steels.

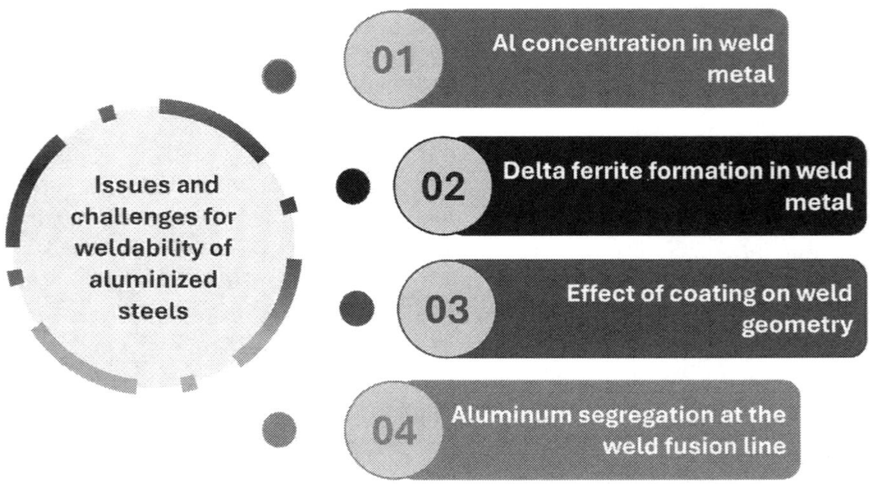

FIGURE 3.1 Issues and challenges for weldability of aluminized steels.

3.2 ISSUES AND CHALLENGES IN WELDABILITY

3.2.1 Al Concentration in the Weld

The aluminium concentration in the substrate or weld zone is very crucial as it is susceptible to cracking. The cracking susceptibility as a function of Al concentration has been investigated for FeCrAl alloys by Field et al. [5]. He reported that during conventional arc welding processes of FeCrAl alloys, the phenomenon of cracking occurs when the aluminium concentration in the weld metal is greater than 8–11% approximately. Hence, aluminium concentration plays a pivotal role in the weldability of aluminized coated steel.

Iron aluminides, which exist in the coating as a diffusion layer and are extremely sensitive to variables related to welding speed and alloy concentration [6], when welded, are brittle and prone to cracking [7, 8]. These intermetalics are difficult to weld and fracture while performing the welding, as they are ductile in nature and have high spatial values as reported by Heinz et al. [9]. Hydrogen embrittlement is one hypothesis for this phenomenon that may exist. Due of their vulnerability to environmental embrittlement, iron aluminides have been found to exhibit low room temperature ductility [10, 11]. The studies of Zacharia et al. reported weld cracking at the centreline due to the presence of Aluminium [12]. The detailed study and scientific explanation have been reported by Regina et al. [13].

3.2.2 Delta Ferrite Formation

Another major concern is the delta ferrite formation in the weld metal during the welding of 9Cr steels. One of the several sub-liquid phases of 9–12% Cr steels depend on the temperature. During the process, the liquid metal solidifies as delta ferrite and transforms into austenite crystals. This process starts at around 1300°C and ends at 1200°C under equilibrium condition. Due to the high cooling rates, this equilibrium transition is not feasible throughout the weld cooling cycle. This means that it is feasible for some delta ferrite to be maintained in the weld metal even at room temperature and to some extent in the heat-affected zone [14]. Figure 3.2 indicates the presence of delta ferrite formation for 9Cr steels.

It has been discovered that the presence of delta ferrite has negative impacts on the toughness of the base metal and welds [15, 16]. Since more delta ferrite in the matrix results in a sharp rise in the ductile brittle transition temperature (DBTT), using delta ferrite in the weld is not recommended since it reduces the notch toughness and creep ductility at high temperatures [17].

Delta ferrite formation has been reported for the weldments of coated steel too. Bertin et al. [18, 19] noticed that after the arc welding of aluminium-coated steel, a phase consisting of delta ferrite formed in the welded joint and reduced the tensile strength. Literature suggests that the concentration of aluminium is also a major concern, as it supports the formation of delta ferrite. Aluminium (>2%) was found in the weld pool, according to Briand et al. [20], which causes delta ferrite to develop and lowers the hardness in the weld zone. In 9–12% Cr metals and welds, delta

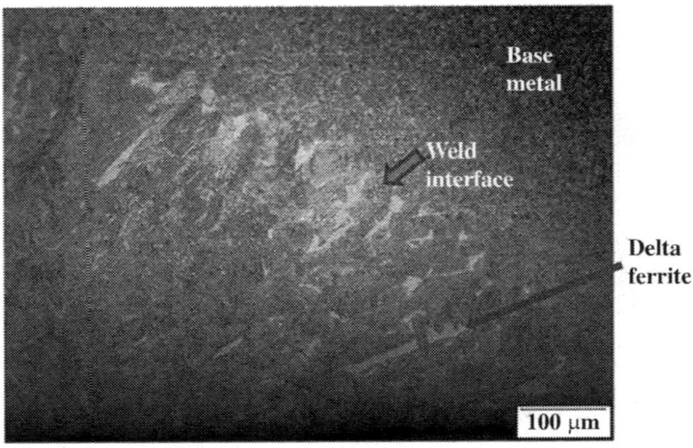

FIGURE 3.2 Delta ferrite formation at the weld interface of 9Cr steel [14].

ferrite development may increase solidification cracking, impair toughness due to the delta ferrite phase's notch sensitivity, reduce creep ductility at high temperatures, and other effects [14]. According to Wang et al. and Anderko et al., delta ferrite impairs the impact characteristics and increases the DBTT of martensitic stainless steel [15, 21].

3.2.3 Effect of Coating on Weld Geometry

The weld geometry includes the depth of penetration (DOP) and weld bead width that are crucial aspects of weldability. The DOP and weld bead width are functions of weld current and travel speed. However, the coating at the top might influence the weld geometry as reported by many researchers. Author Park et al. [22] explained the effect of coatings on weld bead for ferrite stainless steels. They demonstrated the effect of coating with respect to different heat inputs for coated steels and uncoated steels. In their study, the welds were prepared at different weld speeds with a constant current of 90A. Figure 3.3–1a, 1b, 1c, 1d, and 1e shows the cross sections of weld bead at different weld speeds, (a) 0.32 m/min,(b) 0.42 m/min, (c) 0.52 m/min, (d) 0.60 m/min, (e) 0.64 m/min for uncoated steel. On the contrary, Figure 3.3–2a, 2b, 2c, 2d, and 2e shows the depth of penetration and width of the weld bead are different at the same heat input (weld current and weld speed) and are higher and more constricted.

The study concludes that due to the Al-Si coating at the top, different mechanisms such as reverse marangoni conviction are responsible for the higher depths of penetration in the case of coated steels. Similar observations have been reported elsewhere [23, 22] in which authors have investigated the effect of coating on the depth of penetration for Al-10%Si-coated ferritic stainless steel with laser beam welding process.

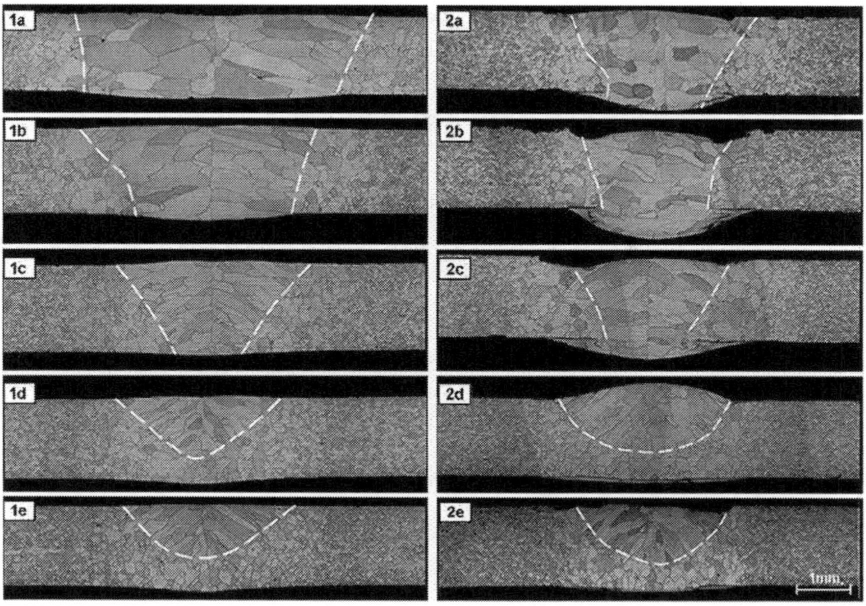

FIGURE 3.3 Cross sections of the bead of the ferritic stainless steel 409L (1) and Al-coated 409L (2) joints welded by various welding speeds at welding current of 90 A: (a) 0.32 m/min, (b) 0.42 m/min, (c) 0.52 m/min, (d) 0.60 m/min, (e) 0.64 m/min [22].

3.3 WELDING PROCESSES FOR 9CR STEELS

Numerous welding methods, such as manual metal arc welding, gas tungsten arc welding (GTAW) or TIG welding, submerged arc welding, and others have been reported for 9Cr–1Mo [24–26]. Among these, TIG welding technique was selected for the present study, as it is the most widely used technique to weld materials such as steels, aluminium, and copper alloys. This welding technique delivers high-quality welds with high purity. In TIG welding, a non-consumable tungsten electrode is connected to the workpiece (base metal) to establish an arc that causes the coalescence of two or more metals. Non-consumable electrodes are normally constructed of pure tungsten or a tungsten alloy and available in a variety of sizes and lengths. TIG welding process has been adopted as the standard welding practice for high-quality engineering jobs related to precision fabrication jobs, including space components, nuclear fission and fusion power plants, and so on. As a result of its weld quality and widespread applicability, the welding experiments were conducted with TIG welding process. For these welding trials, a TIG power supply made of Panasonic was made available at Pandit Deendayal Energy University, Gandhinagar [27, 28]. Welding experiments with various types of TIG welding processes were carried out, i.e., (a) autogenous TIG welding and (b) autogenous activated TIG welding. The welding technique with the most acceptable weld profile was chosen for further testing and assessment.

3.3.1 Autogenous TIG Welding Process

TIG welding can be carried out manually, automatically, or utilizing a hybrid technique if the relevant knowledge and/or tools are used. In-depth welding tests were conducted for this investigation utilizing an autogenous TIG machine. Using bead-on-plate experiments, the weld parameters for coated and bare 9Cr–1Mo samples were optimized. The purpose of the bead-on-plate testing was to measure the depth and width of the weld metal and to assess how coatings affected the microstructure of the weld. The process factors for bead-on-plate testing on coated and uncoated P91 steels are listed in Table 3.1.

3.3.2 Autogenous A-TIG Welding Process

The efficiency and productivity of TIG welding have increased due to several developments. Because the processes claimed maximum weld penetration is between 3.0 mm and 3.5 mm, it cannot be used to weld plates with a thickness of 6 mm or more in a single pass or the autogenous mode [29]. The inclusion of filler metal or double-sided welds was the only option for welding greater thicknesses. However, each of these techniques increased the time needed for welding, increased the costs, and required design considerations, which decreased productivity. The development of the activated TIG (A-TIG) welding method at the Paton Institute of Electric Welding in Ukraine in 1960 provided a partial solution to the problem [30, 31]. A-TIG welding is a variation of the TIG welding method that uses a single chemical powder or a combination of chemical powders (referred to as flux) in the TIG welding process. By adding solvents (such acetone, methanol, and so on), the flux is transformed into

TABLE 3.1
TIG Welding Parameters for Coated P91 and Bare P91 Steel Samples

Welding current	200 A
Creator current	160 A
Travel speed	100, 125, 150 mm/min
Electrode type	Tungsten (2% thoriated)
Electrode diameter	2.9 mm
Electrode angle	18–20° (blunt ground at tip)
Arc gap	2–3 mm
Gas flow rate	10–12 l/min
Welding position	1 G (flat)
Electrode extension	5–6 mm
Nozzle diameter	8 mm

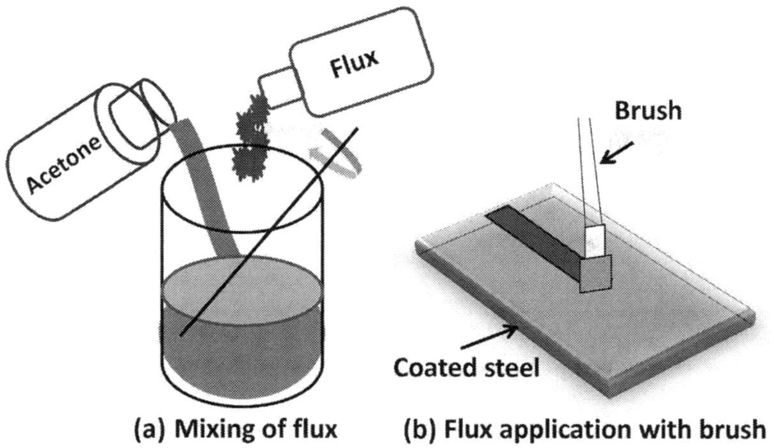

FIGURE 3.4 Schematic of A-TIG welding technique (a) mixing of fluxes with acetone, (b) application of flux-in-paste form with brush.

paste form and applied to the plate's top surface [27, 32]. Then, using the standard A-TIG welding, the flux is applied on top of the sample between the welding arc and the plate. Researchers have observed that by using A-TIG process, the penetration has increased by roughly 300% [33]. The schematic of A-TIG process is explained in Figure 3.4(a) [34]. It contains different steps.

As indicated in the figure, mixing of the flux with a solvent such as acetone/methanol is done before its application on the coated plate. The solvent acts as a carrier when the flux is applied with the help of a brush. Once the acetone evaporates, the TIG welding has been performed. Figure 3.4(b) shows a schematic of A-TIG process after the application of flux on the weld coupon. Once the welding was done, the cross sections of samples were investigated to examine weld metal microstructure as well as the weld's bead width and depth.

3.4 EVALUATION OF WELDED MICROSTRUCTURES

3.4.1 Effect of Coating on Weld Bead

The kinetic energy that is applied to the charged species during TIG welding is used to melt the weld metal (electrons or positively charged ions). The coated and bare P91 samples act as the parent metal. The tungsten electrode was connected as cathode and the parent metal was connected as anode. The anode connection was grounded. Usually, when a welding arc strikes between the tungsten electrode and the metal plate, it spreads in a limited exposed area owing to free electrons in metals showing a conducting path. In the case of aluminized P91 samples, a thin insulating layer of alumina is present at the top surface exposed to the electrode. As a result, unlike in a bare material or uncoated steel, the electron does not have a free path to expand when the arc contacts with the top alumina layer. Consequently, the arc restricts, and the current density rises. Because more energy is deposited in a comparatively smaller area, the DOP improves, and the bead width of the weld narrows in comparison to that generated in the naked P91 steel bead on plate [35]. The arc constriction mechanism is depicted schematically in Figure 3.5(a). As shown in Figure 3.5(b) and (c), with the same heat input, the arc bead (9 mm) in coated P91 steels is narrower than the weld bead (11.5 mm) of naked grade 91 steel. Insulation effect is the name given to the phenomenon of increased penetration depth brought on by coating and was first described by Vilarinho et al. [36]. A similar mechanism is also reported in Refs. [37, 38], where an enhanced DOP was achieved for P91 steels. For coated steels, similar observations on higher DOP were reported by Park et al. [22] using Al–8Si%-based aluminized-based coatings on ferritic stainless steels. The effect of alumina layer can be clearly indentified from the visual inspection of the sample surface.

Esme et al. [39] conducted an analysis of variance (ANOVA) study on the most significant parameter affecting weld bead in the TIG welding process and concluded that the welding speed was the most critical parameter. Based on this, the weld speed was varied, i.e., 100 mm/min, 125 mm/min, 150 mm/min, and the weld current was kept constant at 200 A (see Table 3.1). Samples having weld bead for coated P91 steels were labelled as 1-a,b,c, and the bare P91 steel samples were labelled as 2-a,b,c. It

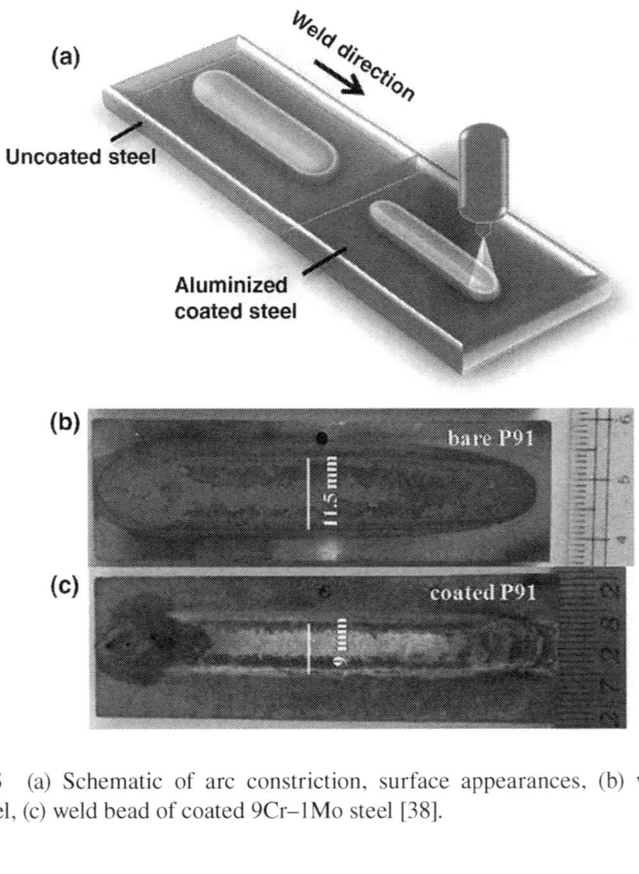

FIGURE 3.5 (a) Schematic of arc constriction, surface appearances, (b) weld bead of 9Cr-1Mo steel, (c) weld bead of coated 9Cr–1Mo steel [38].

should be noted that the welding current (200A) and weld speed (100 mm/min) were kept the same for both coated P91 steel (Sample 1a and 2a) and the uncoated (bare) P91 steel. Similarly, 1b and 2b are welded at 125 mm/min weld speed and 1c and 2c are welded at 125 mm/min.

Metallographic specimens were prepared in accordance with ASTM E3–01 standard by polishing followed by sequential immersions in Villella's solution. The macro and microstructural studies were conducted using an optical microscope (OM) to evaluate the weld bead shape and DOP.

The weld bead macrostructures for coated and bare samples are shown in Figure 3.6, while the information of weld bead geometry (size and shape) are shown in Figure 3.7. Weld beads for coated P91 steels are samples with nomenclatures 1a, 1b, and 1c, whereas bare steels are samples with nomenclatures 2a, 2b, and 2c. When compared to sample 2a, which is uncoated, sample 1a (the coated ones) had deeper penetration. At the same current, 200A, and weld speed, 100 mm/min, both weld beads were created. Other samples which were generated varying the heat input also showed a similar pattern.

It is evident that coatings have a significant impact on arc constriction, which leads to increased DOP and a favourable D/W ratio for coated P91 steel. Samples 1a and 2a were taken into consideration for further characterization since the necessary DOP was seen in each of these samples.

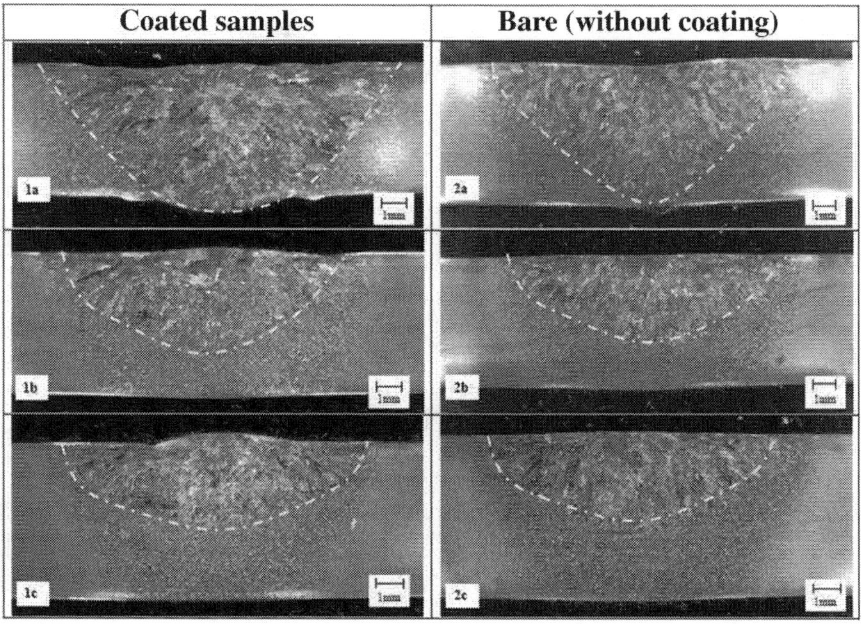

FIGURE 3.6 Cross sections of the bead of the Aluminized-coated 9Cr–1Mo Steel (1) and bare 9Cr–1Mo (2) joints welded by various welding speeds at welding current of 200 A: (1a and 2a) 100 mm/min, (1b and 2b) 125 mm/min, (1c and 2c) 150 mm/min [38].

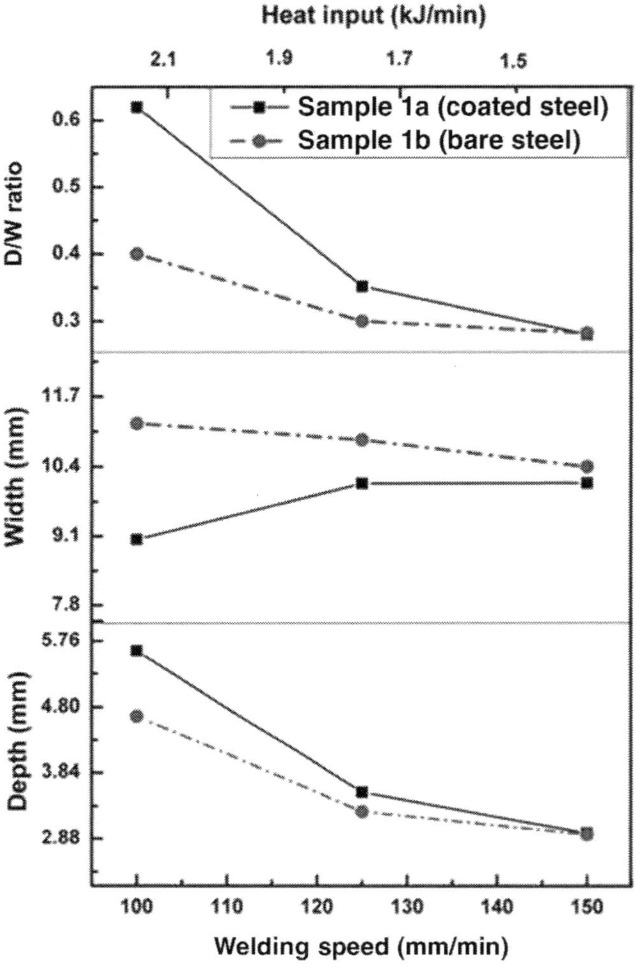

FIGURE 3.7 Variation of the penetration depth (D) and width of the bead (W) and the D/W ratio with heat input for samples 1a and 2a [38].

3.4.2 AlN Inclusions in Weld Microstructures

The cross-sectional examination of coated samples before welding revealed the presence of AlN formation at the interface of FeAl and the substrate. This steel had a nitrogen concentration of 0.07 weight percent (max.). Al diffuses into the base metal during the diffusion coating process, where it combines with the nitrogen at the interface to create AlN [40, 41]. These precipitates were mostly formed during the coating procedure, and when welding took place, they were transmitted into the weld bead. Due to the AlN precipitates' high melting point, they did not dissolve and regenerate during welding and cooling [42]. The presence of the AlN precipitates was confirmed using EDX mapping, as shown in Figure 3.8(a), where the red colour indicates the

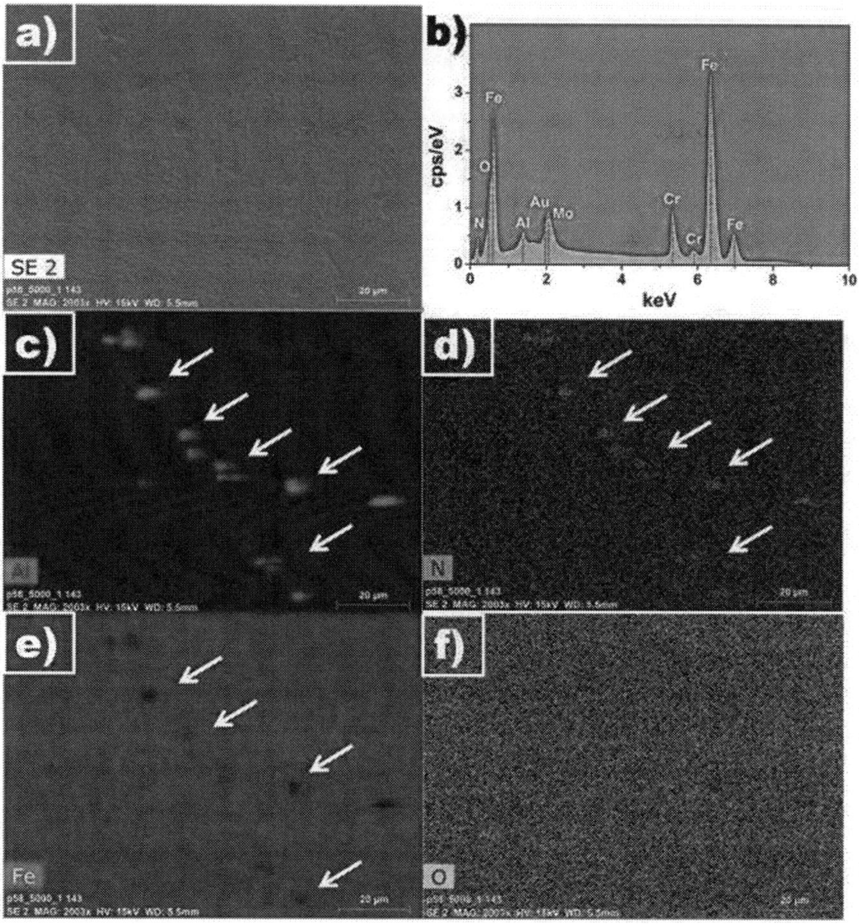

FIGURE 3.8 EDX mapping of AlN precipitates in the weld zone for sample 1a (coated P91 steel). (a) SE image of the weld zone used, (b) EDX spectra of the mapping data, (c) mapping of Al (aluminium) in red colour, (d) mapping of N (Nitrogen), (e) mapping of Fe, (f) mapping of O [38].

presence of Al [denoted with an arrow in 3.8(c)], the shade of green indicates the presence of nitrogen [3.8(d)], the blue shade represents the iron [3.8(e), and the last is oxygen (3.8(f)]. The EDX spectrum of the AlN inclusion is shown in Figure 3.8(b), and the Al:N ratio confirms the composition of AlN.

EDS analysis was used to do elemental depth profiling to look into the Al content in the coating region and weld metal from the surface to the core. On the parent metal (or base metal) side and the weld metal side, the profile of various elements is depicted together with the distance from the surface to the core. The cross section of sample 1a, coated 9Cr–1Mo steel, is shown in Figure 3.9(a) at the parent metal (also referred to as base metal) side, and the elemental depth profile of Al and

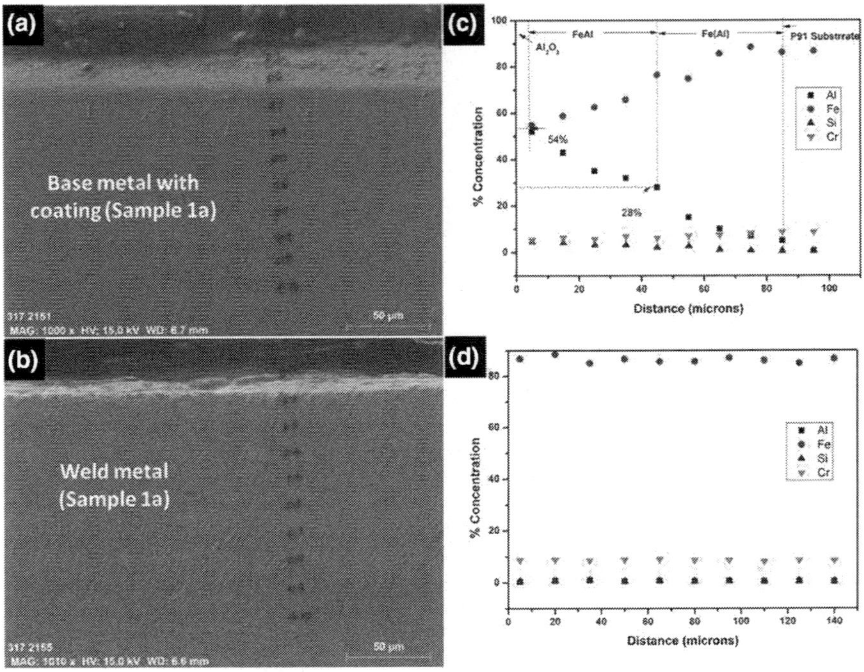

FIGURE 3.9 Elemental depth profiling of cross-section bead of the sample 1a (coated 9Cr–1Mo) (a) at base metal side (b) at weld zone, plot of elemental depth profile (c) at the baseside (d) at the weld zone [38].

other elements, as determined by EDS, is shown in Figure 3.9(c). The oxide layer (Al_2O_3) is visible on the coated sample's base metal side up to a thickness of around ~3–5 microns, followed by an 80–100 micron-thick Fe–Al diffusion zone. The Fe–Al phase diagram states that the existence of FeAl and Fe(Al) is indicated by the gradient in the concentration of Al. On the other hand, aluminium in the weld zone [Figure 3.9(b) and (d)] was found to be nearly the same at different places in the weld zone, indicating that Al is uniformly distributed throughout the weld metal region. Additionally, it confirms to the absence of several phases, including FeAl and Al_2O_3, which were present in the coating before welding. The range of 1–2 wt% was identified for the Al content in the weld zone. Additionally, area EDX was used to evaluate the Al content across the weld, and the same concentration (1–2 wt%) was verified.

The reduction in grain size appears to be significantly influenced by the presence of Al in the weld metal. According to Hong et al. [43], the reduction in grain size was caused by the weld metal's increased Al concentration. This was caused by the fact that greater Al content steels displayed lower "n" values, which impact the grain size as a function of "t" (annealing time). The n values in grains with impurities are often substantially lower than the theoretical estimates ($n = 0.5$) of a system with no defects, according to Bruke et al. [44]. The value of "n" (0.5) would decrease as the number of defects increases [45]. According to some publications [46–48], the

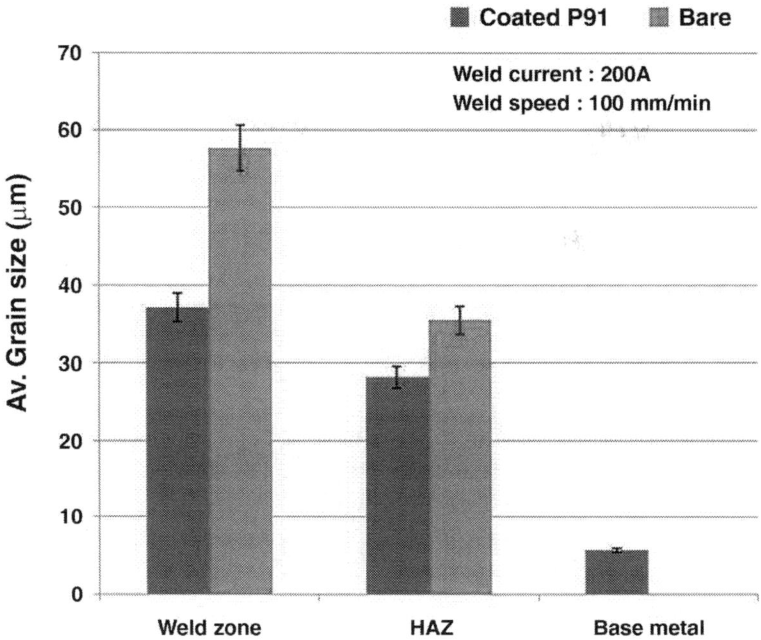

FIGURE 3.10 Effect of Al concentration on the grain size in sample 1a and 2a [38].

presence of AlN precipitates in the weld pool may also have an impact on grain size due to the action of pinning grain boundaries. This corroborates our theory on the impact of aluminium on the weld pool grain size. The precise process of Al dispersion in the weld metal and its impact on the mechanical and microstructural characteristics are the subject of a thorough enquiry. Both samples 1a and 1b's average grain sizes were determined using the ASTM E112–10 standard. Due to the presence of Al in the weld bead, the average grain size in the weld zone and HAZ of coated samples was less than that of the uncoated P91, as shown by the same plot in Figure 3.10.

3.4.3 Delta Ferrite Formation in Weld Metal

The microscopy was carried out in the same state as when it was received, and the microstructures of the weld metal for coated and uncoated welds were then compared. The P91 material used in the experiment was provided in tempered and normalized condition. The tempering process took place at 780°C with a holding duration of 1 hour, and the normalizing treatment took place at 1050°C for 0.5 hours. The tempered martensite of P91 steel in its as-received state after both heat treatments is shown in Figure 3.11. Figure 3.12 depicts the microstructure of the weld metal for uncoated steel. Weld metal was found to be absent of delta ferrite and to consist of entirely martensitic laths and prior austenite grain boundaries.

Figure 3.13 displays the optical and SEM microstructure of the coated-steel weld metal generated by autogenous TIG welding, where heat input was 2.12 kJ/mm. The

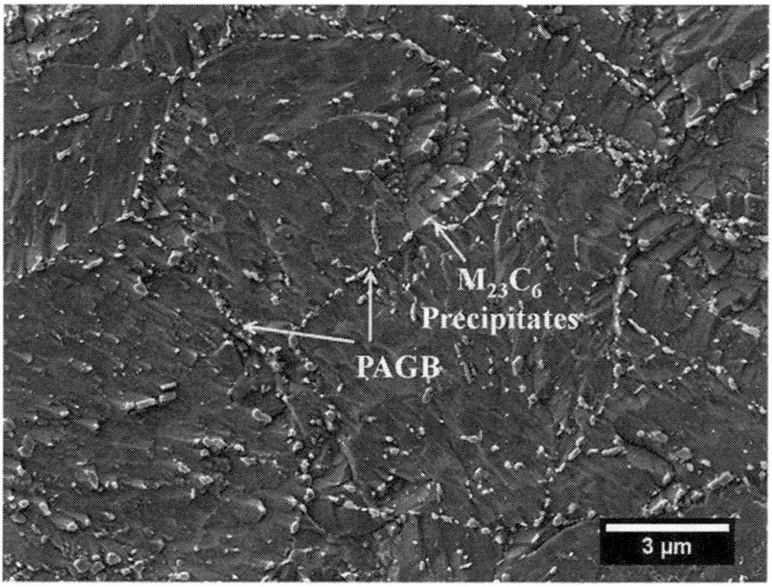

FIGURE 3.11 SE image of P91 steel's microstructure in as-received condition [49].

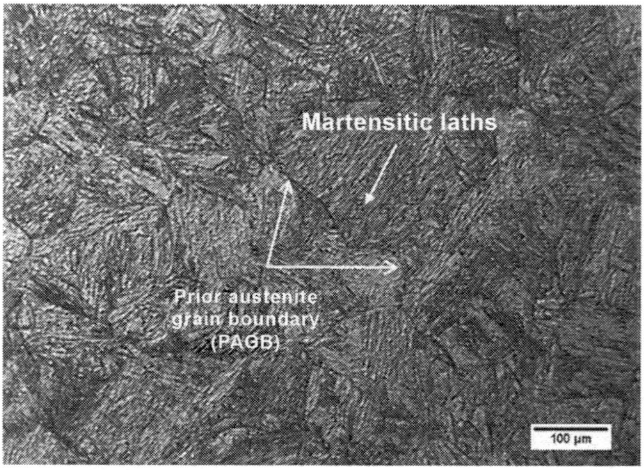

FIGURE 3.12 Weld metal's microstructure of bare P91 steel prepared at 100 mm/min weld speed and 200 A current [49].

shape of the weld metal suggests that delta ferrite is present. A higher magnification picture using electron microscopy [Figure 3.13(b)] was acquired at the location indicated by the circle in Figure 3.13(a) to confirm the same. According to Arivazhagan et al., the delta ferrite is carbide-free [50]. In the delta ferrite area of the SE picture [Figure 13(b)], carbides are completely absent. Additionally, the delta ferrite that was

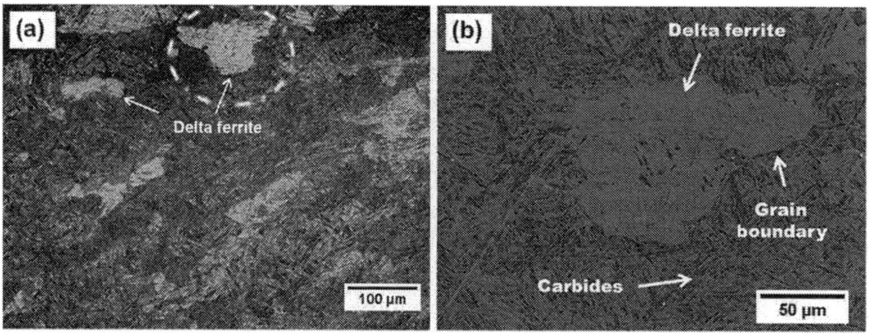

FIGURE 3.13 Weld metal's microstructure of coated P91 prepared at 100 mm/min weld speed and 200 A current. The encircled part in (a) (optical image) is highlighted as (b) (SE image) [49].

found here had a polygonal form. There have been reports in references of polygonal delta ferrite and a delta ferrite island inside a tempered martensite matrix [14, 51]. In Figures 3.13(a) and 3.13(b), where the island of delta ferrite in the martensite matrix structure has been clearly detected, delta ferrite was seen to have a similar type of polygonal form. While the core of the grain exhibited less delta ferrite phase, the delta ferrite areas were more prevalent near grain borders. The fluctuation in the amount of delta ferrite may be caused by the grain border having greater energy than the grain inner area [50]. Using image J software, the volume fraction of delta ferrite in the microstructure of the weld zone was analysed to determine the amount of delta ferrite present and 5.09% of delta ferrite was detected in the weld zone [49].

3.5 CONCLUSIONS

The aim of this chapter was to examine the weldability of P91 steels with an aluminide coating. The 9Cr steel with an aluminide coating has several applications, including in fusion reactor test blanket modules, turbine-driven systems, petrochemical facilities, and automobile manufacturing. However, investigations on the weldability of aluminide-coated structures are seldom ever documented, and a number of problems remain unresolved, resulting in lacunae. As shown from the literature study, issues including the impact of aluminium dissolving from coating into the weld, its impact on delta ferrite formation, the grain size of the weld metal, and inclusion dissolution are required to be addressed. Utilizing autogenous TIG welding, the impact of Al contamination on the weld's microstructural characteristics and weld profile was examined (bead-on-plate trials). The main conclusions of this chapter are summarized here:

Alumina (Al_2O_3) on the top of coated samples enhanced the depth of penetration (DOP) owing to arc constriction, according to bead-on-plate autogenous TIG welding experiments.

The presence of Al-rich coating in bead-on-plate experiments resulted in Al concentration of ~0.19% distributed across the weld zone. This Al content supports the formation of delta ferrite (δ-Fe) in the weld pool, which ultimately deteriorates the mechanical properties of the weld joint. Results with bead-on-plate trials show that the δ-Fe had an average volume fraction of ~5.09% in the weld metal against a maximum permissible value of ~15–17% [52]. The δ-Fe showed a microhardness of 192–198HV$_{0.05}$. This microhardness is approximately half of the reported microhardness for martensitic laths (396–410 HV$_{0.05}$). The Al content in the weld metal was also observed to reduce the grain size. The prior austenite grain size in Al-contaminated weld was found to be ~58 μm, whereas that of uncoated weld metal was ~38 μm. Al has been reported to be a grain reformer leading to finer grain sizes [53]; results of this study are in accordance with the earlier observation.

REFERENCES

1. Rafiee, Ahmad, and Kaveh Rajab Khalilpour. "Renewable hybridization of oil and gas supply chains." In *Polygeneration with polystorage for chemical and energy hubs*, pp. 331–372. Academic Press, 2019.
2. Jamnapara, Nirav I., A. Sarada Sree, E. Rajendra Kumar, S. Mukherjee, and A. S. Khanna. "Compatibility study of plasma grown alumina coating with Pb-17Li under static conditions." *Journal of Nuclear Materials* 455, no. 1–3 (2014): 612–617.
3. Zala, A. B., N. I. Jamnapara, C. S. Sasmal, P. Chaudhuri, and M. Ranjan. "Investigation of alumina film formed over aluminized RAFM steel by plasma assisted heat treatment." *Fusion Engineering and Design* 146 (2019): 2002–2006.
4. Mukherjee, S., and N. I. Jamnapara. "Materials research and development opportunities in fusion reactors." *Proceedings of the Indian National Science Academy* 81, no. 4 (2015): 827–839.
5. Field, Kevin G., Maxim N. Gussev, Yukinori Yamamoto, Richard H. Howard, and Samuel A. Briggs. *Second annual progress report on radiation tolerance of controlled fusion welds in high temperature oxidation resistant FeCrAl alloys*. No. ORNL/TM-2016/770. High Flux Isotope Reactor (HFIR). Oak Ridge National Lab (ORNL), Oak Ridge, TN, 2016.
6. David, S. A., J. A. Horton, C. G. McKamey, T. Zacharia, and R. W. Reed. "Welding of iron aluminides." *Welding Journal* 68, no. 9 (1989): 372–381.
7. Aleshin, N. P., S. A. Gerasimov, Li Yajiang, Ma Haijun, Yu A. Puchkov, and B. F. Yakushin. "Microstructure and composition of welds of Fe 3 Al intermetallic with austenitic corrosion-resistant steels." *Metal Science and Heat Treatment* 52 (2011): 437–441.
8. Stoloff, N. S., A. Castagna, J. Scott, and D. J. Duquette. "Environmental embrittlement of Fe 3 Al alloys under monotonic and cyclic loading." *International Journal of Fatigue* 6, no. 17 (1995): 448–449.
9. Neumann, Heinz, Jaromír Moravec, and J. Bradac. "TIG welding process effect on the iron aluminide Fe3Al." *Metal* 5 (2010): 18–20.
10. Liu, C. T., E. H. Lee, and C. G. McKamey. "An environmental effect as the major cause for room-temperature embrittlement in FeAl." *Scripta Metallurgica* 23, no. 6 (1989): 875–880.
11. Liu, C. T., C. G. McKamey, and E. H. Lee. "Environmental effects on room-temperature ductility and fracture in Fe3Al." *Scripta Metallurgica et Materialia* 24, no. 2 (1990): 385–389.
12. David, S. A., and T. Zacharia. "Weldability of Fe~ 3Al-type aluminide." *Welding Journal, New York* 72 (1993): 201.
13. Regina, J. R., J. N. Dupont, and A. R. Marder. "The effect of chromium on the weldability and microstructure of Fe-Cr-Al weld cladding." *Welding Journal, New York* 86, no. 6 (2007): 170.

14. Onoro, J. "Martensite microstructure of 9–12% Cr steels weld metals." *Journal of Materials Processing Technology* 180, no. 1–3 (2006): 137–142.
15. Anderko, K., L. Schäfer, and E. Materna-Morris. "Effect of the δ-ferrite phase on the impact properties of martensitic chromium steels." *Journal of Nuclear Materials* 179 (1991): 492–495.
16. Sam, Shiju, C. R. Das, V. Ramasubbu, S. K. Albert, A. K. Bhaduri, T. Jayakumar, and E. Rajendra Kumar. "Delta ferrite in the weld metal of reduced activation ferritic martensitic steel." *Journal of Nuclear Materials* 455, no. 1–3 (2014): 343–348.
17. Abe, F., T. Noda, H. Araki, and S. Nakazawa. "Alloy composition selection for improving strength and toughness of reduced activation 9Cr-W steels." *Journal of Nuclear Materials* 179 (1991): 663–666.
18. Bertin, Pascal, Olivier Dubet, Sebastien Gadrey, and Frederic Richard. "Method for arc-welding aluminum-coated metal parts using an inert gas containing nitrogen." U.S. Patent Application 13/809,489, filed May 2, 2013.
19. Bertin, Pascal, Olivier Dubet, Sebastien Gadrey, and Frederic Richard. "Method for arc-welding aluminum-coated metal parts using oxidizing gas." U.S. Patent Application 13/809,531, filed May 9, 2013.
20. Briand, Francis, and Olivier Dubet. "Hybrid arc/laser-welding method for aluminized steel part using a gas including nitrogen and/or oxygen." U.S. Patent Application 13/809,552, filed May 2, 2013.
21. Wang, P., S. P. Lu, N. M. Xiao, D. Z. Li, and Y. Y. Li. "Effect of delta ferrite on impact properties of low carbon 13Cr-4Ni martensitic stainless steel." *Materials Science and Engineering: A* 527, no. 13–14 (2010): 3210–3216.
22. Park, Tae Jun, Jong Pan Kong, Sang Ho Uhm, In Su Woo, Jong Sub Lee, and Chung Yun Kang. "Effect of Al-Si coating layer on the penetration and microstructures of ferritic stainless steel, 409L GTA welds." *Journal of Materials Processing Technology* 211, no. 3 (2011): 415–423.
23. Kong, Jong Pan, Tae Jun Park, Jeong Kil Kim, Sang Ho Uhm, In Su Woo, Jong Sub Lee, Bong Gyu Park, and Chung Yun Kang. "Characterization of laser welds in Al-10 wt.% Si coated ferritic stainless steel." *Materials & Design* 32, no. 2 (2011): 917–925.
24. Pandey, Chandan, Manas Mohan Mahapatra, Pradeep Kumar, and N. Saini. "Comparative study of autogenous tungsten inert gas welding and tungsten arc welding with filler wire for dissimilar P91 and P92 steel weld joint." *Materials Science and Engineering: A* 712 (2018): 720–737.
25. Pandey, C., A. Giri, M. M. Mahapatra, and Pradeep Kumar. "Characterization of microstructure of HAZs in as-welded and service condition of P91 pipe weldments." *Metals and Materials International* 23 (2017): 148–162.
26. Pandey, Chandan, Anoj Giri, and M. M. Mahapatra. "Evolution of phases in P91 steel in various heat treatment conditions and their effect on microstructure stability and mechanical properties." *Materials Science and Engineering: A* 664 (2016): 58–74.
27. Dhandha, Kamal H., and Vishvesh J. Badheka. "Effect of activating fluxes on weld bead morphology of P91 steel bead-on-plate welds by flux assisted tungsten inert gas welding process." *Journal of Manufacturing Processes* 17 (2015): 48–57.
28. Nayee, Sanjay G., and Vishvesh J. Badheka. "Effect of oxide-based fluxes on mechanical and metallurgical properties of dissimilar activating flux assisted-tungsten inert gas welds." *Journal of Manufacturing Processes* 16, no. 1 (2014): 137–143.
29. Tseng, Kuang-Hung, and Chih-Yu Hsu. "Performance of activated TIG process in austenitic stainless steel welds." *Journal of Materials Processing Technology* 211, no. 3 (2011): 503–512.
30. Vora, Jay J., and Vishvesh J. Badheka. "Experimental investigation on microstructure and mechanical properties of activated TIG welded reduced activation ferritic/martensitic steel joints." *Journal of Manufacturing Processes* 25 (2017): 85–93.

31. Vora, Jay J., and Vishvesh J. Badheka. "Experimental investigation on mechanism and weld morphology of activated TIG welded bead-on-plate weldments of reduced activation ferritic/martensitic steel using oxide fluxes." *Journal of Manufacturing Processes* 20 (2015): 224–233.
32. Vora, Jay J., and Vishvesh J. Badheka. "Experimental investigation on effects of carrier solvent and oxide fluxes in activated TIG welding of reduced activation ferritic/martensitic steel." *International Journal of Advances in Mechanical & Automobile Engineering* 3, no. 1 (2016): 75.
33. Sun, Z., and D. Pan. "Welding of titanium alloys with activating flux." *Science and Technology of Welding and Joining* 9, no. 4 (2004): 337–344.
34. Zala, Arunsinh B., N. I. Jamnapara, Vishvesh J. Badheka, C. S. Sasmal, Shiju Sam, and Mukesh Ranjan. "Study of weldability for aluminide coated steels through A-TIG welding process." *Materials Performance and Characterization* 11, no. 2 (2022): 225–238.
35. Zala, A. B., N. I. Jamnapara, V. J. Badheka, S. Sam, and M. Ranjan. "Effect of aluminide coatings on penetration and microstructure of TIG welded 9Cr-1Mo steel for fusion blanket applications." *Fusion Engineering and Design* 144 (2019): 172–179.
36. Vilarinho, Louriel O., Sayee Raghunathan, and Bill Lucas. "Observation of insulation mechanism during FBTIG." In *Proceedings of COBEM 2009: 20th international congress of mechanical engineering (Gramado, RS, Brazil).* ABCM and TWI Ltd. 2009, November 15–20.
37. Jayakrishnan, S., and P. Chakravarthy. "Flux bounded tungsten inert gas welding for enhanced weld performance—a review." *Journal of Manufacturing Processes* 28 (2017): 116–130.
38. Singh, Akhilesh Kumar, Vidyut Dey, and Ram Naresh Rai. "A study to enhance the depth of penetration in grade P91 steel plate using alumina as flux in FBTIG welding." *Arabian Journal for Science and Engineering* 42 (2017): 4959–4970.
39. Greyjevo, O. G. T. V. Z., and Analizo in Taguchijevo Metodo. "Optimization of weld bead geometry in TIG welding process using grey relation analysis and Taguchi method." *Materiali in Tehnologije* 43, no. 3 (2009): 143–149.
40. Agüero, Alina, Raúl Muelas, Ana Pastor, and Steve Osgerby. "Long exposure steam oxidation testing and mechanical properties of slurry aluminide coatings for steam turbine components." *Surface and Coatings Technology* 200, no. 5–6 (2005): 1219–1224.
41. Huttunen-Saarivirta, E., M. Honkanen, S. A. Tsipas, H. Omar, and D. Tsipas. "Microstructure and elevated-temperature erosion-oxidation behaviour of aluminized 9Cr-1Mo steel." *Applied Surface Science* 259 (2012): 674–684.
42. Kalivodová, Jana, David Baxter, Michael Schütze, and Valentin Rohr. "Corrosion behaviour of boiler steels, coatings and welds in flue gas environments." In *Protective systems for high temperature applications EFC 57*, pp. 10–20. CRC Press, 2023.
43. Hong, Jaewan, Hyunseo Choi, Seil Lee, Jae Kwan Kim, and Yang mo Koo. "Effect of Al content on magnetic properties of Fe-Al Non-oriented electrical steel." *Journal of Magnetism and Magnetic Materials* 439 (2017): 343–348.
44. Burke, J. E., and D. Turnbull. *Progress in metal physics*, vol. 3. Pergamon Press, London, 1952.
45. Ghosh, Deepika. "Effects of grain size and nitrogen content on tensile flow properties of 316LN stainless steel at 300 K and 823 K." Master of Technology in Steel Technology Dissertation, Department of Metallurgical Engineering and Materials Science, Indian Institute of Technology, Bombay, 2016.
46. Pous-Romero, H., I. Lonardelli, D. Cogswell, and H. K. D. H. Bhadeshia. "Austenite grain growth in a nuclear pressure vessel steel." *Materials Science and Engineering: A* 567 (2013): 72–79.
47. Oda, Yoshihiko, Yasushi Tanaka, Atsushi Chino, and Katsumi Yamada. "The effects of sulfur on magnetic properties of non-oriented electrical steel sheets." *Journal of Magnetism and Magnetic Materials* 254 (2003): 361–363.

48. Fu, H. M., M.-X. Zhang, D. Qiu, P. M. Kelly, and J. A. Taylor. "Grain refinement by AlN particles in Mg-Al based alloys." *Journal of Alloys and Compounds* 478, no. 1–2 (2009): 809–812.
49. Zala, A. B., N. I. Jamnapara, V. J. Badheka, C. Sasmal, S. Sam, and M. Ranjan. "Delta (δ) ferrite formation in the welds of aluminized 9Cr-1Mo steels." *Metallography, Microstructure, and Analysis* 8 (2019): 256–262.
50. Arivazhagan, B., G. Srinivasan, S. K. Albert, and A. K. Bhaduri. "A study on influence of heat input variation on microstructure of reduced activation ferritic martensitic steel weld metal produced by GTAW process." *Fusion Engineering and Design* 86, no. 2–3 (2011): 192–197.
51. Onoro, J. "Weld metal microstructure analysis of 9–12% Cr steels." *International Journal of Pressure Vessels and Piping* 83, no. 7 (2006): 540–545.
52. Mayr, Peter, Todd A. Palmer, John W. Elmer, Eliot D. Specht, and Samuel M. Allen. "Formation of delta ferrite in 9 Wt pct Cr steel investigated by in-situ X-ray diffraction using synchrotron radiation." *Metallurgical and Materials Transactions A* 41 (2010): 2462–2465.
53. Wallace, John F. "Grain refinement of steels: A general review of grain refinement of carbon, low-, and high-alloy steels." *Jom* 15 (1963): 372–376.

4 Practical Aspects in Welding of Duplex and Super Duplex Stainless Steels

Pradip Goswami

CONTENTS

DOI: 10.1201/9781003352419-4

4.1 INTRODUCTION

Duplex stainless steels are engineering alloys combining good corrosion resistance with high strength and ease of fabrication. Duplex stainless steels (DSS) are two-phase alloys consisting of austenite and ferrite. These alloys have one of the highest levels of proven engineering integrity in all types of engineering industries [1–4].

These alloys are based on the Fe–Cr–Ni–N system and are characterized by low carbon contents with additions of molybdenum, nitrogen, and copper. They exhibit higher strength and better resistance to stress corrosion cracking (SCC) and pitting than austenitic stainless steels; also, they provide better toughness compared to ferritic stainless steels. Nowadays, these steels are widely used in the oil and gas industries in both upstream and downstream applications, downstream petrochemical industries, chemical industries, mining and metal industries, and many more. Common components fabricated from these steels are pressure vessels, reactors, exchangers, pumps, valves, process piping, pipelines systems, offshore manifolds, and risers, as well as they are used in all types of piping.

All grades of duplex stainless steels belong to a family of engineering alloys combining extremely good corrosion resistance with proven high strength and ease of fabrication. Their physical properties are between those of the austenitic and ferritic stainless steels but tend to be closer to those of the ferritic alloys. The chloride pitting and crevice corrosion resistance of the duplex stainless steels are a function of chromium, molybdenum, tungsten, and nitrogen content. These properties are far superior than common austenitic grades, such as SS 316/316L and similar to those of super austenitic (6 Mo) stainless steels. All duplex stainless steels have chloride stress corrosion cracking resistance significantly greater than that of the 300-series austenitic stainless steels. They all provide significantly greater strength than the austenitic grades while exhibiting good ductility and toughness. There are many similarities in the fabrication of austenitic and duplex stainless steels, but there are important differences. The high alloy content and the high strength of the duplex grades require some changes in fabrication practice. This chapter is for fabricators and for end users with fabrication responsibility. It presents, in a single source, practical information for the successful fabrication of duplex stainless steels. This chapter assumes the reader already has experience with the fabrication of stainless steels; therefore, it provides data comparing the properties and fabrication practices of duplex stainless steels to those of the 300-series austenitic stainless steels and of carbon steel.

The following discussion on duplex stainless steels will provide the readers a detailed understanding on the fabrication aspects of the structures and components made from duplex stainless steel, and as well as the knowledge that fabrication of duplex stainless steels is different than other steels but not difficult at all. Duplex stainless steels form several embrittling precipitates at relatively low temperatures. Hence, as a rule of thumb they are not recommended for service exceeding about 315°C (600°F). They are more expensive than austenitic stainless steels, not so much due to the cost of alloy elements but primarily due to the cost of processing involved for the as-cast steel to finished plate, sheet, or tubular form. All duplex grades offer distinct corrosion advantages and weight savings. They may be used in lieu of some Ni-base alloys in many aggressive environments, such as in oil and gas, petrochemicals, minerals processing, and more at a fraction of the material cost.

Duplex stainless steels can be broadly divided into a few categories such as lean duplex, 22% Cr duplex, and 25% Cr super duplex groups. The recently developed hyper duplex group involves typically 27% Cr and above. The classification is based primarily on the alloying level in terms of pitting resistance equivalent number (PREN) which measures the alloy's resistance for pitting corrosion. PREN is calculated using a simple formula: PREN = % Cr + 3.3% Mo +16% N; an allowance for W is sometimes made, having a factor of 1.65. Typically, duplex steel has a PREN less than 40; a super duplex PREN is typically between 40 and 45, a hyper duplex is above 45, whilst the lean grades typically have lower nickel content and hence lower price.

All grades of duplex stainless steels are significantly stronger than austenitic stainless steels. Typical yield strengths of these alloys are above 425 MPa (60 ksi) compared to 210 Mpa (30 ksi) for austenitic steels. Since the duplex stainless steels are stronger, they are also harder, making them more attractive where abrasion is a concern as well as corrosion. Most of today's duplex stainless steels have good toughness and ductility, however, they do undergo a ductile-to-brittle transition at low temperature, so that they are generally not suitable for service at cryogenic temperatures. The service temperature range of duplex stainless steels is limited to a range of about –40°C to 280°C (–40–535°F).

Metallurgical aspects of the duplex stainless steel family are complex and require very close control of weld metal composition and heat treatment regime so as to ensure that mechanical properties and/or corrosion resistance would not get adversely affected during fabrication [5–8]. For obtaining the optimum mechanical properties and corrosion resistance, the microstructure along with the phase balance of both the parent and weld metal should be approximately 50% ferrite and 50% austenite. Usually, this precise value is impossible to achieve with accuracy, but a range of phase balances between 35–60% ferrite and the rest being austenite phases is acceptable in the various industry standards.

Though the composition and, perhaps more importantly, heat treatment parameters are relatively easy to control theoretically, this is not the case during welding. The amount of ferrite phases formed is dependent not only on composition but also on the cooling rate. Faster cooling rates retain more of the ferrite that forms at elevated temperature. Therefore, to minimize the risk of producing very high ferrite levels in the weld metal, it is necessary to ensure that there is a minimum heat input and, therefore, a maximum cooling rate. A common rule of thumb is that heat input for duplex and super duplex steels should not be less than 0.5 kJ/mm, although thick sections will need this lower limit to be increased.

4.2 HISTORY OF DUPLEX STAINLESS STEELS

Duplex stainless steels had been in existence since the 1930s, however, the real thrust behind the developments and usage started in the 1970s and beyond with the huge development in offshore oil and gas explorations.

Back in the early 1980s, second-generation (nitrogen-alloyed) duplex stainless steels were developed rapidly. With very limited understanding of the formation of the intermetallic phases, early views of welding duplex grades were focused only on limiting or controlling the heat input. This approach was typically applied to special and common austenitic grades for controlling intergranular corrosion. With such severe limitations on heat input being imposed, many of the more economical welding methods with high deposition rates, such as submerged arc welding, were thought to be inappropriate for duplex stainless steels. However, the final properties and the economics behind choices of the duplex stainless steels are of such interest that much effort was directed to learning how to use the more economical processes.

In the early 1990s, the International Institute of Welding (IIW) published the document titled "Guide for Industry: Welding Duplex and Super Duplex Stainless Steels". This was published in the IIW publication "*Welding in the World*". Material engineers in the petrochemical industry and chemical tankers as well as in shipbuilding became convinced that the combination of higher strength and increased corrosion properties as compared to common austenitic stainless steel grades offers a real advantage to duplex stainless steels.

Currently, many variants of duplex stainless steel, from "lean" to "hyper" grades, are commercially available, providing a wide range of strength levels, corrosion resistance, and cost-efficient solutions. Duplex and super duplex stainless steels have found their ways in various applications such as the chemical, oil, and gas industries, petrochemical process plants, pulp and paper industries, pollution control equipment, transportation, and, last of all, in general engineering.

All arc welding processes are recommended for welding duplex and super duplex stainless steel grades. Today, arc welding processes like gas tungsten arc welding (GTAW), gas metal arc welding (GMAW), shielded metal arc welding (SMAW), flux core arc welding (FCAW), PAW, and submerged arc welding (SAW) have all seen practical applications and are recommended for welding of duplex stainless steels.

The demand for duplex grades would always remain high due to its versatile and growing applications in various major industries like in oil and gas, desalination facilities, air pollution control, flue gas desulfurization units, chemical processes, paper and pulp industries, and also in architecture and infrastructure facilities.

Duplex stainless steels (DSS) are alloys characterized by low carbon content and a duplex microstructure comprising both body-centered-cubic (BCC) ferrite and face-centered-cubic (FCC) austenite phases. This group of stainless steels is based on Fe–Cr–Ni–N alloys with additions of molybdenum, tungsten, and copper, depending on the alloy classification. Nitrogen is a key alloying element, used in addition to nickel to improve the mechanical and corrosion (pitting) properties of DSS. It is important to note that nitrogen is a gaseous alloying element, which is added to liquid DSS by a high-pressure processing process only. Consequently, nitrogen can partially evolve from the liquid weld pool to the environment, depending on the welding procedure utilized (specifically the shielding gas composition and the size of the weld pool). The technical interest in DSS alloys as compared to the Series 300 austenitic stainless steel alloys *is derived* from their resistance to stress corrosion cracking and pitting and the fact that their yield strengths are twice those of the Series 300 alloys. DSS alloys combine some of the advantages of ferritic and austenitic stainless steels, but they also have some disadvantages of each. Above all, the good resistance to pitting corrosion and chloride-induced stress corrosion cracking makes them so useful for these industries.

Table 4.1 lists all the commercial grades of duplex stainless steels, which are broken up into four categories. The compositions of austenitic and super austenitic stainless steels are given in the same analysis to provide a relative comparison of the chemical composition and the PREN values. PREN is an index developed to use the resistance of duplex stainless steels and other austenitic alloys used as an index to quantify the resistance to pitting and other forms of corrosion.

Table 4.2 lists all the applicable ASME and ASTM (American Society for Testing and Materials) specifications for various DSS and SDSS grades; Table 4.3 lists all the mechanical properties.

In the "sweet as well as sour" corrosive environments of upstream oil and gas environments, in many instances, DSS are considered to be the standard design basis materials. Regarding the downstream side for many engineering applications in the petroleum and refining industry, various DSS grades are currently the preferred material over austenitic stainless steels and even nickel-based alloys. There has been considerable application experience with DSS and super duplex stainless steels (SDSS) for items such as upstream downhole and well head tubular, well heads, flow lines, pipelines, manifolds, coolers, heat exchangers, valves, and pumps.

A newer grade of duplex alloys termed as "Hyper Duplex" is the latest addition to the duplex family and is well accepted by the offshore oil and gas industry. Sandvik SAF 3207 HD (UNS S32707, C-max–0.03, Cr-32, Ni-7, Mo–3.5, N–0.5, PREN-50

TABLE 4.1
Typical Gas Metal Arc Welding (GMAW) Parameters for Short-Circuiting Arc Transfer and for Spray Arc Transfer for Welding Duplex Stainless Steels with Various Wire Sizes

Short Circuit Arc Transfer			
Weld wire diameter		Current	Voltage
mm	inch	Amperes (A)	Volts (V)
1.0	0.035	90–120	19–21
1.2	0.045	110–140	20–22
Spray Arc Transfer			
1.0	0.035	170–200	25–28
1.2	0.045	210–280	26–29
1.6	0.063	270–330	27–30

(*Source*: Avesta/Outokumpu Welding).

TABLE 4.2
Typical Submerged Arc Welding (SAW) Parameters for Welding Duplex Stainless Steels with Various-Size Wire

Diameter of Welding Filler Wire		Current	Voltage
Millimeter (mm)	Inch (in.)	Amps (A)	Volts (V)
2.5	0.094	250–350	28–31
3.25	0.125	300–500	29–34
4.0	0.156	400–600	30–35
5.0	0.203	500–700	30–35

(*Source*: Avesta/Outokumpu)

TABLE 4.3
Examples of Various Welding Parameters for Different Types of Joints

Weld Process	Electrode/Filler AWS Classification	Diameter (in.)	Welding Position	Bead	Recommended Current (A)	Voltage (V)	Travel Speed (inch/min)
SMAW	E-2209	3/32	3G-upwards	Root	50–60	20–24	4–6/7–9
		&		Cap	80–95		
		1/8					
GTAW	ER-2209	3/32	1G	Root	100–120	16–18	5–8
GTAW	ER-2209	3/32	1G	Root	100–120	16–18	5–8
&	+	&		Cap	190–210	28–30	17–22
FCAW	E2209T1–4	1/16					
SAW	ER-2209+ P100 Flux	3/32	1G	Cap	350–400	30–32	30–40

and above) is the alloy which is commonly termed as hyper duplex stainless steel. This new material has a yield strength **20% higher than that of the** SDSS and is recommended for a service temperature up to 90°C as an umbilical tubing.

Although this alloy is referred to in this chapter, more details are not provided as it is a fairly recent alloy and is in the process of establishing credentials in oil and gas industry.

4.3 WELDING METALLURGY

4.3.1 Solidification Behavior

The elevated temperature region of a pseudo binary phase diagram for DSS compositions is shown in Figure 4.1, where the shaded region represents the range for common commercial duplex alloys.

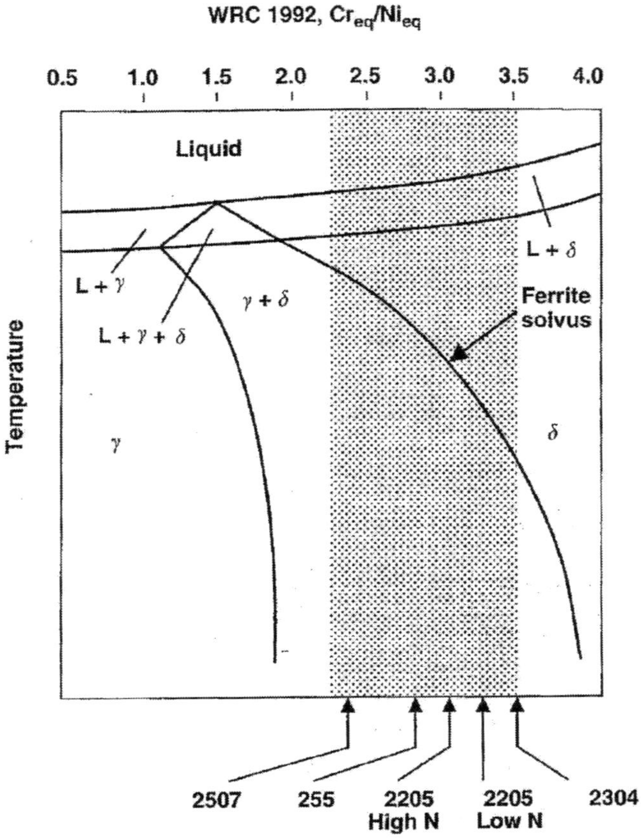

FIGURE 4.1 Elevated temperature region of a pseudo binary phase diagram for duplex stainless steel compositions. The shaded region represents the range for common commercial duplex alloys.

Virtually all the DSS solidify as ferrite and are fully ferritic at the end of solidification. Based on the compositions of the alloys involved, the ferrite phase is stable over some range of elevated temperature before it falls below the ferrite solvus temperature and the metallurgical transformation to austenite begins. The nature of ferrite-to-austenite transformation is dependent on both composition and cooling rate. It is this transformation that determines the ultimate ferrite–austenite balance and austenite distribution in the weld metal. The transformation sequence for duplex stainless steels is as follows:

- L L+F F F+A

Here, **L = Liquid, F = Ferrite, A = Austenite**.

The iron–chromium–nickel ternary phase diagram is a roadmap of the metallurgical behavior of DSS. A section through the ternary at 68% iron (Figure 4.2)

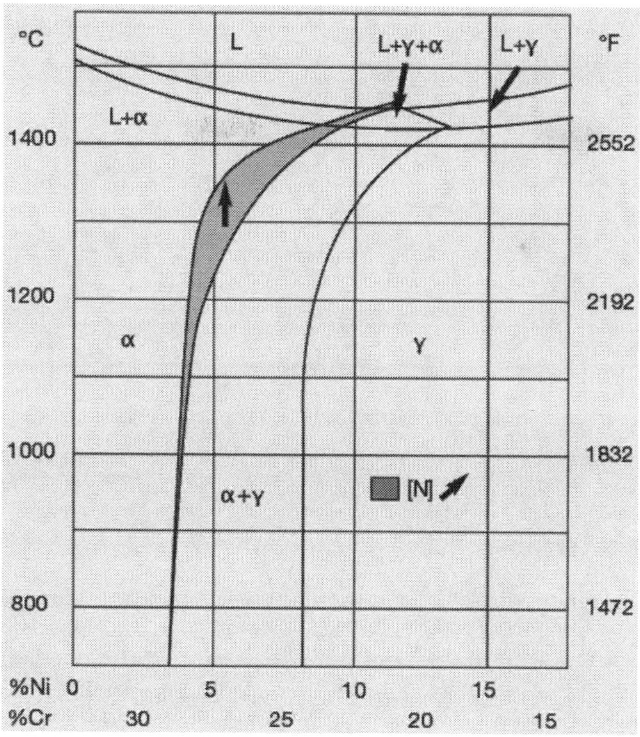

FIGURE 4.2 Solidification diagram of duplex stainless steels Fe–Cr–Ni system at 68% iron.

illustrates that these alloys solidify as ferrite (α), some of which then transform to austenite (γ) as the temperature falls to about 1000°C (1832°F) depending on alloy composition. There is little further change in the equilibrium ferrite–austenite balance at lower temperatures. Upon further cooling to lower temperatures, carbides, nitrides, sigma, and other intermetallic phases are formed and are precipitated as various microstructural constituents.

The ferrite–austenite phase balance in the microstructure can be predicted with multivariable linear regression as follows:

- $Cr\ eq = \%Cr + 1.73\ \%Si + 0.88\ \%Mo$
- $Ni\ eq = \%Ni + 24.55\ \%C + 21.75\ \%N + 0.4\ \%Cu$
- $\%\ Ferrite = -20.93 + 4.01\ Cr_{eq} - 5.6\ Nieq + 0.016\ T$

Where T (in degrees Celsius) is the annealing temperature ranging from 1050°C to 1150°C, and the elemental compositions are in wt.%.

The desired phase balance in a DSS is primarily achieved by adjusting the chromium, molybdenum, nickel, and nitrogen contents and then by controlling the thermal history during metal processing. However, the exposures to cooling rates also influence the phase balance. Because fast cooling rates favor retention of ferrite. For example, low heat input welding of a heavy section might result in excessive ferrite in the heat-affected zone (HAZ).

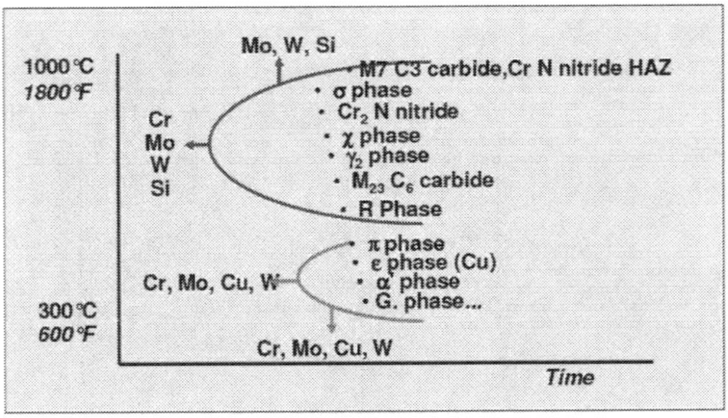

FIGURE 4.3 Typical phase precipitations which may occur in duplex grades.

4.3.2 Role of Nitrogen

Modern DSS contains nitrogen. Intentional additions of nitrogen are made to improve the mechanical strength and pitting corrosion resistance. The range of nitrogen content is between 0.08 wt.% and 0.35 wt.%. These levels of nitrogen are well above the solubility limit in ferrite at temperatures below about 1000°C (1830°F). By contrast, the solubility of nitrogen in austenite is much higher. This differential in nitrogen solubility has important implications with respect to nitride precipitation in these alloys. When the microstructure is balanced (nearly a 50/50 mixture achieved by an elevated temperature hold followed by rapid cooling), the nitrogen partitions between the ferrite and austenite remain predominantly in solid solution.

If the ferrite content is high, such as in the weld metal and HAZ under rapid cooling conditions, an intense nitride precipitation reaction occurs upon cooling since the solubility limit of nitrogen in the ferrite is exceeded, and the nitrogen has insufficient time to partition to the austenite. In most cases, these nitrides are Cr-rich and are thought to be primarily Cr_2N. The consequences of this intense nitride precipitation are most notably, loss of ductility, toughness, and corrosion resistance. In most of the second-generation DSS (which are very commonly used nowadays), the optimum levels of nitrogen addition reduce the problem of excess ferrite in the HAZ.

The *sigma phase,* shown in Figure 4.4, precipitates at temperatures below austenite formation from the ferrite on cooling. The goal of avoiding sigma phase in mill products is achieved by controlling the annealing temperature and ensuring that the steel is quenched as rapidly as possible from the annealing temperature. This reduces the possibilities of sigma phase formation during cooling. The required cooling rate is quite rapid, allowing the use of water quenching. Only when welding widely differing section sizes or when welding heavy sections with very low heat inputs may excessively cooling rates be encountered during actual fabrication.

Alpha prime is also a stable phase in duplex alloys, forming in the *ferrite phase below about 525°C* (950°F) in the same manner as it forms in fully ferritic alloys. Though not as prominent as sigma phase, alpha prime causes the *loss of ambient*

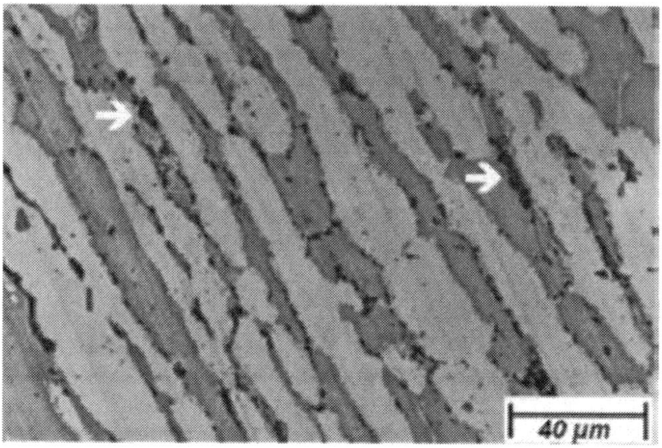

FIGURE 4.4 Microstructure of a 2205 sample aged at 850°C (1560°F) for 40 minutes showing sigma phase precipitation (arrows) on the austenite–ferrite grain boundaries. The ferrite (F) phase appears darker than the austenite (A) phase in the micrograph.

temperature toughness in ferritic stainless steel after extended exposure to temperatures around 475°C (885°F); this behavior is known as *475°C/885°F embrittlement.*

Nitrogen is preferentially used as an alloying element in these stainless steels. This causes the formation of *chromium nitrides*, which may be present on ferrite–ferrite grain boundaries and on austenite–ferrite boundaries in the HAZ of welds. *Higher nitrogen promotes austenite*, which has a high solubility for nitrogen. The second-generation DSS *seldom contain significant amounts of chromium nitrides*. Furthermore, the second-generation DSS are made with *very low carbon* content so that carbide formation to a detrimental extent is not usually a practical concern.

Detrimental sigma, alpha prime, carbides, and nitrides can form in a matter of minutes at certain temperatures. Consequently, the thermal treatments required for processing and fabrication, as well as the service cycles, must take the reaction kinetics of phase formation into account. This ensures that the desired corrosion resistance and mechanical properties are obtained. Modern duplex grades have been developed to maximize corrosion resistance and retard the effects of such precipitation reactions sufficiently to allow successful fabrication.

Further illustrations on Fe–C–Ni systems as discussed earlier in Figure 4.2 are explained as follows.

According to the established and published results:

- Alloys with above or with approximately 10% Ni solidify as ferrite but subsequently transform largely to austenite.
- Alloys near 5% Ni solidify as ferrite but only partially transform to austenite.
- Solidification as 100% ferrite is the fundamental principle underlying DSS, where Ni content varies between 5% and 10% in the base metal and the weld metal.

Generally, the austenite–ferrite ratio depends on the chemical composition of the alloys and the heat treatments adopted. However, most alloys are designed to contain similar amounts of each phase at room temperature. The austenite–ferrite matrix is attainable by combining various phase-stabilizing elements. Chromium and molybdenum are effective ferrite stabilizers, producing a wide ferrite field in phase diagrams. In general, stainless steels containing ferrite as the predominant phase have excellent corrosion resistance due to the high solubility of Cr in ferrite. Regarding the high temperature stability of the duplex structure, it is important to note the gamma (austenite) stabilizing action of nitrogen, which is intentionally added in DSS.

In the 1980s, higher alloyed DSS grades started getting manufactured and were developed into SDSS grades. They are made to withstand more aggressive environments but also bear a higher risk of unfavourable precipitation phases due to the higher alloying element content. In the making of SDSS, Cr- and Ni-forming elements are balanced, and more nitrogen was added. SDSS are usually characterized by having a PREN greater than 40. The higher the PREN, the better the predicted corrosion properties of a DSS. The application of SDSS is increasingly more commonplace for certain offshore applications.

However, PREN numbers provide only an approximate grading of alloys and do not account for the microstructure of the material. An acceptance corrosion test on

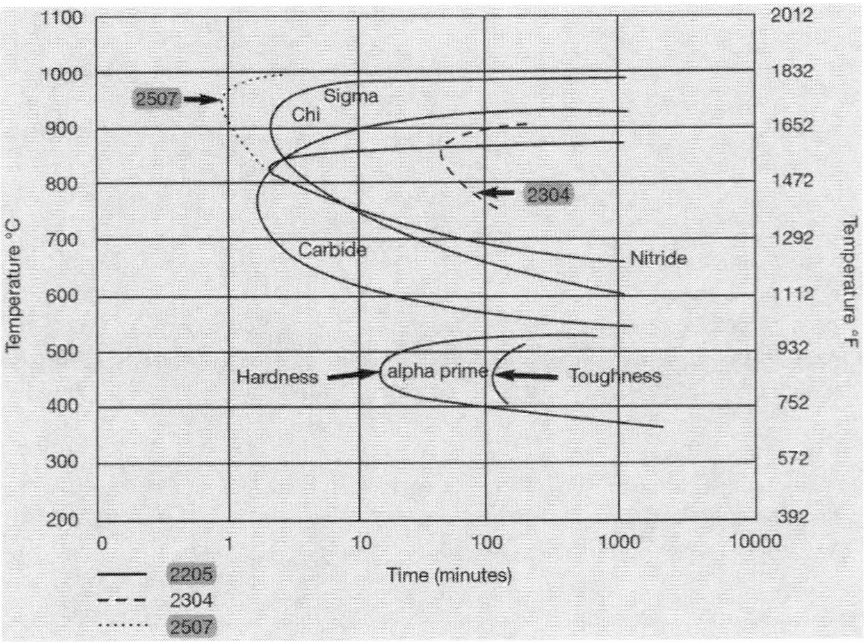

FIGURE 4.5 Isothermal precipitation diagram for 2205 duplex stainless steel, annealed at 1050°C (1920°F). (Duplex grades 2304 and 2507 are shown for comparison). It's worth noting the phase contents for various alloys concerned.

material in the supply condition is much more meaningful and a necessity in many offshore applications. The most common way of ranking all stainless steels for their PREN is according to the relation between the amount of the essential alloying elements and the corrosion properties.

4.3.3 Alloying Elements in Duplex Stainless Steels

a) *Chromium (Ferrite Stabilizer):* A minimum of approximately 10.5% chromium is necessary in DSS to form a stable chromium passive film for protection against atmospheric corrosion. Typically, the corrosion resistance of a stainless steel increases with increasing chromium content. The duplex grades are clearly designed for significantly better corrosion resistance than the classical ferritic or 300 series austenitic grades, e.g., 304/316. Some duplex grades have additional copper additions known to have very positive effects in sulfuric acid solutions. Tungsten is known to further enhance the resistance of the stainless steel to abrasion-corrosion resistance phenomenon. Moreover, duplex grades have high nitrogen contents, which are known to be beneficial for localized corrosion resistance (pitting and crevice).
b) *Molybdenum (Ferrite Stabilizer):* Molybdenum acts to support chromium in providing resistance against chloride pitting corrosion to stainless steels. When the chromium content of a stainless steel is at least 18%, additions of molybdenum become approximately three times as effective as chromium additions against pitting and crevice corrosion in chloride-containing environments. Molybdenum is a ferrite former and increases the tendency of stainless steel to form detrimental intermetallic (sigma and other) phases. Therefore, it is usually restricted to approximately 4% maximum in DSS.
c) *Nickel (Austenite Stabilizer):* The addition of nickel to iron-based alloys promotes a change of the crystal structure of the stainless steel from body-centered cubic (ferritic) to face-centered cubic (austenitic). DSS contain an intermediate amount of nickel such as 4– 7%. The addition of nickel delays the formation of detrimental intermetallic phases in austenitic stainless steels but is far less effective than nitrogen in delaying their formation in DSS. The face-centered cubic (FCC) structure is responsible for the excellent toughness of the austenitic stainless steels. Its presence in approximately half of the microstructure of duplex grades greatly increases their toughness relative to ferritic stainless steels.
d) *Nitrogen: (Austenite Stabilizer):* Nitrogen is added to DSS containing high chromium and molybdenum contents to offset their tendency to form sigma phase. The ferrite-forming alloying elements, chromium and molybdenum, are balanced by the austenite formers, nickel and nitrogen, to obtain the duplex structure. Nitrogen increases the resistance of DSS to pitting and crevice formation and substantially increases their mechanical strengths. It is the most effective solid solution-strengthening element.

Because of their higher strength, the nitrogen-enhanced DSS also have increased notch toughness. Nitrogen delays the formation of intermetallic phases typically occurring during thermal processing and fabrication of the duplex grades. In DSS, nitrogen is typically added almost to its solubility limit, and accordingly the amount of nickel is adjusted to achieve the desired phase balance.

e) *Manganese:* The addition of manganese has a controversial effect; they are known to increase the nitrogen solubility and, in some respect, to stabilize the austenitic phase directly or indirectly. On the contrary, most of the localized, mainly pitting corrosion resistance investigations have underlined a negative effect of manganese additions.
f) *Copper (Cu):* Copper is normally present in stainless steel only as a residual element. However, it is added to selected alloys to enhance their corrosion resistance, especially in seawater and sulfuric acid solutions. Adding copper to stainless steels has an interesting beneficial effect around these inclusion-initiated pitting corrosion sites. Dissolution of copper at the surface results in the precipitation of an insoluble copper sulfide (Cu_2S) compound. This, in turn, prevents sulfur species from adsorbing on to the surface of the steel, thereby inhibiting further pit growth. In some ways, copper provides a self-healing effect for stainless steels, which will not prevent pitting corrosion from initiating at points of weakness in the passive layer, but it will slow or halt further corrosion at these sites preventing more harmful corrosion.
g) Among the detrimental alloying elements, sulfur has the worst effect. Sulfur generally combines with manganese or several oxides to form sulfide precipitates. The worst case is a large oxide inclusion surrounded by sulfur species. Locally, the passive film would be unable to resist, and severe local corrosion may get started.
h) Chromium carbides, and to some extent chromium nitrides, are also known to reduce corrosion resistance mainly due to the formation of chromium depleted areas around the precipitates, which take place mainly at the grain boundaries.

4.4 DETRIMENTAL PHASES DURING WELDING

Ideally, DSS should have a microstructure consisting of 50% ferrite (α) and 50% austenite (γ). These alloys solidify as ferrite, which then partially transforms to austenite as the temperature decreases during solidification. The 50%:50% microstructure is achieved at room temperature after water quenching from the solution at annealing temperature.

This two-phase microstructure provides the DSS with attractive mechanical and corrosion-resistant properties. As mentioned previously, these properties are strongly dependent on the heat treatment and chemical composition (appropriate balance between austenite-formers nickel and nitrogen and ferrite-formers chromium and molybdenum).

Temperature during welding is a key parameter which drives the microstructure in both HAZ and the weld metal. An inappropriate temperature control can lead to the following issues: (i) the formation of detrimental phases and (ii) the formation of an unbalanced microstructure (ferrite and austenite).

4.4.1 Detrimental Phases

The formation domains of these phases are illustrated in the TTT diagram shown in Figure 4.6. These phases can have a detrimental influence on the corrosion and toughness properties of the duplex materials.

- **Sigma phase** can precipitate in the temperature range from 700°C to 1000°C. Its formation is strongly promoted by the molybdenum and chromium content of the material. The main effect of the formation of sigma phase is the decrease of the pitting corrosion resistance. Such low resistance is due to the formation of a chromium-depleted zone in the surrounding areas of the sigma phase. Also excess of sigma phase (> 5%) typically is known to reduce notch toughness.

  Sigma is a hard brittle intermetallic phase which is expected to contain iron, chromium, and molybdenum. In duplex alloys, sigma phase generally can be formed between about 600°C and 950°C, with the most rapid formation occurring between 700°C and 900°C.
- **Nitrides** can precipitate in the temperature range from 600°C to 900°C, most often *in lean DSS*. They can also precipitate in HAZ and weld metal, especially if the ferrite content near the fusion line is too high leading to the formation of chromium carbides upon cooling.

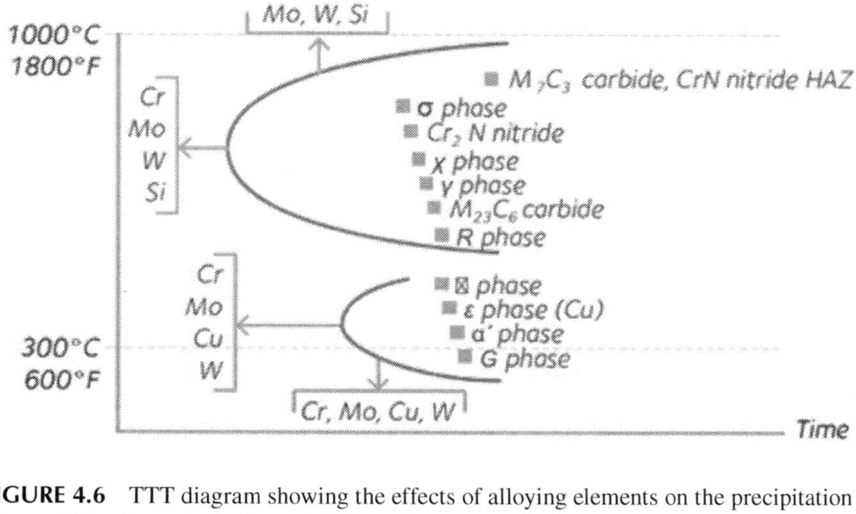

FIGURE 4.6 TTT diagram showing the effects of alloying elements on the precipitation of intermetallic phases in duplex stainless steels.

- **Alpha prime** can form in the ferrite phase in the temperature range from 350°C to 550°C, resulting in the decrease of toughness properties of the material.

  The cumulative time in these temperature ranges due to the welding operation can lead to the formation of these phases. They can be avoided or controlled largely by following the welding recommendations provided for various duplex grades by controlling the *cooling rate, heat input, and interpass temperature.*

4.5 WELDING OF DUPLEX STAINLESS STEELS, THE CHALLENGES

4.5.1 Welding Consumables

In the early 1980s, second-generation (nitrogen-alloyed) DSS developed rapidly. With only limited understanding of the formation of intermetallic phases, the early views of welding duplex grades focused on limiting heat input, possibly because this approach is typically applied to special austenitic grades. With such severe limitations on heat input, many of the more economical welding methods with high deposition rates, such as SAW, were thought to be inappropriate for DSS. However, the final properties of the DSS are of such interest that much effort was directed to learning how to use the more economical processes. Nowadays, virtually all welding processes, except for oxyacetylene with its associated carbon contamination of the weld, are applied to DSS. Today's arc welding processes like GTAW, GMAW, SMAW, FCAW, and SAW have all seen practical applications.

The welding characteristics of DSS are extremely sensitive to minor within-grade variations in chemistry or processing than austenitic stainless steels. For example, the importance of having sufficient nitrogen in the DSS base metal has been repeatedly emphasized. Air cooling of a plate, even when rapid, through the 980°C to 705°C range will limit the applicability. The metallurgical condition of the material used in actual fabrication should be of the same quality, with regard to composition and production practices, as the material used for qualification of welding procedure.

Consumables grades/standards

For consumables, the applicable standards are:

- ISO3581/AWS A5.4 for covered electrodes, SMAW
- ISO 17633/AWSA5.22 for tubular cored wires, e.g., FCAW
- ISO 14343/AWSA5.9 for solid wires, e.g., GTAW, GMAW, SAW

In Tables 4.4 and 4.5, an overview of each process is listed to meet both ISO/AWS requirements. Typically, consumables are over-alloyed with 2–4% Ni to encourage austenite formation and to maintain phase balances of ferrite and austenite.

In addition to ambient strength requirements, there have been projects where the strengths at elevated temperatures for super duplex have been specified. For coated electrodes, impact toughness and chemistry might differ substantially between manufacturers as well as based on the classification type (basic or rutile).

TABLE 4.4
Recommended Shielding Gases for GMAW, GTAW, and FCAW for Duplex and Super Duplex Stainless Steels

Welding Process	Shielding Gases	Remarks
GMAW	Ar + 30He + 1O_2	Short arc welding gives very convex beads
	Ar + 30He + 1O_2 (22Cr duplex) Ar +2CO_2 (Super duplex	Spray arc welding
	Ar + 30He +1O_2 (22Cr duplex) Ar (99,996%) for super duplex	Pulsed arc welding
GTAW	• Shielding -Ar + 2% N_2 • Purging Gas- Ar	Purge to maintain 0.5% oxygen max Ar + 2% N_2 shielding gas is recommended for the root run. For subsequent runs, Ar may be used as shielding gas. Root is mandatory for the first two runs. **Note:**—Per ASME Sec-IX, a change for purging gas is a non-essential variable. Formier gas (90% nitrogen and 10% hydrogen) is a cheaper and sometimes a tried alternative for root purging. However, the acceptance is subject to satisfactory procedure qualification only.
FCAW	• N	Typical gas flow rate 20–25 l/min wire sickout length 15–20 mm

Rutile-basic electrodes have been in use traditionally, but fully basic electrodes gain preference due to low defect rates, good weld pool control, and better mechanical properties. For cored wires, the available range in both downhand and positional grades are available from lean 2304 grades to super duplex grades with copper (Cu) and tungsten (W).

4.5.2 Welding Practice, Process, and Procedures

4.5.2.1 Joint Design

Welding DSS always requires good joint preparation (Figure 4.7). Weld joint design must facilitate complete penetration and avoid autogenous regions in the weld solidification. It's considered to be a good practice to machine rather than grind the weld edge or bevel preparation to provide uniformity of the weld thickness as well as the root gap. Special attention should be given to the uniformity of the welding preparation and the fit-up, when grinding is performed. Any grinding burr should be removed to maintain complete fusion and penetration. The assembly must be designed in such a manner that the penetration weld is performed without an excessive dilution of the base metal. The groove type must be configured to have a good gaseous shielding protection (for GTAW, GMAW, PAW, and FCAW), with good accessibility for welding torches to reach the bottom of the joint. As for all other stainless steels, the

TABLE 4.5
Welding Consumables for Various Duplex Stainless Steels (Refer to ASME BPV Code, Sec-II, Part C, 2010) [(1)]

Welding Process/SFA No	AWS Classification	Tensile Strength, min, Ksi/MPa	Elongation min. Percent	C	Cr	Ni	Mo	Nb (Cb) Plus Ta	Mn	N	Cu	Others
SMAW—SFA–5.4, Duplex	E2209-XX	100/690	20	0.04	21.5 – 23.5	8.5 – 10.5	2.5 – 3.5	—	0.5 – 2.0	0.08 – 0.20	0.75	—
SMAW—SFA–5.4, Super Duplex	E2553-XX	110/760	15	0.06	24.0 – 27.0	6.5 – 8.5	2.9 – 3.9	—	0.5 – 1.5	0.10 – 0.25	1.5 – 2.5	—
	E2593-XX	110/760	15	0.04	24.0 – 27.0	8.5 – 10.5	2.9 – 3.9	—	0.5 – 1.5	0.08 – 0.25	1.5 – 3.0	—
	E2594-XX	110/760	15	0.04	24.0 – 27.0	8.0 – 10.5	3.5 – 4.5	—	0.5 – 2.0	0.20 – 0.30	0.75	
	E2595-XX	110/760	15	0.04	24.0 – 27.0	8.0 – 10.5	2.5 – 4.5	—	2.5	0.20 – 0.30	0.4 – 1.5	W = 0.4 – 1.0
GTAW-GMAW-SAW- SFA–5.9	ER2209	100/690	20	0.03	21.5 – 23.5	7.5 – 9.5	2.5 – 3.5	—	0.50 – 2.00	0.08 – 0.20	0.75	—
	ER2553	110/760	15	0.04	24.0 – 27.0	4.5 – 6.5	2.9 – 3.9		1.5	0.10 – 0.25	1.5 – 2.5	
	ER2594	110/760	15	0.03	24.0 – 27.0	8.0 – 10.5	2.5 – 4.5	—	2.5	0.20 – 0.30	1.5	W= 1.0

Brief Descriptions Per ASME Sec-II, Pt-C Code

E2209 The nominal composition (wt.%) of this weld metal is 22.5 Cr, 9.5 Ni, 3 Mo, and 0.15 N. Electrodes of this composition are used primarily to weld DSS which contain approximately 22% of chromium.

E2553 The nominal composition (wt.%) of this weld metal is 25.5 Cr, 7.5 Ni, 3.4 Mo, 2 Cu, and 0.17 N. These electrodes are used primarily to weld DSS which contain approximately 25% of chromium.

E2593 The nominal composition (wt.%) of this weld metal is 25 Cr, 9.5 Ni, 3.4 Mo, 2.5 Cu, and 0.2 N. These electrodes are used primarily to weld DSS which contain approximately 25% chromium.

E2594 The nominal composition (wt.%) of this weld metal is 25.5 Cr, 10 Ni, 4 Mo, and 0.25 N. The sum of the Cr + 3.3 (Mo + 0.5 W) + 16 N, known as the pitting resistance equivalent number (PREN), is at least 40, thereby allowing the weld metal to be called a "superduplex stainless steel." This number is a semi-quantitative indicator of resistance to pitting in aqueous chloride-containing environments. It is designed for the welding of Type 2507 super duplex stainless steels UNS S32750 (wrought) and UNS J93404 (cast) and similar compositions. It can also be used for the welding of carbon and low-alloy steels to DSS as well as to weld "standard" DSS such as Type 2205 although the weld metal impact toughness may be inferior to that from E2209 electrodes. If post-weld annealing is required, this weld metal will require a higher annealing temperature than that required by the duplex base metal.

E2595 The nominal composition (wt.%) of this weld metal is 25.5 Cr, 9 Ni, 3.8 Mo, 0.7 Cu, 0.7 W, and 0.25 N. The sum of the Cr + 3.3 (Mo + 0.5 W) + 16 N, that is PREN, is at least 40, thereby allowing the weld metal to be called a "superduplex stainless steel." This number is a semi-quantitative indicator of resistance to pitting in aqueous chloride-containing environments. It is designed for the welding of superduplex stainless steelsUNS S32550, S32750, and S32760 (wrought), and UNS J93370, J93380, J93404, CD4MCuN (cast), and similar compositions. It can also be used for the welding of carbon and low-alloy steels to DSS as well as to weld "standard duplex stainless steel" such as UNS S31803 and UNS S32205.

ER2209 The nominal composition (wt.%) of this classification is 22.5 Cr, 8.5 Ni, 3 Mo, and 0.15 N. Filler metal of this classification is used primarily to weld DSS which contain approximately 22% chromium such as UNS S31803 and S32205.

ER2553 The nominal composition (wt.%) of this classification is 25.5 Cr, 5.5 Ni, 3.4 Mo, 2 Cu, and 0.2 N. Filler metal of this classification is used primarily to weld duplex stainless steels UNS S32550 which contain approximately 25% chromium.

ER2594 The nominal composition (wt.%) of this classification is 25.5 Cr, 9.2 Ni, 3.5 Mo, and 0.25 N. The sum of the Cr + 3.3(Mo + 0.5 W) + 16 N, that is, PREN, is at least 40, thereby allowing the weld metal to be called a "superduplex stainless steel." This number is a semi-quantitative indicator of resistance to pitting in aqueous chloride-containing environments. It is designed for the welding of superduplex stainless steels UNS S32750 and 32760 (wrought), and UNS J93380, J93404 (cast). It can also be used for the welding of UNS S32550, J93370, J93372 when not subject to sulfurous or sulfuric acids in service. It can also be used for the welding of carbon and low-alloy steels to DSS as well as to weld "standard' DSS such as UNS S32205 and J92205 especially for root runs in pipe.

[1] International Molybdenum Association, 2014. *Practical guidelines for the fabrication of duplex stainless steels*. 3rd ed. London: IMOA.

Groove	Process	Thickn. t (mm)	Gap d (mm)	Root k (mm)	Bevel α (°)
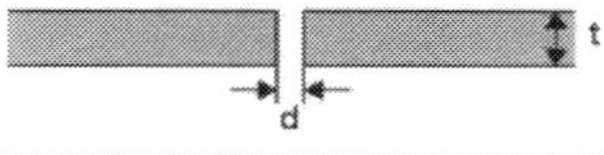	GTAW	3–5	1–3	–	–
	GMAW	3–6	1–3	–	–
	SMAW	3–4	1–3	–	–
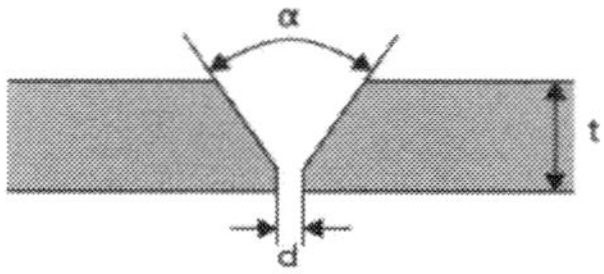	SMAW	4–15	1–3	1–2	55–65
	GTAW	3–8	1–3	1–2	60–70
	GMAW	5–12	1–3	1–2	60–70
	SAW	9–12	0	5	80
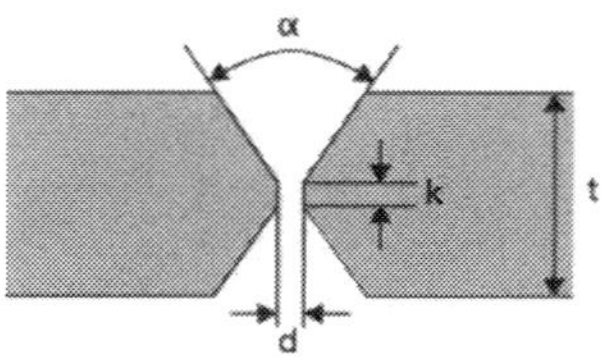	SMAW	>10	1.5–3	1–3	55–65
	GMAW	>10	1.5–3	1–3	60–70
	SAW	>10	0	3–5	90
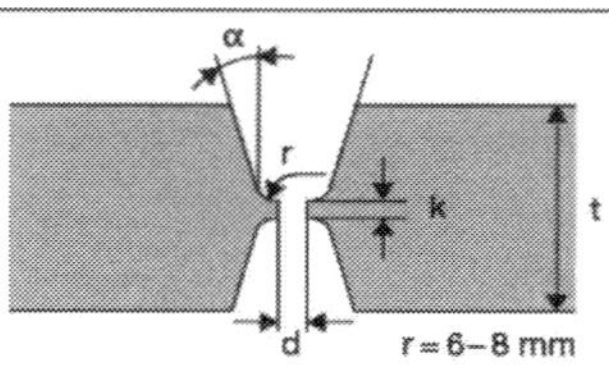	SMAW	>25	1–3	1–3	10–15
	GMAW	>25	1–3	1–3	10–15
	SAW	>25	0	3–5	10–15
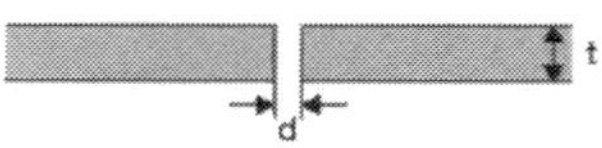	GTAW	>3	0–2	–	–
	GMAW	>3	0–2	–	–
	SMAW	>3	0–2	–	–
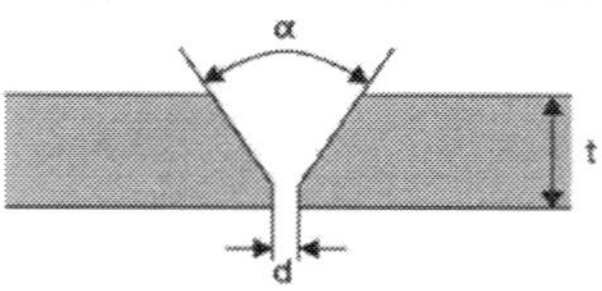	SMAW	3–15	2–3	1–2	60–70
	GTAW	2.5–8	2–3	1–2	60–70
	GMAW	3–12	2–3	1–2	60–70
	SAW	4–12	2–3	1–2	70–80
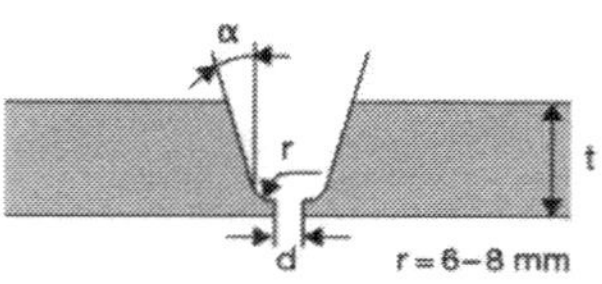	SMAW	12–60	1–2	2–3	10–15
	GTAW	>8	1–2	1–2	10–15
	GMAW	>12	1–2	2–3	10–15
	SAW	>10	1–2	1–3	10–15

FIGURE 4.7 Typical weld joint designs used for duplex stainless steels.

Courtesy: Arcelor Mittal.

welding zone must be carefully cleaned with a chlorine-free solvent to eliminate any grease or paint marks.

While welding an austenitic stainless steel weld, *a skilled welder can overcome some deficiencies in joint preparation by manipulation of the torch.* For a DSS, *these techniques (excessive manipulation) can cause a longer exposure in the harmful temperature range, exposure to higher heat input, formation of detrimental phases, thus leading to results outside of the qualified procedure. Hence, stringer bead technique or controlled weaving (Max 2x core wire diameter) is advised.*

Some examples of welded joint preparations are presented in the following tables, which can be used for butt or groove welding of DSS, involving various plate thicknesses.

4.5.2.2 Welding Heat Input

Typically, a heat input range of 0.5 KJ/mm to 2.5 KJ/mm is recommended for welding DSS, and a range of 0.5 KJ/mm to 1.5 KJ/mm is recommended for SDSS. These recommendations are typically as per all the design and fabrication codes and project-specific recommended practices, as well as various applicable specifications and generic guidelines for welding DSS.

When choosing a suitable heat input, it should be noted that a *lower heat input may cause an increase in ferrite content, which may impair the mechanical properties and corrosion resistance.* In the same way, *higher or excessive heat input increases the cooling time, and consequently intermetallic phases, e.g., sigma phase, may precipitate during the welding– cooling cycle.*

Hence, the weld parameter selection as well as the selection and control of heat input range based on the section thickness involved is absolutely important for obtaining the best/optimum microstructure, mechanical properties, and corrosion resistance.

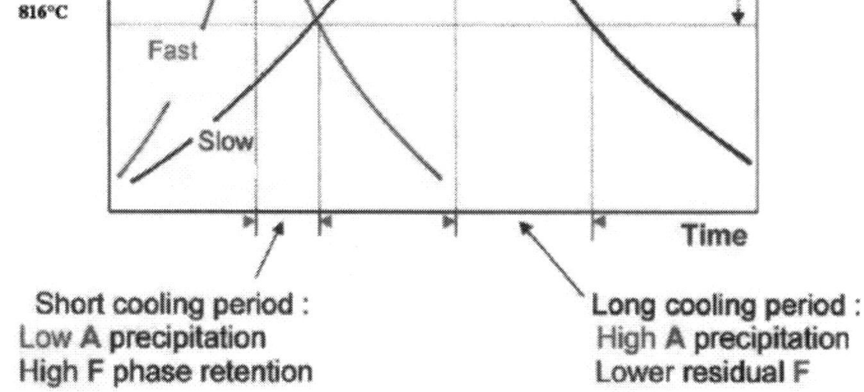

FIGURE 4.8 The effects of cooling rates on the ferrite contents on duplex stainless steel welds.

4.5.2.3 Post-Weld Heat Treatment

Post-weld stress relief heat treatment is not necessary nor recommended due to the probability of intermetallic phase precipitation and the 475°C (885°F) embrittlement phenomenon (alpha prime), which causes a significant loss in corrosion resistance and toughness of the weldments.

Post-weld heat treatment (PWHT), when applied to balance the duplex microstructure, is a full-solution anneal, followed immediately by water quenching. When recommended, the PWHT should be carried out in a furnace with a protective atmosphere, because the scaling temperature in air for DSS and SDSS is around 1000°C (1830°F), which is lower than the minimum solution-annealing temperature for most DSS and SDSS. Also, at a PWHT temperature of about 1000°C (1830°F), DSS and SDSS may undergo plastic deformation in the structure by its own weight.

Localized PWHT is also not recommended because it will create a region in the base metal exposed to 975°C to 700°C (1800°F to 1300°F), where sigma phase precipitation can occur causing embrittlement.

Solution annealing duplex stainless weld metal presents a special problem for weldment. The higher nickel content (9%) of the weld metal raises its sigma solvus compared to that of the base metal. Whereas a minimum of 1040°C (1900°F) is sufficient for proper annealing of base metal, 9% Ni weld metal may require annealing at temperatures as high as 1120°C (2050°F) to dissolve all sigma. Hence, for duplex stainless welds the solution annealing heat treatment in the temperature range 1040–1100°C is generally recommended.

4.5.2.4 Measuring and Controlling Ferrite Content

Control of the volume fraction of ferrite is fundamental and very essential to obtain the desired properties of weldments. Some measurement techniques present the results as a percentage of ferrite (%), while others give results in terms of the ferrite number (FN).

The ferrite percentage in a DSS or SDSS microstructure is measured by a quantitative metallographic methodology (ASTM E 562), which uses either image analysis or point grid measurements of the area of interest in the microstructure. The ferrite number is the measured result of a magnetic equivalent of a quantity of ferrite in the material. For lower amounts of ferrite (less than 10%), the two techniques yield nearly equivalent values.

In accordance with IIW II 1196–92, the relation between ferrite content (F%) and ferrite number (FN) for ferritic-austenitic steels is expressed by the following equation:

$$F\% = 0.54\ FN + 9.7 \quad (5.3)$$

Another relation between the amount of ferrite and the FN for duplex alloys has been developed as follows:

$$F\% = 0.7\ FN$$

The amount of ferrite can be estimated from constitution diagrams such as the Espy diagram (with the results in F%) or WRC-92 diagram (with the result in FN). Other diagrams, commonly adopted by the welding industry (the Schaeffler and DeLong diagrams), are not adequate for DSS that contain nitrogen and copper.

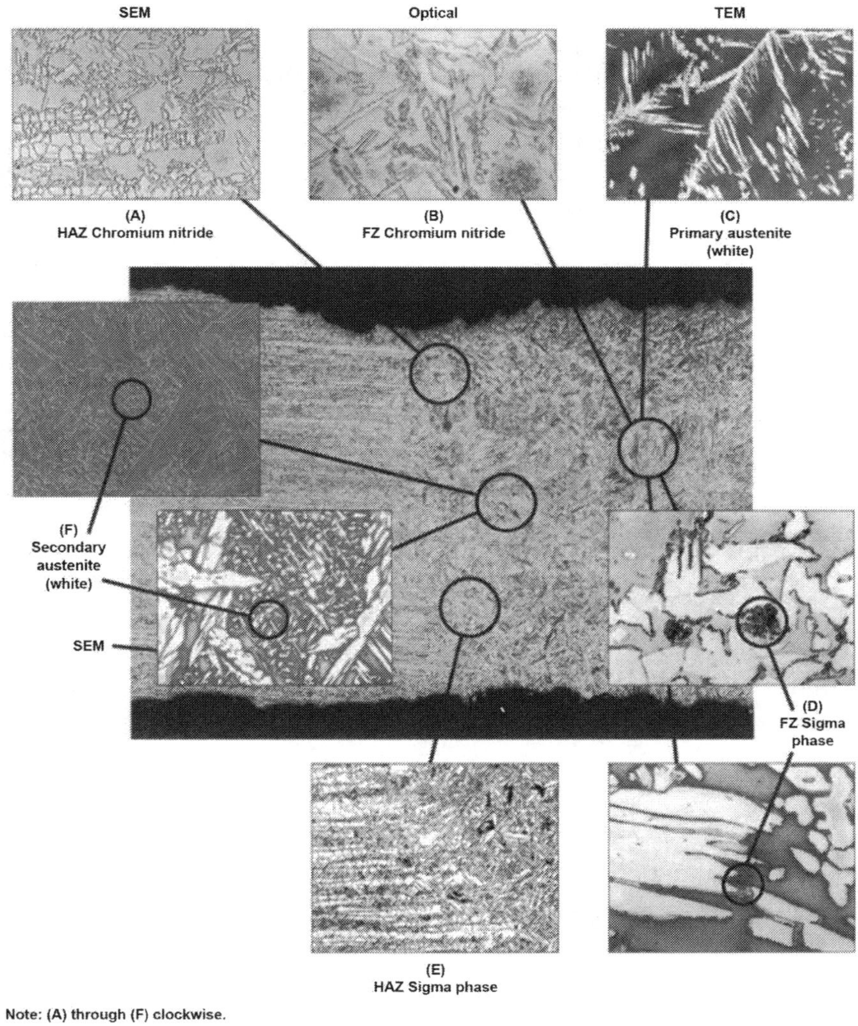

FIGURE 4.9 Microstructures of a welded joint of a duplex stainless steel.

The ferrite content can be measured by a grid method, according to ASTM E 562, Standard Test Method for Determining Volume Fraction by Systematic Manual Point Count, or by image analysis, using ASTM E 1245, Standard Practice for Determining the Inclusion or Second-Phase Constituent Content of Metals by Automatic Image Analysis.

4.5.2.5 Welding and Joining Processes for Duplex and Super Duplex Stainless Steels

Duplex and super duplex stainless steels are weldable by all conventional arc welding processes such as shielded metal arc welding (SMAW) (covered electrodes), gas metal arc welding (GMAW), gas tungsten arc welding (GTAW),

SAW, and flux core arc welding (FCAW), as well as by electron beam welding (EBW), laser beam welding (LBW), and resistance welding (RW). Welding without filler metals is permitted only where subsequent heat treatment (solution heat treatment) is permitted. If heat treatment is not carried out, there is a great risk that the ferrite content in the weld metal will be too high. As a result, ductility and corrosion resistance will be lower. Welding property requirements and the ease of positional weldability and productivity typically drive the selection of welding method.

General recommendations for filler metal for arc welding of DSS and SDSS usually call for a matching composition but over-alloyed in nickel to guarantee the balance of austenite in the weld metal. Other filler metal compositions may be used, depending on the requirements of a properly qualified welding procedure.

It's a kind of mandatory requirement that all tack welds are performed by qualified welders, using proper protection on both sides (shielding and purge gases) for all welding processes. A good technique would involve eliminating tack welds at the start point of the root pass. During welding, tack welds should be removed before continuing the root pass. Argon is preferred over pure nitrogen as the backing gas because the absorption of nitrogen in the weld pool affects compositional balance, although the use of nitrogen is acceptable in some specific cases. Some literature advocates using the argon–nitrogen mixture as a purge gas. Molten DSS filler materials are more viscous than the standard 300-Series (austenitic S.S) filler metals. Therefore, a slightly wider root opening and included angle than those for standard stainless steel joints should be used. The use of helium in the shielding gas can also enhance weld metal flow properties and penetration. When using single-sided weld joint designs, the best results are obtained by welding the root pass with GTAW, utilizing a back purge. The purging should be continued until the base metal temperature is below 260°C (500°F).

Stringer beads may be used with a slight weave. Excessive weaving, with dwell at the turning points, should be avoided; otherwise, the heat input can locally exceed the recommended maximum. However, a technique using a weave bead in an open root with an opening of 3 mm to 4 mm (0.12–0.16 in.) has been performed successfully.

The technique of back gouging the root and welding from the reverse side of the joint can be used when the weld design permits. As in standard stainless steel fabrication practice, contact with mild steel and low-alloy steel must be avoided.

- **SMAW** welding is particularly excellent for positional welding, single-sided welding, and welding joints where access is limited. With all products, direct current positive (DC+) gives the best welding results. Nonetheless, all rutile-acid electrodes can also be used with alternating current. However, weldability is clearly inferior to that with direct current (DC+). A short arc is always recommended for welding. This gives the best stability and reduces the risk associated with nitrogen pick-up, which is common with long weld arc. The latter (nitrogen pick-up) can lead to the formation of porosities, increased surface oxidation, and can alter the ferrite–austenite ratio. Electrodes should be properly stored in an oven heated to 95°C (200°F) to prevent absorption of moisture, which may lead to weld porosity or hydrogen-induced cracking.

- **GMAW** is a preferred welding process when long welding runs with higher weld deposition rates are desired. This process may be mechanized, semi-automated, or automated. GMAW may be used in the short arc mode (for workpieces of sheet metal-thickness welding), spray arc, or pulsed arc modes. Each arc mode has its typical thermal characteristics, and excessive temperature should be avoided by using proper welding techniques, such as controlling the interpass temperature.

Shielding gases for GMAW of duplex and super duplex stainless steels are somewhat more complicated than those used for GTAW. Shielding gases used for the GMAW process may be argon, argon–oxygen, argon–helium–xygen, argon with a small amount of carbon dioxide, and argon–nitrogen. Helium additions to the gas mixture are helpful for improving the flow characteristics of the molten weld metal and joint penetration. The same GTAW recommendations are valid for preparation and mounting the weld joint.

Wire sizes of 1.0 mm, 1.2 mm, and 1.6 mm (1/32, 3/64, and 1/16 inch) are commonly used for GMAW. Filler wire should be clean and dry and should be stored in a covered container until used. The guide tube should be kept clean and dry. The best results are obtained when the work is done in the flat position. The torch should be maintained as vertical as possible to minimize aspiration of air into the shielding gas. Some references are provided in the following Table 4.1.

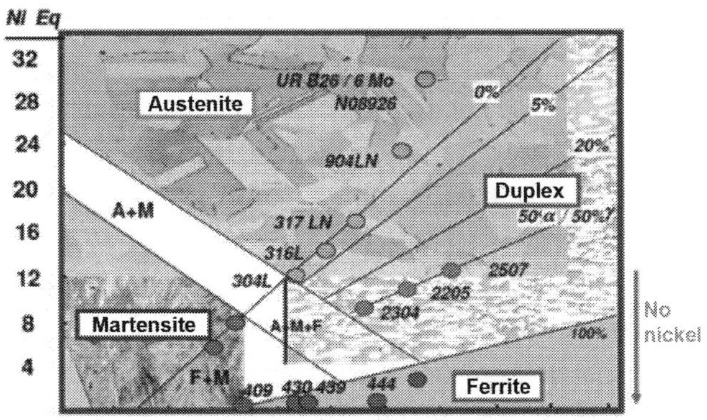

FIGURE 4.10 A modified version of Schaeffler diagram: the predicted ferrite content of various duplex weld metals is shown. All conventional duplex alloy grades, such as 2304, 2205, and 2507 fall in 50% ferrite and 50% austenite range.

4.5.2.5.1 Gas Tungsten Arc Welding

GTAW is generally used to produce short runs of manual welding, with or without filler metal. It may be automated or mechanized depending on the equipment geometry and the application. GTAW is normally used for thin (up to around 4 mm) workpieces. The method is also highly suitable for welding single-sided root beads (both with and without root backing) and is strongly recommended as a process with very high integrity. This welding process is specifically common in the welding of pipe joints. The GTAW process is usually selected for single-sided root-pass welding due to its capability for achieving complete penetration as well as higher weld quality compared to other welding processes. GTAW process may be adopted for welding thin sections (typically 3 mm or 1/8″ and below) by the autogenous welding process. In autogenous welding, the amount of austenite in the weld metal and in the high-temperature HAZ should be closely controlled to avoid deterioration of properties in these regions. Following autogenous welding, a solution-annealing heat treatment should be applied to restore the balance of the microstructure and the mechanical and corrosion properties of the weld joint.

Shielding gases for GTAW process may be argon or a mixture of argon and nitrogen (2% to 3% N_2) to aid in nitrogen retention and Ferrite–Austenite balance in the weld pool. Joint preparation and alignment should produce a consistent root opening.

Flux Cored Arc Welding: FCAW is a process that provides high deposition rates, like GMAW. Gas-shielded and self-shielded welding wires are commercially available for DSS and SDSS welding. As in SMAW, these filler metals produce a deposit with more austenite than the nearly balanced base metal microstructure, which improves its low-temperature properties. When shielding is necessary, the shielding gases typically utilized for FCAW are mixtures of 75% argon–25% CO_2 or 100% CO_2.

FCAW is suitable for material thicknesses above approximately 2.5 mm. Due to the slag that is formed, positional weldability is very good. When FCAW is used, the arc and weld pools are protected by both the slag and the shielding gas. The drop transfer is even, and finishes are extremely smooth and fine. FCAW can advantageously be used for single-sided welding against a ceramic backing; this is fast and efficient. At the same time, the surface properties on the root side are very good.

Submerged Arc Welding: SAW is widely used with duplex steels. Its high productivity and good surface finishes are attractive options for welding DSS. SAW allows the deposition of relatively large volumes of welds with less time at above the transformation temperature for the HAZ. This is the opposite of other welding processes which require a large number of passes with less deposition per pass, resulting in more thermal/transformation cycles for the HAZ. DSS can be welded by SAW process with minimal risk of hot cracking. This is due to the ferritic solidification and duplex transformation over the solidification cycle of the weld metal. SAW welds made at very fast travel speeds combined with unfavorable groove designs may lead to centerline cracks, but lowering the travel speed to the normal range (to match the heat input control) normally solves the cracking problem.

Furthermore, the SAW work environment is considerably better than that of other methods. Both fume generation and radiation are minimal. The disadvantages of SAW are that the heat input is relatively more, and this requires very fine manipulations of welding parameters, especially *the travel speed*. Consequently, small objects present problems. Typical parameters for SAW of duplex stainless steels are summarized in Table 4.2.

For large constructions and for large straight runs of weld, SAW is a cost-efficient and technically satisfactory approach to welding DSS. SAW is commonly used to manufacture heavy wall DSS pipes. For SAW, the usual matching duplex filler metal is appropriate. However, it is important to select the correct flux to achieve the desired properties. It is reported that highly basic fluxes give the best impact toughness for the DSS. For example, a basic agglomerated flux, such as Avesta 805 or Lincoln/Metrode SSB or Sandvik 15W flux has shown proven performance in SAW of DSS/SDSS.

Welding DSS and SDSS to design code/specification requirements are always challenging. All of the common arc welding processes are suitable for joining DSS and SDSS depending on process and economy-related conditions as discussed in Tables 4.3, 4.4, and 4.5. Efforts hae been made to discuss about all the relevant welding parameters in details.

4.5.2.5.2 Shielding Gases

Welding of duplex steels is possible using the conventional shielding gases used with stainless steels. Normally, argon is used with an addition of 2% O_2. Addition of oxygen in the shielding gas increases wetting of the welding pool. An addition of around 30% helium is advantageous; it increases arc energy which, in turn, increases weld pool fluidity and enables higher welding speeds and is preferred for thicker sections, e.g., typically ¾″ (19.00 mm) upwards. Arc stability varies greatly between different arc types, different steel grades, and even between different welding machines. Table 4.4 provides general recommendations for the MIG welding of various duplex grades.

Tungsten inert gas (TIG)/GTAW welding is typically performed with pure argon as the shielding gas. In particular resistance to pitting and crevice corrosion can be considerably raised by the addition of up to 2% nitrogen in the shielding. However, because the risk of weld porosities increases with increased nitrogen content, the latter should not exceed 2%.

Selection of shielding gas for GMAW is somewhat more complex than for GTAW. The GMAW shielding gasses range from pure argon to about 80% argon with additions of helium, nitrogen, and oxygen in various proportions to enhance weldability and final properties of the welded structure. The flow rates depend on the transfer mode, travel speed, and wire diameter but are typically in the range of 12–16 l/min (0.4–0.6 cfm) for 1–1.6 mm (0.035–0.063 inch) diameter wire.

The shielding gases most typically used for FCAW are 75% Ar–25% CO_2 and or 90% Ar–10% O_2 for flat and vertical welding positions, respectively. The flow rate for either gas or position is 20–25 l/min (0.7–0.9 cfm). Welding with 100% CO_2 shielding should preferably be avoided.

4.7.2.5.3 Commonalities for DSS/SDSS

For the root passes, GTAW is most widely used welding process; for the filling and capping runs FCAW, pulsed GMAW, SMAW, SAW are widely used, The aim for welding capping passes is to fill the joint as quickly as possible whilst maintaining the procedural controls necessary to obtain the required mechanical and corrosion properties. All of these welding processes except the SAW are suitable for positional welding.

The weld metal compositions are like the base material compositions given in Table 4.5, except for nickel which is typically approximately 3% higher, compared to the base metal. The relevant ASME/ASTM specifications for the various base materials are also provided in Table 4.5. For manual GTAW, wire is supplied in cut lengths of 1.6 mm, 2.4 mm, and 3.2 mm diameters. The most used size is 2.4 mm; the largest diameter is used to increase productivity when filling thicker joints; 1.6 mm is used only for thin wall tube. For GMAW, and mechanized GTAW, spooled wires are used in 0.8 mm, 1.0 mm, and 1.2 mm diameters.

The most adaptable process for welding duplex stainless steel, SMAW, is widely used for welding DSS with electrodes of 2.5–5.0 mm diameter. Basic coated electrodes *(Type 15 coatings in American Welding Society (AWS) terminology) are utilized for positional welding and for optimum toughness; for optimum operability in less critical applications, rutile electrodes can be used (16 or 17 types in AWS).* As a matter fact, Type-17 rutile-basic electrodes are extremely popular these days. The basic character of the electrode coating ensures smooth and reliable welding operations.

Flux-cored wires are used almost exclusively in 1.2 mm diameter as either flat and horizontal or vertical up positions (with restrictions on weaving). Only wires with excellent slag detachibility and bead profile and all-positional suitability are recommended. *Typically, flux-cored wires are formulated to utilize a rutile flux system and mixed Ar-20% CO_2 shielding gas.*

Submerged arc welding makes use of the same wire as GTAW and GMAW *with the wire diameter normally restricted to 2.4 mm maximum to control heat input. A fully basic agglomerated flux is used in combination with the wire when submerged arc is welded to get the best combination of properties and operability.*

When welding the DSS, the heat input and the cooling rate are important parameters. At too rapid cooling, chromium nitrides are formed, since at high temperatures the solubility of nitrogen in the ferrite is increased. Chromium nitrides are formed when at rapid cooling, the solubility drops. A limited amount of chromium nitrides does not have any effect on the mechanical and corrosion resistance properties of the weld unless they are located on the grains close to the surface. In that case, the corrosion resistance will be decreased because of the depletion of chromium. *Thus, welding of heavy wall thicknesses with too low heat input must be avoided.*

An understanding of the metallurgical structure of DSS and the effects of various treatments on that structure is needed. This is required to address the interaction between proper specification of the materials and proper fabrication/welding practices in obtaining the expected toughness and corrosion resistance in the WM and HAZ.

TABLE 4.6
Chemical Composition of Commonly Used DDS and Other Alloys (1) & (2)

		Composition, Mass % [a e]											
UNS #	Type [b]	Cr	Mo	Ni	N	Cu	C	Mn	P	S	Si	Min PREN [d]	Other
First–Generation DSS													
S32900	Type 329	23.0 – 28.0	1.0 2.0	2.5 – 5.0	–	–	0.080	1.00	0.040	0.030	0.75	26.3	–
S31500	"3RE60"	18.0 – 19.0	2.5 – 3.0	4.25 – 5.25	0.05 – 0.10		0.030	1.20 – 2.00	0.030	0.030	1.40 – 2.00	27.1	
Lean and Semi–Lean DSS													
S32304	2304 [c]	21.5 – 24.5	0.05 – 0.60	3.0 – 5.5	0.05 – 0.20	0.05 – 0.60	0.030	2.50	0.040	0.030	1.00	22.5	–
S32101	2101	21.0 – 22.0	0.10 – 0.80	1.35 – 1.70	0.20 – 0.25	0.10 – 0.80	0.040	4.0 – 6.0	0.040	0.030	1.00	24.5	–
S32003	2003	19.5 – 22.5	1.50 – 2.00	3.0 – 4.0	0.14 – 0.20	–	0.030	2.00	0.030	0.020	1.00	26.7	–
22% Cr Duplex Grades													
S31803	2205	21.0–23.0	2.5–3.5	4.5–6.5	0.08–0.20		0.030					34.1	
S32205	2205, UR45N	22.0–23.0	3.0–3.5	4.5–6.5	0.14–0.20		0.030					35.9	
25% Cr and SDSS													
S31200	–	24.0 – 26.0	1.2 to 2.0	5.5 to 6.5	0.14 – 0.20	–	0.030	2.00	0.045	0.030	1.00	30.2	
S31260	–	24.0 – 26.0	2.5 to 3.5	5.5 to 7.5	0.10 – 0.30	0.20 – 0.80	0.030	1.00	0.030	0.030	0.75	34.0	W: 0.1 – 0.5
S32520	–	24.0 – 26.0	3.0 to 5.0	5.5 to 8.0	0.20 – 0.35	0.50 – 3.00	0.030	1.50	0.035	0.020	0.80	37.1	
S32550	255 [c]	24.0 – 27.0	2.9 to 3.9	4.5 to 6.5	0.10 – 0.25	1.50 – 2.50	0.040	1.50	0.040	0.030	1.00	35.2	–
S32750	2507 [c]	24.0 – 26.0	3.0 to 5.0	6.0 to 8.0	0.24 – 0.32	0.50	0.030	1.20	0.035	0.020	0.80	37.7	–
S32760	–	24.0 – 26.0	3.0 to 4.0	6.0 to 8.0	0.20 – 0.30	0.50 – 1.00	0.030	1.00	0.030	0.010	1.00	37.9 [d]	W: 0.5 – 1.0
S32950	–	26.0 – 29.0	1.0 to 2.5	3.5 to 5.2	0.15 – 0.35	–	0.030	2.00	0.035	0.010	0.60	31.7	

(Continued)

TABLE 4.6
Chemical Composition of Commonly Used DDS and Other Alloys [(1) & (2)] *(Continued)*

UNS #	Type [b]	Composition, Mass % [a e]										Min PREN [d]	Other
		Cr	Mo	Ni	N	Cu	C	Mn	P	S	Si		
S39274	–	24.0 – 26.0	2.5 to 3.5	6.0 to 8.0	0.24 – 0.32	0.20 – 0.80	0.030	1.00	0.030	0.020	0.80	38.6	W: 1.5 – 2.5
S39277	–	24.0 – 26.0	3.0 to 4.0	6.5 to 8.0	0.23 – 0.33	1.20 – 2.00	0.025	0.80	0.025	0.002	0.80	38.9	W: 0.8 – 1.2
Austenitic SS													
S31603	Type 316L	16.0 – 18.0	2.0 – 3.0	10.0 – 14.0	0.10	–	0.030	2.00	0.045	0.030	0.75	22.6	–
S31703	Type 317L	18.0 – 20.0	3.0 – 4.0	11.0 – 15.0	0.10	–	0.030	2.00	0.045	0.030	0.75	29.5	–
N08020	"Alloy 20"	19.0 – 21.0	2.0 – 3.0	32.0 – 38.0	–	3.00 – 4.00	0.070	2.00	0.045	0.035	1.00	25.6	Cb + Ta: 8× C—1.00
N08904	904L	19.0 – 23.0	4.0 – 5.0	23.0 – 28.0	0.10	1.00 – 2.00	0.020	2.00	0.040	0.030	1.00	33.8	
6% Mo Super Austenitic SS													
N08367	–	20.0 – 22.0	6.0 – 7.0	23.5 – 25.5	0.18 – 0.25	0.75	0.030	2.00	0.040	0.030	1.00	42.7	–
S31254	–	19.5 – 20.5	6.0 – 6.5	17.5 – 18.5	0.18 – 0.22	0.50 – 1.00	0.020	1.00	0.030	0.010	0.80	42.2	–

[1] International Molybdenum Association, 2014. *Practical guidelines for the fabrication of duplex stainless steels*. 3rd ed. London: IMOA.

[2] Industeel, 2025. Duplex stainless steels welding guidelines. Technical support document. Industeel (ArcelorMittal).

[a] Single values indicate maximum content unless otherwise specified. The number of significant figures reflects the ASTM recommended practices as shown in ASTM A959 and in ASTM A240, but these rules have not yet been universally adopted for all product forms and all specifications systems.

[b] Unless otherwise indicated, a grade designation originally assigned by the American Iron and Steel Institute (AISI). Names shown in quotation marks are not listed in ASTM specifications.

[c] As listed by ASTM, a widely used common name (not a trademark and not associated with any one producer).

[d] Minimum PREN is calculated based on the minimum chemistry requirements. Note that UNS S32760 which has a minimum PREN of 40 required by the ASTM/ASME material specifications.

[e] The chemistry may vary slightly between product forms, and the specifications often change with time. Hence, for the latest chemistry requirements, the product specifications should be reviewed.

TABLE 4.7
ASME and ASTM Specifications for DSS [(1)]

Product Form	ASME or ASTM Specifications
Plate, sheet	SA-240
Bar products	SA-479, A276
Pipe	SA-790, A928
Tubing	SA-789
Fittings	SA-815
Forgings	SA-182
Castings	SA-351, A890, A995
Testing	ASTM A923

[1] International Molybdenum Association, 2014. *Practical guidelines for the fabrication of duplex stainless steels*. 3rd ed. London: IMOA.

TABLE 4.8
Mechanical Properties of Various Duplex and 316L SS [(1)]

		Tensile Strength, min		Yield Strength, min			Hardness, max	
UNS Number	Type	MPa	ksi	MPa	ksi	Elongation min %	Brinell	Rockwell C
S32304	*2304*	600	87	400	58	25	290	—
S32101	*2101*	650	95	450	65	30	290	—
S32003	*2003*	620	90	450	65	25	293	31
S31803	—	620	90	450	65	25.0	293	31
S32205	2205	655	95	450	65	25.0	293	31
S32304	2304	600	87	400	58	25.0	290	32
S32550	255	760	110	550	80	15.0	302	32
S32750	2507	795	116	550	80	15.0	310	32
S32760	—	750	108	550	80	25.0	270	—
S31603	316L	485	70	170	25	40.0	217	95 R_b

[1] International Molybdenum Association, 2014. *Practical guidelines for the fabrication of duplex stainless steels*. 3rd ed. London: IMOA.

[a] The values shown are for ASME SA-240 plate grades and may vary slightly between product forms. Also, specifications often change with time; hence, for the latest requirements, the product specifications should be reviewed.**UNS S31803** field experience with no failures is as follows: production tubing (110 ksi [760 MPa] YS), 280 psi (1.9 MPa), H_2S, 520 psi (3.6 MPa), CO_2, 20 psi (138 kPa), CO_2, 5% Cl-, 140°C (284°F), eight (8) years; disposal well tubing (75 ksi [520 MPa] YS), 1 psi (7 kPa) H_2S, high % Cl-, 80–100°C (176–212°F), one (1) year.18 [86NCC00331]**UNS S31803** field experience (no failures) is as follows: stinger, gas well (130 ksi [900 MPa] YS), 0.3 psi (2 kPa) H_2S, 1,500 psi (10 MPa) CO_2, 4% Cl-,190°C (374°F), 1.5 years.18 [86NCC00331]**UNS S31803** field experience (no failures) includes the following conditions: gas flowline (65 ksi [448 MPa] YS), 0.06 psi (0.4 kPa) H_2S, 50 psi (344 kPa) CO_2 5% Cl-, 140°C (284°F), two (2) years; gas flowline (65 ksi [448 MPa] YS), 0.04 psi (0.3 kPa) H_2S, 70 psi (483 kPa) CO2? %Cl-, 120°C (248°F), one (1) year. 18 [86NCC00331] Field experience (no failures) with annealed **UNS S31803** included flowlines (three (3) cases); field conditions ranged from 0.05 psi to 125 psi (0.3–862 kPa)H2S, 50°C to 150°C (122 to 302°F), and up to 50,000 ppm Cl–.20 [90NCC00064]

4.6 CORROSION RESISTANCE

All the varieties of DSS have excellent corrosion resistance properties, including resistance to stress corrosion cracking. Such properties are possible, thanks to their austenite–ferrite (duplex) microstructure. In addition to offshore oil and gas industries, where they're the backbones, DSS are finding increasing use in the downstream industries, such as refining, petrochemicals, chemicals and more. This is primarily because they often offer an economical combination of strength and corrosion resistance. Most upstream oil and gas facilities, downstream refinery, petrochemical, chemicals, and other applications where DSS are used are highly corrosive, and DSS or super duplex alloys are required for extremely high corrosion resistance and for resistance against environmental cracking at fairly higher operating temperatures. These days DSS/SDSS are being considered as baseline material for many processes equipment, piping, and many more components. They are using it in applications where carbon steel or austenitic stainless steels had been used in the past. However, DSS/SDSS alloys have been proven to show better long-term integrity considering their higher strength and better long-term reliability. DSS are often used in lieu of austenitic SS, in services where the common austenitic would have problems with chloride pitting or chloride stress corrosion cracking (CSCC).

Most refinery services where DSS are currently being used or being considered in the refining industry contain:

a) a wet, sour, H_2S-containing environment, which may also contain hydrogen, ammonia, carbon dioxide, chlorides, and/or hydrocarbons.
b) water, containing chlorides, with or without hydrocarbons; or
c) hydrocarbons with naphthenic acids at greater than 200°C (400°F) but below the maximum allowable temperatures in the ASME Codes for DSS 260°C to 343°C [500°F to 650°F] depending on the grade and code section involved.

4.6.1 Critical Pitting Temperature (CPT)

Pitting corrosion has to be taken into account for applications involving chloride ions. CPT is defined as the minimum temperature required to produce pitting attack, generally measured in a ferric chloride solution according to ASTM A. The lean duplex UR™ 2304 is more resistant to pitting corrosion than the standard austenitic grades 304L and 316L. UR™ 2205 has an equivalent resistance to 904L, while the 25% Cr super duplex grades present the same pitting resistance as some of the more alloyed 6% Mo super-austenitic materials. Refer to Figure 4.11, for guidance.

Note: The ASTM G48 medium is not representative of the in-service conditions of most applications but gives a ranking of the corrosion resistance of the materials.

4.6.2 Critical Crevice Temperature (CCT)

Crevice corrosion has to be taken into account when confined areas are likely to be formed (for instance, under deposits, flanges, gaskets etc.). UR™ 2205 offers

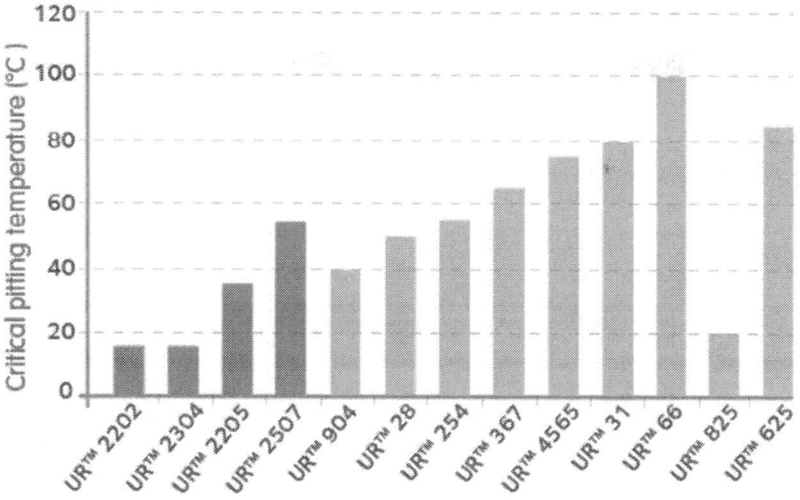

FIGURE 4.11 Critical Pitting Temperatures for various grades of duplex stainless steel. **(Courtesy: Arcelor, Uranus).**

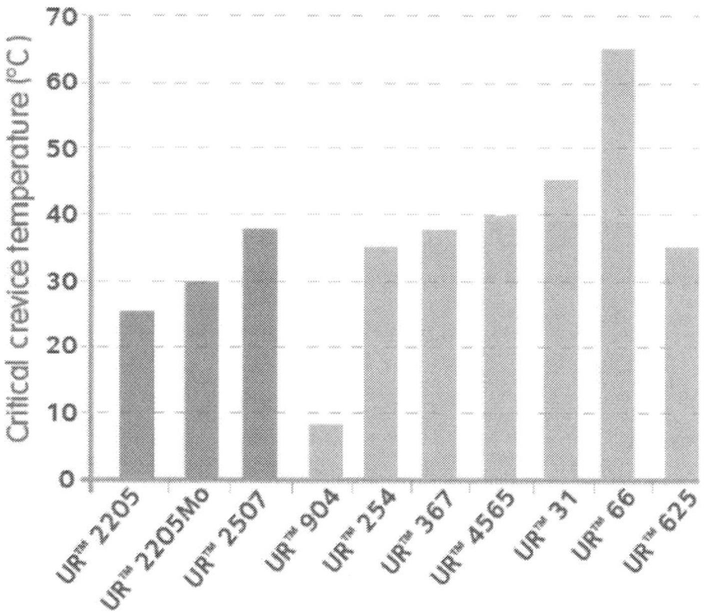

FIGURE 4.12 Critical crevice temperatures for various grades of duplex stainless steel. **(Courtesy: Arcelor, Uranus).**

a significantly higher resistance to crevice corrosion than the standard austenitic steels including 904L. URTM 2507 presents an equivalent resistance to some 6% Mo super-austenitic grades. Refer to Figure 4.12 for guidance. Crevice corrosion tests for DSS/SDSS alloys are usually performed as per testing practice, ASTM G48 B.

4.6.3 Stress Corrosion Cracking (SCC)

Stress corrosion cracking (SCC) is due to the combined influence of a corrosive environment and stress (applied or residual). Standard austenitic stainless steel grades are highly sensitive to SCC in chloride-containing environments, especially above 50°C. The combination of ferrite and austenite phases makes duplex steels significantly more resistant to SCC.

4.6.4 Intergranular Corrosion

Thanks to their two-phase microstructure and their low carbon content, DSS possess an excellent resistance to intergranular corrosion. DSS exhibit a high level of corrosion resistance in most corrosive environments. DSS are often used in lieu of austenitic SS in services where the common austenitic would have problems with chloride pitting or CSCC (chloride SCC).

Duplex alloys are decidedly superior in certain corrosive medias, such as both oxidizing and reducing acids. Such properties result from their high chromium content, which is beneficial in oxidizing acids, along with sufficient molybdenum and nickel to provide resistance in mildly reducing acid environments. The relatively high chromium, molybdenum, and nitrogen also give them very good resistance to chloride-induced pitting and crevice corrosion. Their duplex structure is an advantage in potential CSCC (chloride-induced SCC) environments.

If the microstructure contains at least 25% or 30% ferrite, DSS are far more resistant to CSCC than Types 304/304L or 316/316L. Ferrite is, however, susceptible to hydrogen embrittlement; thus, the DSS do not have high resistance in environments or applications where hydrogen charging is possible into the metal. For example, a cathodic protection (CP) system in offshore super duplex pipelines often may initiate hydrogen-induced stress cracking (HISC). Hydrogen is always generated under the CP system. Cracking potential exists when the hardness is high, typically above 350 HV.

Duplex grades have a minimal chromium content of 20%, which is significantly higher than most of the ferritic and austenitic stainless steels. Furthermore, most of them have Mo additions, as well as added complementary W, Cu elements. All of

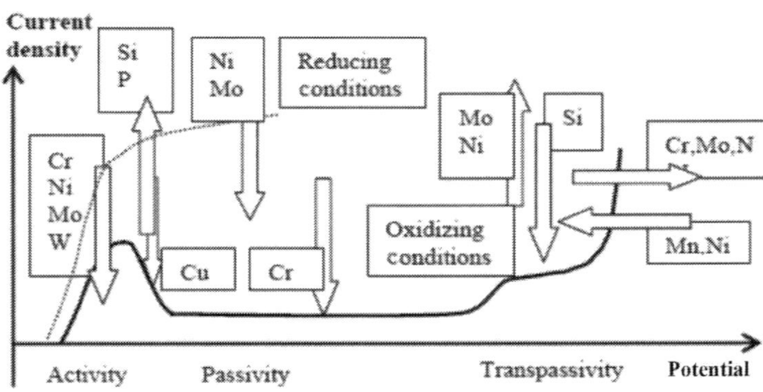

FIGURE 4.13 Influence of alloying elements on corrosion resistance.

these alloying elements are known as beneficial to enhance the corrosion resistance of various duplex grades in most of the in-service conditions, some as stated earlier. Refer to Figure 4.13.

4.6.5 Resistance to Acids

Various industry experiences shows that SDSS containing "Cu" and "W" have proven good corrosion resistance in all types of acids. The presence of copper in SDSS is able to stifle incipient pit growth and thus has been shown to impart significant corrosion resistance improvements. Acid leaching is an important hydrometallurgical process for the extraction of critical materials such as "Nickel." Duplex especially SDSS have proven their worth in such environments.

- FERRALIUM 255 ® (UNS-S32550, 25Cr, 6Ni, 3.5Mo, 1.75Cu, 0.20N) has been designed to harness together this particularly advantageous aspect of copper with the passivating elements chromium and molybdenum to produce an SDSS with proven enhanced corrosion resistance in sulfuric, phosphoric, and nitric acid environments.
- ZERON 100 ® (UNS-S32760, 25Cr, 7.0Ni, 3.5Mo, 0.75W, 0.75Cu, 0.15N) is highly resistant to corrosion in a wide range of organic and inorganic acids. The copper content gives excellent resistance to corrosion in many non-oxidizing acids. Commercial acid applications often contain chlorides and other impurities which can cause corrosion of some stainless steels. ZERON 100 offers much improved corrosion performance in these environments.

4.6.6 Resistance to Caustics

The high chromium content and presence of a large percentage of ferrite provide for the excellent performance of DSS in caustic environments. At moderate temperatures, corrosion rates are much lower than those of the standard austenitic grades. For demanding applications, high nickel alloys and super austenitic grades are recommended. Based on the industry experience some examples are cited here.

- ZERON 100 ® (UNS-S32760, 25Cr, 7.0Ni, 3.5Mo, 0.75W, 0.75Cu, 0.15N) is highly resistant to strong alkalis. The production process of caustic soda results in hot, strong solutions and even in high concentration, e.g., 60 wt% caustic soda, Zeron 100 has very low corrosion rates (<0.1 mm/y). Caustic soda is often found with chlorides in extraction processes and even with 10g/l (10,000 ppm) chloride, Zeron 100 has excellent corrosion resistance. Three years of service experience of fabricated Zeron 100 pipework in 2M caustic soda with chlorides at 230°C has been excellent.
- An iso-corrosion curve for FERRALIUM 255-SD50 in high-temperature solutions of sodium hydroxide indicates typical corrosion rate of 0.5 mm/year, in environments particularly encountered for bauxite purification through the Bayer's process.

4.6.7 Pitting and Crevice Corrosion Resistance

Each grade of stainless steel can be characterized by a temperature above which pitting corrosion will initiate and propagate within approximately 24 hours to a visibly detectable extent. Pitting initiation below this temperature will not occur if left exposed to indefinitely long times. This temperature is known as the critical pitting temperature (CPT). It is a characteristic of stainless steel, which varies with the composition and specific service environment and may occur over a range of temperatures. This is true, as pit initiation is random, and the CPT is sensitive to minor within product/grade as well as compositional variations. However, with a new research tool described in ASTM G 150, CPT can now be accurately and reliably measured by electropotential measurements.

Similarly, critical temperature for initiation of crevice corrosion is called the critical crevice temperature (CCT). The CCT is dependent on the individual composition of stainless steel, chloride environment, and nature (i.e., tightness, length etc.) of the crevice. Typically, the CCT could be 15–20°C (27–36°F) lower than the CPT for the same steel and the corrosion environment.

DSS and SDSS are generally designed with higher Cr levels than the austenitic stainless grades. Since the duplex grades are purposefully alloyed with molybdenum and nitrogen, they usually have significantly better chloride pitting and crevice corrosion resistance than the standard austenitic SS grades such as 304L and 316L (UNS S30403 and S31603). The most common tools for predicting the chloride pitting resistance of corrosion-resistant alloys are the PREN and CPT (as discussed later). A comparison of pitting and crevice corrosion resistance for several stainless steels in the solution annealed condition as measured by the ASTM G 48 procedures (10% ferric chloride) is shown in Figure 4.14.

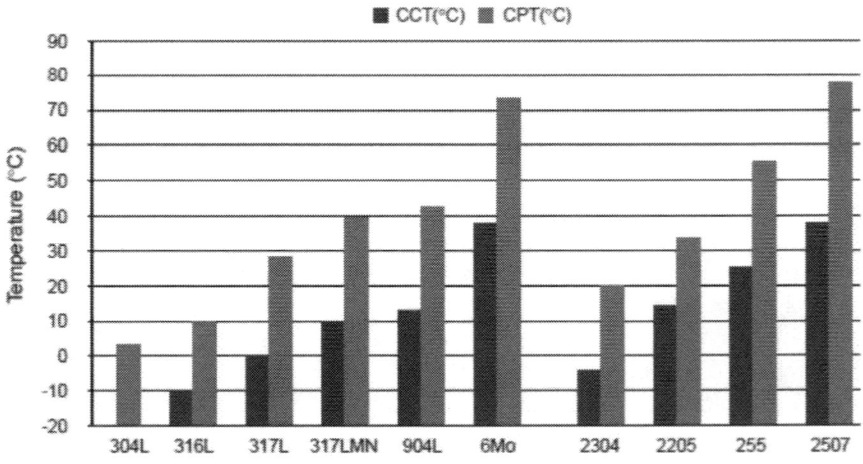

FIGURE 4.14 Comparative values of the CCT and CPT temperatures (°C) in various duplex stainless grades and other alloys.

It is true that PREN is useful in approximate ranking of alloys against chloride-induced pitting and other similar corrosion. Other material factors important for chloride pitting resistance are surface finish, welding quality, and other fabrication details. Generally, higher critical pitting or crevice corrosion temperatures for a specific alloy indicate greater resistance to the initiation of these forms of corrosion. Due to the welding heat input cycle, CPT and CCT in welded or stress-relieved conditions would be expected to be somewhat lower than the similar grade of base metal.

The CPT and CCT of DSS 2205 and all other grades of DSS are well above those of Type 316 (e.g., refer to Figure 4.14). This makes 2205 and other DSS/SDSS as most versatile and preferred materials for applications involving chlorides/salinity, such as brackish waters and deaerated brines and even for critical equipment involved in seawater environment which is extremely a hostile environment. Duplex and super duplex grades are extremely popular and highly competitive against higher alloyed super austenitic stainless steels and nickel-based alloys.

$$\text{CPT} = \text{Constant} + \%\text{Cr} + 3.3 \times \%\text{Mo} + 16 \times \%\text{N}$$

4.6.8 Resistance to Stress Corrosion Cracking

Initial applications of DSS for offshore oil and gas environments were based heavily on their resistance to chloride SCC, which is significantly better than the 300 series austenitic stainless steels with similar chloride pitting and crevice corrosion resistance. In many downstream refining, petrochemical process industries, DSS are replacements for austenitic grades. Such replacement significantly lowers the risk of stress corrosion cracking (both chloride SCC and sulphide SCC). However, DSS may be susceptible to cracking under certain conditions such as in high temperature with chloride-containing environments or when conditions favour hydrogen-induced cracking.

4.6.9 Requirements of Design Codes and Specifications

With the proper solution-annealed microstructure, DSS can have good notch toughness for low (arctic) ambient temperatures but not for cryogenic applications. The minimum allowable temperatures are –51°C (–60°F) in the B31.3 Code and –29°C (–20°F) for some cases in the ASME Section VIII Code. These limits put restrictions on design thickness, among others, and the lower temperatures which can be assured by impact testing of the material. Hence, actual limits are determined by the applicable design and construction codes and project specifications.

The ASME Section VIII requirements for impact testing for DSS base and weld metals are given in UHA-51(d) (3). It requires impact testing of all DSS thicker than 10 mm (3/8 in.) or those with a minimum design metal temperature (MDMT) less than –29°C (–20 °F). The maximum operating temperatures are limited by the susceptibility of the ferritic phase to intermetallic transformation, i.e., leading

TABLE 4.9
ASME Code Maximum Allowable Temperatures, °C (°F) (1)

Grade	ASME Section VIII (Div. 1)	ASME B31.3
S32304	316 (600)	316 (600)
S32101	316 (600) Code Case 2418	NL
S32003	343 (650) Code Case 2503	343 (650) Code Case
S31803 (Note 1)	316 (600)	316 (600)
S31200	316 (600)	NL
S31260	343 (650)	NL
S32550	260 (500)	NL
S32750	316 (600)	316 (600)
S32760	316 (600) Code Case 2245	316 (600)

Note: NL = Not listed

[1] Grade S32205 can use the design allowable for S31803, and the material should be dual-certified.

to 475°C (885°F) embrittlement. To avoid this problem, most design codes applicable to downstream refinery and oil and gas processing plants limit the upper application temperature of various DSS grades to between 260°C and 340°C (500°F–650°F). This code limit applies to the risks associated with continuous long-term exposures above the limiting temperature. This requirement applies to actual metal temperature and not to the process stream temperature. Brief infrequent excursions of the actual metal temperature into the embrittlement range may be tolerated without significant loss of toughness properties; however, the damage from overheating is cumulative. The code does not address the issue of excursions to higher temperatures.

The typical design limits for DSS and SDSS are as specified in ASME Section VIII (Div. 1), which is the reference code for pressure vessels while ASME B31.3 is the reference code for process piping and are listed in Table 4.9.

4.6.10 Design Considerations for Duplex Stainless Steels

As per some aforementioned criteria, Table 4.10 lists all of the mechanical properties as per ASME Section VIII (Div.1)/ASME Sec II, Part D Code. (*Note*: a change from SS 316 to duplex UNS 31803 causes a change of YS from 25 ksi to 65 ksi and for super duplex 32750 to 80 ksi.) Table 4.10 gives a comparison of some allowable design stress values following several codes for the design and construction of pressure vessels. Those results clearly indicate that thickness reductions may be considered with duplex grades when vessels are properly designed.

TABLE 4.10
Allowable Stress Values for Various Grades of Duplex Stainless Steels, Various Pressure Vessel Design Codes

Country	Codes	Allowable Stresses e > 5 mm—20°C/MPa				Savings	
		316L	UR45N	URB6	UR52N	UR45N/ 316L	UR52N+/ UR B6
USA	ASME VIII	115	155	123	190	26%	35%
France	CODAP90	170	275	176	287	38%	38%
UK	BS 5500	150	289	173	294	48%	42%
EU	PED	150	300	167	327	50%	49%

TABLE 4.11
Upper Temperature Limits for DSS for Maximum Allowable Stress Values in Pressure Vessel Design Codes

Grade	Condition	ASME		TÜV	
		°C	°F	°C	°F
2304	Unwelded	315	600	300	570
2304	Welded, matching filler	315	600	300	570
2304	Welded with 2205/2209	315	600	250	480
2205	Unwelded	315	600	280	535
2205	Welded	315	600	250	480
2507	Seamless tubes	315	600	250	480
Alloy 255	Welded or unwelded	315	600	–	–

TABLE 4.12
Typical Temperatures for Precipitation Reactions and Other Characteristic Reactions in DSS (1)

	2205 (Duplex)		2507 (Super duplex)	
	°C	°F	°C	°F
Solidification range	1445–1385	2630–2525	1450–1390	2640–2535
Scaling temperature in air	1000	1830	1000	1830
Sigma phase formation	700–975	1300–1800	700–975	1300–1800
Carbide precipitation	450–800	840–1470	450–800	840–1470
475°C/885°F Embrittlement	350–525	650–980	350–525	650–980

1 International Molybdenum Association, 2014. *Practical guidelines for the fabrication of duplex stainless steels*. 3rd ed. London: IMOA.

TABLE 4.13
Applications of DSS and SDSS in Upstream Oil & Gas Industries. Service Experiences with Duplex/Superduplex Stainless Steels, Some References, NACE, 1F192,2000, Use of Corrosion-Resistant Alloys in Oilfield Environments

Alloy/Location	Operating Conditions	Years in Service	Remarks
UNS S31803 Production tubing	H_2S, 520 psi (3.6 MPa) CO_2, 20 psi (138 kPa) CO_2, 5% Cl-, 140°C (284°F)	8	Field experience with no failures as is.
UNSS31803: Stinger, gas well,	0.3 psi (2 kPa) H_2S, 1,500 psi (10 MPa) CO_2, 4% Cl-, 190°C (374°F)	1.5 years	Field experience (no failures)
UNS S32760 used in offshore (North Sea), installations piping, manifold, and fittings.	3,300 psi (23 MPa) total pressure, 2 psi (0.001 MPa) H2S, 110 psi (0.8 MPa), CO_2, 20,000 ppm Cl-, 104°C (219°F)	Sour gas field, less than 2 years	No inspection information is available
UNS S32550 manifolds, UNS S32760 piping North Sea installa-tions.UNSS32760 valves and piping were used at North Sea installations.	Total gas pressure of 155 psi (1.00 MPa) with 5 ppm H_2S, 0.12% CO_2, 71°C (160°F) total gas pressure of 700 psi (5 MPa) with 33 ppm H_2S, 0.8% CO_2, 46°C (115°F)	2 years or more	Piping, manifold, and fittings of UNS S32760 and UNS 32550 had been in service in a sour gas field. No inspection information is available. However, according to the design conditions and service experiences, these duplex stainless grades are considered to be the best used in the industries.
UNS S32760 Flow lines and manifolds, Vesslefrikkk platform in the North Sea.	3,500 psi (24 MPa) total pressure, 0.2 psi (0.001 MPa) H_2S, 90 psi (0.6 MPa) CO_2, 10,000 ppm Cl-, 110°C (230°F) Sand production was 3 lb/1,000 bbl (1 mg/L) Fluid velocities in the piping were up to 30 m/s (100 ft/s)	3.5 years	Were in excellent condition
Thirteen subsea flow lines, Alloy UNS S31803 North Sea	NA	some since 1978	Shown to be cost-effective compared with carbon steel for distances up to 8–12 km (5–7.5 miles)

TABLE 4.13 (*Continued*)
Applications of DSS and SDSS in Upstream Oil & Gas Industries. Service Experiences with Duplex/Superduplex Stainless Steels, Some References, NACE, 1F192,2000, Use of Corrosion-Resistant Alloys in Oilfield Environments

Alloy/Location	Operating Conditions	Years in Service	Remarks
UNS S31803 manifold piping and well jumper systems	NA	NA	Used in the sweet gas subsea production system in the Popeye Field (Gulf of Mexico)
UNS S31803, Gas flow line	0.06 psi (0.4 kPa) H_2S, 50 psi (344 kPa) CO_2 5%Cl-, 140°C (284°F)	2 years	field experience (no failures)

TABLE 4.14
Environmental and Materials Limits for the Uses of Duplex Stainless Steels. Some References, NACE MR-0175, ISO 15156–3:2003

Materials Type/ Individual UNS Number	Temperature Max, °C (°F)	Partial Pressure H_2S, pH_2S, Max, kPa (psi)	Chloride Conc, Max. mg/l	pH	Sulfur-Resistant?	Remarks
30≤ F_{PREN} ≤40, Mo ≥ 1.5%	232(450)	10(1.5)	See remarks	See remarks	No data submitted	Any combinations of chloride concentration and *in situ* pH occurring in production environments are acceptable
S31803(HIP)	232(450)	10(1.5)	See remarks	See remarks	No	
40< F_{PREN} < 45	232(450)	20(3)	See remarks	See remarks	No data submitted	

Wrought and cast duplex stainless steels shall

- be solution-annealed and liquid-quenched,
- have a ferrite content (volume fraction) of between 35% and 65%, and
- not have undergone ageing heat treatments.

Hot-isostatic pressure-produced (HIP) duplex stainless steel UNS S31803 (30 ≤ F_{PREN} ≤ 40, Mo ≥ 1.5%) shall have a maximum hardness of 25 HRC and shall

- be in the solution-annealed and water-quenched condition,
- have a ferrite content (volume fraction) of between 35% and 65%, and
- not have undergone ageing heat treatments.

Note: Higher values of F_{PREN} provide higher corrosion resistance, however they also lead to increased risk of sigma- and alpha-prime phase formation in the material's ferrite phase, during manufacture, depending on product thickness and achievable quench rate.
The ranges of F_{PREN} quoted are typical of those found to minimize the problem of sigma- and alpha-prime phase formation.

TABLE 4.15
Environmental and Materials Limits for Duplex Stainless Steels used as Downhole Tubular Components and as Packers and Other Subsurface Equipment

Materials Type/ Individual UNS Number	Temperature Max, °C (°F)	Partial Pressure H_2S, pH_2S, Max, kPa (psi)	Chloride Conc, Max. mg/l	pH	Sulfur-Resistant?	Remarks
30≤ F_{PREN} ≤40, Mo≥ 1.5%	232(450)	2(0.3)	See remarks	See remarks	No data submitted	Any combinations of chloride concentration and *in situ* pH occurring in production environments are acceptable.
40< F_{PREN} < 45	232(450)	20(3)	120,000	See remarks	No data submitted	Any combinations of chloride concentration and *in situ* pH occurring in production environments are acceptable. Chloride limits have been found to be strongly dependent upon yield strength and the level of cold work.

For these applications, these materials shall:

- be in the solution-annealed, liquid-quenched, and cold-worked condition,
- have a ferrite content (volume fraction) of between 35% and 65%, and
- have a maximum hardness of 36 HRC.

Note: Higher values of F_{PREN} provide higher corrosion resistance, however they also lead to increased risk of sigma- and alpha-prime phase formation in the materials ferrite phase, during manufacture, depending on product thickness and achievable quench rate.
The ranges of F_{PREN} quoted are typical of those found to minimize the problem of sigma- and alpha-prime phase formation.

4.7 APPLICATIONS OF DUPLEX STAINLESS STEELS, EXAMPLES

High corrosion resistance, resistance to stress corrosion cracking, durability, formability, ease of fabrication, welding, and low maintenance requirements—these are just a few of the characteristics that make DSS the material of choice for various industrial applications.

4.7.1 Desalination and Water Treatment

Desalination and water treatment applications demand a durable material that can withstand the corrosiveness of water, liquids, and salt for at least 30 years. Such a material is DSS. Thanks to its excellent strength, corrosion resistance, long lifespan, and cost-effectiveness as well as other aspects, the benefits of using duplex stainless steel far exceed the initial costs of investment.

FIGURE 4.15 Seawater manifold fabricated from UNS 32760.

FIGURE 4.16 Large super duplex seawater pump for handling high suspended solids in service.

FIGURE 4.17 Banks of evaporators in a desalination plant manufactured from various grades of duplex stainless steel.

FIGURE 4.18 A desalination plant with welded super duplex stainless heat exchanger to handle seawater with high sand contents.

Evaporators for desalination plants are a very key equipment for any desalination plant. Lean Duplex DX2304 or conventional duplex 2205 are very popular grades used for such equipment. Figure 4.17 shows a set of ten evaporators in a Middle East desalination plant. The material chosen here was Lean Duplex DX2304 and other various grades of duplex stainless steels.

4.8 LIST OF ABBREVIATIONS

- ASME—American Society of Mechanical Engineers
- DSS—Duplex Stainless Steel
- SDSS—Super Duplex Stainless Steel
- PREN—Pitting Resistance Equivalent Number
- SCC—Stress Corrosion Cracking
- CSSC—Chloride Stress Corrosion Cracking
- CPT—Critical Pitting Temperature
- CCT—Critical Crevice Temperature

REFERENCES

1. International Molybdenum Association, 2014. *Practical Guidelines for the Fabrication of Duplex Stainless Steels.* 3rd ed. London: IMOA.
2. Industeel, 2025. *Duplex Stainless Steels Welding Guidelines.* Technical support document. Industeel (ArcelorMittal).
3. Damian J. Kotecki, *Lecture Series on Stainless Steel Welding in India*, 24 November–9 December 2006.
4. J. Charles and P. Chemelle, *The History of Duplex Developments*, 8th Duplex Stainless Steels Conference, Beaune, France, 13–15 October 2010.
5. J.O. Nilsson and G. Chai-Sandvik, *The Physical Metallurgy of Duplex Stainless Steels*, Materials Technology. In Proceedings of the Conf. Duplex Stainless Steels, Beaune, France, 13–15 October 2010; KCI: Zutphen, The Netherlands, pp. 369–390.
6. L. van Nassau, H. Meelker and J. Hilkes, *Welding Duplex and Super-Duplex Stainless Steels.* Welding in the World, 31(5), 322–334.
7. NACE International Publication 1F192 (2000 Revision), *Use of Corrosion-Resistant Alloys in Oilfield Environments.* (Publication 1F192, 2000 Revision). Houston, TX: NACE International.
8. ISO & NACE International. (2009). Petroleum and Natural Gas Industries—Materials for Use in H2S-Containing Environments in Oil and Gas Production. (ISO 15156 / NACE MR0175). Geneva: ISO; Houston, TX: NACE International.

5 Practical Aspects of Hard Facing

Pradip Goswami

CONTENTS

5.1 INTRODUCTION

According to practical definitions, hard facing is defined as a surfacing process. Surfacing is an industrial fabricating process defined as the application of a layer or layers of metal to a surface by welding, brazing, or thermal spraying to obtain the desired properties or dimensions. Surfacing is used to meet service requirements,

DOI: 10.1201/9781003352419-5

such as corrosion resistance, wear resistance, and dimensional control, or to provide necessary metallurgical properties. Surfacing process can be carried out by shielded metal arc welding (SMAW), gas metal arc welding (GMAW), gas tungsten arc welding (GTAW), plasma transferred arc welding (PTAW), flux-cored arc welding (FCAW), submerged arc welding (SAW), laser beam welding (LBW), oxyfuel gas welding (OFW) processes, brazing (B), and the thermal spraying (THSP) processes.

Hard facing is a variation of the surfacing process, in which the surfacing material is deposited to *reduce wear by* increasing the resistance of a metal surface to abrasion, impact, erosion, galling, or cavitation. Based on the service requirements, hard facing may or may not be considered in the original design of the component.

Hard facing is usually performed with manual, semiautomatic, mechanized, automatic, or robotic welding techniques. The common welding processes are listed earlier. Hard facing metals are available in the form of bare welding rods, covered electrodes, cored wire, pastes, and powders. Specifications for surfacing metals are difficult to develop, as many of the product compositions tend to be proprietary, not standardized. The latest versions of ASME Sec-II, Pt-C code together with ASME Sec IX code, provide some guidance on how to procure hard facing consumables and how to qualify hard facing procedures to the requirements of various design services.

AWS A5.13/ASME SFA 5.13 (available in ASME Sec-II, Pt-C) is the specification for surfacing electrodes shielded metal arc welding process dealing with multiple types of alloy types. About 26 classifications of iron-base alloy-covered electrodes, five classifications of cobalt-base alloy-covered electrodes, two classifications of nickel-base alloy-covered electrodes, 11 classifications of copper-base alloy-covered electrodes, and five classifications of covered electrodes containing tungsten carbide are described in this specification. Of these AWS classifications, the most commercially used are those for the cobalt-base and nickel-base alloys.

5.2 IMPORTANCE OF HARD FACING

The important properties associated with hard facing are hardness, abrasion resistance, impact resistance, erosive wear resistance, heat resistance, and corrosion resistance. The explanations of these above stated important properties are as follows.

a) **The following levels of hardness can be measured on any hard faced weld deposit:**
   1. Microhardness (gross hardness);
   2. Microhardness (hardness of individual constituents/phases in a heterogeneous structure);
   3. Hot hardness (resistance to the weakening effect of service at elevated temperatures during short-term loading); and
   4. Creep resistance (resistance to plastic deformation when loaded at elevated temperatures for relatively long periods of time).
b) **Impact resistance is described as follows:**
   1. Resistance to plastic deformation under repeated impact loading (related to yield strength and fatigue strength),

2. Resistance to cracking under impact loading, i.e, work-hardening considerations, which is avoidable in alloys with higher ductility.

c) **The properties of metal-to-metal wear resistance are the following:**
1. Friction coefficients, lower friction values mean better wear resistance.
2. Galling resistance. This property helps to avoid any localized fusion and bonding.
3. Surface films (protective oxide layers), thicker oxide layers cause higher wear resistance.
4. Surface toughness, greater the toughness better is the wear resistance.
5. Plasticity (ability to deform).

d) **Several types of abrasion can occur under the following conditions:**
1. Scratching wear under low stress.
2. Grinding wear under high stress; and
3. Gouging wear under high stress and impact.

e) **The types of heat resistance include the following:**
1. Resistance to tempering (softening with time at temperature);
2. Retention of strength when hot (including hot hardness);
3. Creep resistance (time factor added to hot strength);
4. Resistance to oxidation or corrosion by hot gases; and
5. Resistance to thermal fatigue (thermal cycling during service).

Note: All or some of the properties are desired in the hard facing deposits based on the service requirements.

5.3 BENEFITS OF HARD FACING

"Hard facing" is the deposition of a surface layer by welding, which is harder than the base material. Its purpose is to give wear resistance. Hard-faced layers may also be characterized by the following properties:

- Soundness (cracks are acceptable in some cases)
- Toughness, depending on the need to resist impacts
- Resistance to environmental stresses such as corrosion and high temperatures

Hard facing may involve depositing one or multiple layers of the weld metal. Some alloy types are designed to be applied in one layer only, while others can be applied without limit. "Preventive hard facing" is the application of hard facing techniques in the production of a brand-new component. In this case, the nature of the base metal may be less relevant, apart from cost considerations. Most types of equipment, such as crushers and earth-moving equipment, incorporate such hard facing deposits on the new components during manufacturing.

"Remedial hard facing" involves reconstitution of an already worn part to the original or near-original requirements. Hence, compatibility with the material of the part to be hard faced needs to be considered.

By hard facing the equipment, one will obtain the following benefits:

- Reduced maintenance
- Reduced operation costs
- Lower repair costs
- Extended equipment lifetime

While considering the hard facing alloys for specific application-oriented services, the following factors are worth considering.

a) **Abrasion Resistance:** Such as
    - Under low stress (scratching wear)
    - Under high stress (grinding)
    - Under high stress and impact (gouging)
b) **Impact Resistance:** Such as
    - Resistance to plastic deformation under repeated impact loading (related to yield strength and fatigue strength)
    - Resistance to cracking under impact loading (related to ductility but including work-hardening considerations)
c) **Heat Resistance:** Such as
    - Resistance to tempering (softening with time at temperature)
    - Retention of strength when hot (including hot hardness)
    - Creep resistance (time factor added to hot strength)
    - Resistance to oxidation or corrosion by hot gases
d) **Metal-to-Metal Resistance:** Such as
    - Friction coefficients (relative ease of sliding)
    - Galling tendency (localized welding)
    - Surface films (oxide layers)
    - Lubricity (slipperiness)
    - Plasticity (ability to deform)
e) **Erosive Wear Resistance:** Such as
    - Under high-angle solid particle impingement
    - Under low-angle solid particle impingement
    - Under liquid droplet erosion
    - Under cavitation conditions

5.4 HARD FACING ADVANTAGES

When compared with other surfacing processes, hard facing has the following advantages:

1. Additional resistance to wear or corrosion exactly where it is needed
2. Ready application in the field
3. Economical use of expensive alloys
4. A hard surface layer to resist wear that is supported by a tough substrate to carry the load

5.5 APPLICATION OF HARD FACING BY ARC WELDING PROCESSES

Table 5.1 shows a snapshot of common hard facing processes used in industrial applications. It summarizes the commonly used hard facing processes and metal forms. A number of welding processes as explained here are applied for deposition in the hard facing process. Hard facing deposits can be applied with a variety of joining processes, including welding, brazing, and thermal spraying. Most processes involve fusion welding, where the surfacing metal and the surface of the base metal melt, join, and then re-solidify to provide a continuous metallurgical bond. Chemical reactions between the molten metal and the atmosphere can be prevented by using a flux that covers the weld pool or by shielding it with gas or by including protective elements (e.g., deoxidizers) within the surfacing metal.

Note: The scope of this chapter is to cover the hard facing techniques applicable through arc welding processes only.

5.6 HARD FACING BY ARC WELDING PROCESSES

A number of arc welding processes are used for this purpose, as explained here.

5.6.1 Gas Tungsten Arc Welding

The gas tungsten arc welding (GTAW) process employs a non-consumable tungsten electrode shielded by argon, helium, or both. A wrought, tubular, or cast welding rod is fed into the arc area, where heat for melting the surfacing metal is supplied by an arc between the tungsten electrode and the workpiece. Fluxes are not used with this

TABLE 5.1
Common Hard facing Processes and Materials Used

Process	Mode of Application	Hard Facing Metal Form
Shielded metal arc (SMAW)	Manual	Covered electrodes, solid cast electrodes
Flux-cored arc (FCAW)	Semiautomatic or automatic	Composite electrode of metallic sheath
Gas metal arc (GMAW)	Semiautomatic or automatic	Bare solid and tubular electrodes
Submerged arc (SAW)	Automatic	Bare solid and tubular wires and strip added through flux
Gas tungsten arc (GTAW)	Manual or automatic	Powder, bare solid and tubular wires
Plasma arc (PAW)	Automatic	Powder and bare and tubular wires
Thermal Spray Processes		
Flame spraying (FLSP)	Flame spraying (FLSP)	Flame spraying (FLSP)
Plasma spraying (PSP)	Plasma spraying (PSP)	Plasma spraying (PSP)
Arc spraying (ASP)	Arc spraying (ASP)	Arc spraying (ASP)
High-velocity flame	High-velocity flame	High-velocity flame

process. Although slower than other arc welding processes, GTAW provides hard facing of excellent quality. *A common use of this technique is in the application of cobalt alloys on intricate components such as valves.*

Dilution in GTAW process is somewhat greater than other consumable wire processes, such as GMAW. This method requires very little preheating and results in less heat build-up, minimizes distortion, affects the base metal to a lesser degree, and last of all deposits cleaner welds.

Almost every *major ferrous and non-ferrous alloy* can be hard faced with the gas tungsten arc process. The thickness of the base metal typically ranges from a minimum of *6.4 mm (1/4 in.) to much thicker sections.*

The hard facing materials used with this process include the *nickel and nickel alloys, cobalt and cobalt alloys, and tungsten-based alloys also.* Most of the materials are used in wire or rod form, but some are deposited using *metal powders.* Hard facing with *tungsten carbide* typically calls for the use of metal powders, *whereas wire and rod surfacing metals* are used for all other hard facing applications. For the powered carbide products, the size of surface material granules that gives the desired quality of deposit (according to the OEM specifications) should be selected. Welding wires from 0.9 mm (0.035 in.) to 6.4 mm (1/4 in.) typically are used with the gas tungsten arc surfacing process.

Argon, helium, and mixtures of these gases are used as shielding gases for gas tungsten arc surfacing. While most surfacing is performed with *argon, which generally results in lower dilution,* other gases and gas mixtures should be considered. The type of gas used can greatly affect the bead shape and surface quality.

Gas tungsten arc surfacing procedures are essentially the same as those used for actual GTAW welding, with due consideration for *dilution control and deposit thickness.* This gas tungsten arc process is basically a low-deposition rate process that produces a high-quality deposit with minimum dilution and minimum workpiece distortion. A summary of the effects of welding variables on gas tungsten arc surfacing is provided in Table 5.2.

TABLE 5.2
Gas Tungsten Arc Process Variables—Independent Effects on Key Surfacing Characteristics and Process Variables

Variable	Variable Type	Dilution	Deposition	Deposit Thickness
Current type	AC	Average	Average	Average
	DC	Lower or higher	Lower or higher	Lower or higher
Polarity	DCEN	High	High	Thick
	DCEP	Low	Low	Thin
Shielding gas	Argon	Lowest	Lowest	Thinnest
	Helium	Highest	Highest	Highest
Amperage	High	High	High	Thick
	Low	Low	Low	Thin

(Continued)

TABLE 5.2 ***(Continued)***
Gas Tungsten Arc Process Variables—Independent Effects on Key Surfacing Characteristics and Process Variables

Variable	Variable Type	Dilution	Deposition	Deposit Thickness
Bead deposition Technique	Stringe	High	No Effect	Thick
	Weave	Low	No effect	Thin
Bead spacing	Narrow	Low	No effect	Thick
	Wide	**High**	No effect	Thin
Electrode extension	Short	No effect	No effect	No effect
	Long	No effect	No effect	No effect
Surfacing wire	Small	High	Low	Thin
	Large	Low	High	Thick
Voltage	High	Low	No effect	Thin
	Low	High	No effect	Thick
Travel speed	Fast	High	No effect	Thin
	Slow	Low	No effect	Thick
Auxiliary wire(s)	—	Low	High	Thicker

5.6.2 Gas Metal Arc Surfacing

In gas metal arc surfacing, the arc and the weld pool are shielded from the atmosphere by a flow of gas from an external source. *Alloying elements used to produce the desired deposit requirements are contained in the electrode, which may be either solid or tubular. Shielding gases, including argon, carbon dioxide, and helium may be used singly, in various mixtures, or with small amounts of oxygen.* The type of shielding may have a significant effect on the composition, dilution, and other characteristics of the deposit.

Numerous types of base metals can be surfaced with the gas metal arc process, including carbon steels, low-alloy steels, high-alloy steels, and many non-ferrous metals. The thickness of the base metals normally may range from about 6 mm (1/4 in.) and up. The hard facing electrodes are too numerous to list except by broad alloy classification. The common classifications for the welding wires are the nickel and nickel alloys and the tungsten carbide and cobalt and cobalt alloys. Gas metal arc surfacing basically is a *high-dilution process when operating in the spray transfer mode* and can be expected to deposit *4 kg/h to 7 kg/h (9 lb/h to 15 lb/h)* at dilution levels of about 30% to 50%.

The equipment used for gas metal arc hard facing is the same as that used for conventional gas metal arc welding process. In numerous hard facing applications, the equipment needed to oscillate (or weave) the electrode, so as to control the bead shape (with alloyed weld puddles) and the dilution patterns as needed. Oscillators may be controlled by mechanical or electronic means.

TABLE 5.3
Gas Metal Arc and Flux-Cored Arc Process Variables—Independent Effects on Key Surfacing Characteristics

Variable	Variable Type	Dilution	Deposition	Deposit Thickness
Current type	AC	Average	Average	Average
	DC	Lower or higher	Lower or higher	Lower or higher
Polarity	DCEN	High	High	Thick
	DCEP	Low	Low	Thin
Shielding Gas	Argon	Lowest	Lowest	Thinnest
	Helium	Highest	Highest	Thickest
	Carbon dioxide	Intermediate	Intermediate	Intermediate
Arc transfer	Spray	1 (Highest)	1 (Highest)	1 (Thickest)
	Globular	3 (Med to Low)	3 (Med to Low)	3 (Med to Low)
	Short circuit	4 (Lowest)	4 (Lowest)	4 (Thinnest)
	Pulsed	2 (Med to Low)	2(Med to Low)	2 (Med to Low)
Amperage	High	High	High	Thick
	Low	Low	Low	Thin
Bead deposition technique	Stringe	High	No effect	Thick
	Weave	Low	No effect	Thin
Bead spacing	Narrow	Low	No effect	Thick
	Wide	High	No effect	Thin
Electrode Extension	Short	No effect	No effect	No effect
	Long	No Effect	No effect	No effect
Surfacing wire	Small	High	Low	Thin
	Large	Low	High	Thick
Voltage	High	Low	No effect	Thin
	Low	High	No effect	Thick
Travel speed	Fast	High	No effect	Thin
	Slow	Low	No effect	Thick
Auxiliary wire(s)	—	Low	High	Thicker

5.6.3 Flux-Cored Arc Surfacing

Flux-cored arc welding (FCAW) is a modification of the gas metal arc welding process. Therefore, the base metals on which this process can be used for surfacing essentially are identical to those used for gas metal arc surfacing. Flux-cored electrodes are available in a much broader range of alloy classifications when compared to solid electrodes. More varieties of flux-cored electrodes are available within each class.

For many hard facing alloys, flux-cored electrodes are the only available option as *the alloyed solid wires are very stiff and difficult to form. Hence, getting alloys through the solid wires is extremely difficult.* Alloy powders can easily be added to the flux core with an appropriate sheath alloy. Sheaths of low-carbon steel, 300-Series and 400-Series stainless steel, nickel, and cobalt can be used to produce flux-cored surfacing alloys. *Flux-cored electrodes may or may not require auxiliary shielding gas, depending on the specific wire and core formulation and the related AWS OR other industry specifications.* In general, all the process variables for flux-cored arc surfacing are almost similar as those for gas metal arc surfacing. When an auxiliary shielding gas is employed, it is usually carbon dioxide or a mixture of carbon dioxide and another gas, such as argon. With carbon dioxide shielding, the transfer mode across the arc is usually globular and not a spray transfer. In general, the FCAW method produces more dilution and higher deposition rates than gas metal arc surfacing. Unlike gas-shielded processes in field welding applications, flux-cored arc surfacing could provide continuous operation without disruption of gas flow by wind. Self-shielded FCAW wires are proven to be useful for in situ or field applications.

Compared to gas metal arc surfacing, the flux-cored arc surfacing process has the following disadvantages:

1. It may produce a slag, like shielded metal arc surfacing, that may have to be removed before subsequent beads can be deposited.
2. It is much more sensitive to electrode extension, especially the self-shielded variation.
3. The cored electrode cannot be fed around as small a radius as is possible with an equivalent solid electrode.

Although most cored wire deposits contain a slag, there are some electrode wires that contain very little or no fluxing ingredients and are more appropriately described as metal-cored wires. These wires are classified as GMAW filler metals. They typically do not require slag removal between subsequent beads and are more appropriate for automatic operation. An advantage of flux-cored, and metal-cored electrodes is that the manufacturer can customize the chemical composition of the electrode more easily than the solid electrode compositions.

5.6.4 Plasma Transferred Arc (PTA) Hard Facing

Plasma transferred arc (PTA) hard facing is a versatile method of depositing high-quality metallurgically fused deposits on low-cost surfaces. Soft alloys, medium- and high-hardness materials, and carbide composites can be deposited on a variety of substrates to achieve diverse properties such as mechanical strength, wear and corrosion resistance, and creep. PTA hard facing has major significant advantages over traditional welding processes such as oxyfuel (OFW) and as tungsten arc welding (GTAW).

Applications:

The perpetual demand for oil and gas pushes companies to optimize the extraction process to be as efficient as possible. Various components of drilling hammers used in oil extraction experience abrasion from mud, metal-to-metal wear, and erosion.

To combat excessive wear, drilling hammers are hard-faced with composite alloys of tungsten carbide, dispersed in a nickel- or cobalt-based matrix. Plasma-transferred arc (PTA) deposits of Ni–Cr–B–Si alloys are used to combat wear on plungers, sleeves, sucker rods, and seals of mud pumps and submersible pumps. In the refining sector, cobalt-based alloys applied by PTA are used extensively to combat wear, erosion, abrasion, and corrosion. The hard facing of valve seats, which is a high-volume process, was originally done using oxyfuel welding (OFW) and GTAW processes. Since the 1980s, application of hard facing on valves have moved steadily to *PTA due to consistently repeatable quality, productivity and enhanced deposit characteristics.*

In the offshore oil and gas environment, valve seats experience a variety of wear modes such as erosion, adhesion, galling, corrosion, and fatigue. *Cobalt-based alloys, such as Stellite® F and Stellite® 6, have proven effective under these circumstances.* Nowadays, a host of cobalt-based alloys are used in the automotive industry for wear resistance. In addition to cobalt-based alloys, several nickel-based alloys that depend on *borides and carbides for hardness* are also used for hard facing engine valves.

The steam cycle of *power generators contains generators, turbines, and pumps that handle steam and water.* The most significant *wear occurs in steam and water valves,* which experience *high-pressure, high temperatures, and metal-to-metal wear* at seating areas. For efficiency, valve seats and spindles need to inhibit oxidation on the surface, which otherwise causes sticking. PTA deposits of *cobalt-based alloys (e.g., Stellite-6 & Stellite-12)* are used extensively on *valve seats and valve faces* due to their high resistance to *metal-to-metal adhesive and erosive wear. Cobalt alloys also retain hardness at temperatures ranging from 550°C or higher.* Cobalt alloys are deposited on the leading edges of turbine blades to combat the effects of liquid droplet erosion due to condensing steam. Cobalt-based alloys perform well on high-pressure steam turbine nozzles and induced draft fan blades.

Benefits of Plasma Transferred Arc (PTA):

- Easily automated, providing a high degree of reproducibility.
- Allows precise metering of metallic powder feedstocks, resulting in lesser material quantity used.
- Precise control of weld parameters (i.e., powder feed rates, gas flow rates, amperage, voltage, and heat input), ensuring consistency.
- Produces alloy deposits that are tougher and more corrosion resistant than similar deposits laid down by GTAW or OFW processes.
- Produces smooth deposits that significantly reduce required post-weld machining.
- Variety of deposits in thicknesses from 1.2 mm to 2.5 mm (0.05–0.10 in.) or higher. Can be deposited by a single pass at a rate of 1 kg/h up to 13 kg/h depending upon torch, powder, and application.

5.7 METALLURGY OF HARD FACING ALLOYS

A range of hard facing alloys are popular for applications in various industries. Four main groups of hard facing alloys are deposited by welding. These are alloys of iron, cobalt, nickel, and copper. Of these, iron alloys make up by far the greatest tonnage of hard facing application.

Cobalt- and nickel-based alloys are in demand in the most aggressive environments in the oil, gas, nuclear, and other critical applications in the related industries and are the most advanced type of hard facing alloys. This chapter mainly highlights these alloys and their applications.

5.7.1 Cobalt-Based Alloys[1, 2]

Cobalt alloys are resistant to high oxidation, corrosion, and heat. High hardness and creep resistance are retained at temperatures above 540°C (1000°F); these alloys are useful for some types of service at temperatures as high as 980°C (1800°F). The cobalt base combined with chromium provides good corrosion resistance and very good resistance to metal-to-metal wear. Unlike steel, the cobalt alloys are not subject to hardening transformations and have negligible response to heat treatment. Occasionally, a stress relief treatment may be advisable to minimize the risk of cracking. As shown in Table 5.4, four commonly used grades are available, with hardness, abrasion resistance, and crack sensitivity increasing as the carbon and tungsten contents increase. The cobalt-based Stellite™ alloys are the most well-known and successful alloys in the world with the best "all-round" properties. They combine excellent mechanical wear resistance, especially at high temperatures, with very good corrosion resistance. The Stellite™ alloys are mostly cobalt-based with additions of Cr, C, W, and/or Mo. They are resistant to cavitation, corrosion, erosion, abrasion, and galling. The lower-carbon alloys are generally recommended for cavitation, sliding wear, or moderate galling. The higher-carbon alloys are usually selected for abrasion, severe galling, or low-angle erosion.

Stellite 6 as being the most popular alloy in oil, gas, nuclear, and other industries provides a good balance of all of these properties. The Stellite group of alloys retain their properties at high temperatures where they also have excellent oxidation resistance. They are typically used in the temperature range 315–600°C (600–1112°F). They can be finished to exceptional levels of surface finish with a low coefficient of friction to give good sliding wear.

A summarized version of the various properties of these cobalt-based alloys are explained in Table 5.5.

TABLE 5.4
Commonly used Cobalt Surfacing Alloys

		Typical Chemical Composition, %						
AWS Class	Commercial Grade	C	Cr	W	Mo	Ni	Co	Deposit Hardness, HRC
CoCr-A	Stellite-6	1.0	29	4.5	—	—	63	32–42
CoCr-B	Stellite 12	1.5	29	8.0	—	—	60	40–47
CoCr-C	Stellite 1	2.5	29	12.5	—	2—	55	45–55
CoCr-E	Stellite 21	0.2	27	—	5.5	2.5	63	20–35

TABLE 5.5
Various Properties of Cobalt-based Hard Facing Alloys

Properties	Characteristics
Hardness	The usual hardness ranges for these alloys are dependent upon the specific alloy and welding process selected. • For example, CoCr-A may range from C 38 to C 47 when welded with the oxyfuel process and from C 23 to C 47 when an arc welding process is used. Similarly, CoCr-B can range from C 45 to C 49 with oxyfuel and from C 34 to C 47 with arc welding, and alloy CoCr-C may range from C 48 to C 58 with oxyfuel and C 43 to C 58 with arc welding processes. • The cobalt-based alloys are exceptions to the norm in that although they exhibit lower hardness while hot, they return to approximately their original hardness upon cooling and can be considered immune to tempering.
Hot hardness	Elevated temperature strength and hardness are outstanding properties of CoCr filler metals. • These materials are generally considered superior to other surfacing alloys where these properties are required above 1200°F (650°C). • Additionally, at temperatures above 1000°F to 1200°F (540–650°C), weld deposits of these filler metals have greater resistance to creep than other available surfacing alloys for which data is available. • This distinction and their hardness at 1200°F (650°C) and above are the primary reasons for their selection for use in many applications.
Impact	Resistance to flow under impact increases with carbon content in CoCr filler metals. • CoCr-C weld deposits are quite brittle and crack readily when impact flow does occur. • CoCr-A deposits, while easily deformed, can withstand some plastic flow under compression before cracking.
Oxidation resistance	• The presence of more than 25% chromium in CoCr filler metals promotes the formation of a thin, tightly adherent protective scale under oxidizing conditions. • Scaling resistance to combustion products of internal combustion engines is also generally adequate, even in the presence of lead compounds.
Corrosion resistance	• CoCr filler metals are frequently useful where abrasion and corrosion are involved. The corrosion resistance is due to large contents of Cr in the alloys. • They can be considered corrosion resistant in the less severe media, in foods, and in air; and they may even have good resistance in some corrosives. • However, an application that involves corrosion should be confirmed by a field test.
Abrasion	Carbon content has much to do with the response of CoCr filler metals to abrasion. • At 1% carbon (CoCr-A), the performance is inferior to that of carbon steel; at 2.5% carbon (CoCr-C), the resistance to high-stress grinding abrasion is good.
Metal-to-metal wear	• The CoCr filler metals are well suited for metal-to-metal wear because of their ability to take a high polish and their low coefficient of friction.
Mechanical properties in compression	• The compressive yield strength ranges from 64 ksi to 76 ksi (441–524 MPa) for CoCr-A and 85 ksi to110 ksi (586–758 MPa) for CoCr-C. • The ultimate compressive strength similarly ranges from150 ksi to 230 ksi (1,034–1,586 MPa) for CoCr-A and 250 ksi to 270 ksi (1,724–1,861 MPa) for CoCr-C.

(Continued)

TABLE 5.5 (*Continued*)
Various Properties of Cobalt-based Hard Facing Alloys

Properties	Characteristics
Machinability	• None of the deposits from CoCr filler metals are easily machinable, and the difficulty increases along with the increased carbon content.
Identification	• These materials usually may be distinguished by their relative hardness and brittleness. • They are non-magnetic. • A spark test may be used to differentiate them from austenitic manganese steel. • However, the austenitic chromium ions are so similar that an acid test may be required to differentiate between these materials.

5.7.2 Stellite Alloys

Stellite® are predominantly cobalt-base alloys consisting of complex mixed carbides in a CoCr-based solid solution-strengthened alloyed matrix. These alloys are called cobalt-based alloys, as cobalt is the major alloying element. They are resistant to corrosion, erosion, abrasion, and sliding wear and retain these properties at high temperatures, where they also resist oxidation. They have outstanding anti-galling properties. The wear resistance of Stellite alloys is due to the unique inherent wear characteristics properties imparted by a hard carbide phase dispersed in a cobalt alloy matrix. Stellite alloys are often used as a weld hard facing on valves, spindles, and other wearing parts.

5.7.3 Nickel Alloys

Nickel hard facing alloys were developed as substitutes for certain cobalt alloys to reduce alloy cost or to avoid the presence of cobalt altogether in areas of high radioactivity, where the high radiation intensity of the cobalt-60 isotope is objectionable. Typically, limitations of cobalt-based hard facing alloys are common in nuclear power plant due to retained radioactivity of cobalt. Hence, nickel-based alloys are preferred for hard facing applications in radioactive services.

A few types of nickel-based alloys are used: those containing about 2–3% boron, which are hardened by chromium and nickel borides (often with chromium carbides), and those employing Laves phases.

At low-chromium content (Cr in the order of 5%), nickel boride (Ni_3B) is the dominant hard phase. With increasing chromium, the nickel boride gives way to chromium borides (CrB and, at still higher chromium levels, Cr_5B_3). The borides provide resistance to severe abrasion similar to abrasion resistance provided by carbides in the iron and cobalt-based alloys. Nickel provides better matrix strength at high temperatures than iron, so these alloys must be used in applications similar to those of cobalt alloys. The general comments previously made on the cobalt alloy hard facing metals also apply to this group, although hot hardness and resistance to high-stress abrasion are lower for nickel alloys.

A few grades of nickel-based hard facing alloys are listed in Table 5.8.

TABLE 5.6
Various Stellite Alloy as Base Metals

Alloy	Composition, Wt%	Hardness, HRC	Melting Range °C	Typical Applications
Stellite-1	C–2.5, Cr-32, W-13, Co-Bal	55	1250–1290	• Very high resistance to abrasion • Retains hardness even above 600°C • Used for pump sleeves, rotary seat rings, wear pads, and bearing sleeves • Also available as welding filler rod and electrode
Stellite-6	C-1.0, Cr-27, W-5, Co-Bal	42	1280–1390	• Tough erosion-resistant alloy, widely used, good all-round performances • Lower tendency to crack as compared to Stellite 12 in multiple layers but more wear resistant than stellite 21 • One of the most popular stellite alloys used in the industry • Also available as welding filler rod and electrode
Stellite-12	C-1.8, Cr-30, W-9, Co-Bal	48	1280–1315	• Properties between those of Stellite-1 and Stellite-6 • More abrasion resistant than Stellite-6 but still has good impact resistance • Also available as welding filler rod and electrode
Stellite -20	C-2.5, Cr-32, W-17, Co-Bal	58	1260–1320	• The most abrasion-resistant standard cobalt-based alloy • Low shock resistance • Stellite 20 is commonly used in slurry environments involving pumps, pump sleeves, rotary seal rings, wear pads, and bearing sleeves. • Also available as welding filler rod and electrode
Stellite -21	C-0.2, Cr-27 Mo-5.0, Co-Bal	32	1180–1360	• Excellent resistance to thermal and mechanical shock • Excellent high-temperature strength • Applications involve erosion and high-temperature corrosion • Valve trims for high-pressure steam, oil, gas, and petrochemical industries • Also available as welding filler rod and electrode

TABLE 5.7
Various Stellite Alloys Available as Welding Consumables

Alloy	Nominal Composition of Welding Consumables Wt%	UNS	ASME/AWS Specification	Weld Hardness, HRC
Stellite alloy-1	Co:Bal Cr:32 W:12 C:2.45 Ni:< 3.0 Mo:<1.0	R30001	SFA 5.21-ERCoCr-C	51–56
Stellite alloy-6	Co:Bal Cr:30 W:4.0–5.0 C:1.2 Ni:< 3.0 Mo:<1.0	R30006	SFA 5.21-ERCoCr-A	40–45
Stellite alloy-12	Co:Bal Cr:30 W:8.0 C:1.55 Ni:< 3.0 Mo:<1.0	R30012	SFA 5.21-ERCoCr-B	46–51
Stellite alloy-20	Co:Bal Cr:33 W:16 C:2.45 Ni:< 3.0 Mo:<1.0	–	–	53–59
Stellite alloy-21	Co:Bal Cr:28 W:0.25 C:1.55 Ni:3.0 Mo:5.2	R30021	SFA 5.21-ERCoCr-E	28–40

TABLE 5.8
Represents Various Nickel-based Alloys, Bare Welding Filler Metals

Alloy	Nominal Composition of Welding Consumables Wt%	UNS	ASME/AWS	Hardness, HRC
Nestelle Alloy C	Cr:17 W:5.0 C:0.1 Ni: Bal	N30002	—	17–27

TABLE 5.8 (*Continued*)
Represents Various Nickel-based Alloys, Bare Welding Filler Metals

Alloy	Nominal Composition of Welding Consumables Wt%	UNS	ASME/AWS	Hardness, HRC
	Mo:17 Others:0.3%V			
Delloro Alloy40	Cr:12 W:0.0 C:0.4 Ni: Bal Mo:0.0 Others: 1.6% B	N99644	SFA 5.21 ERNiCr-A	36–42
Delloro Alloy50	Cr:12 W:0.0 C:0.5 Ni: Bal Others:2.2%B	N99645	SFA 5.21 ERNiCr-B	48–55
Delloro Alloy 55	Cr:12 W:0.0 C:0.6 Ni: Bal Others:2.6%B	—	—	52–57
Delloro Alloy 60	Cr:13 W:0.0 C:0.7 Ni: Bal Others:3.0%B	N99646	SFA 5.21 ERNiCr-C	57–62

5.8 STELLITE HARD FACING DEPOSITION—EXAMPLES AND PRACTICAL ISSUES

5.8.1 Deposition on Carbon and Low-Alloy Steels—Preheating

The weldability of these steels is directly related to their hardenability, which results in the formation of a brittle heat-affected zone (HAZ) below the weld if they are cooled too fast from the welding temperature.

Effectively, if the part is cooled too fast or if the preheat before welding is too low, then the steel in the HAZ is quenched from an austenitic structure to a brittle semi-martensitic structure directly after welding. The final structure is dependent on the cooling rate from the peak temperature. It is therefore standard practice to reduce the cooling rate by preheating the part to be hard-faced before welding. For larger components, the bulk of the steel can "quench" the overlay once welding is completed. A certain amount of preheat is usually necessary even for small parts

FIGURE 5.1 Stellite despositon on valve spindle; process PTAW.

FIGURE 5.2 Stellite deposition on valve spindle; preheating and posting are the requirements, when the substrate is alloy steels, martensitic steels, tool steels.

FIGURE 5.3 Stellite despositon on valve spindle; process GTAW.

FIGURE 5.4 Stellite alloy overlay, process GTAW.

FIGURE 5.5 Stellite deposition on valve spindle, finished look.

with low hardenability, as it facilitates the hard facing process itself. The hardness and brittleness of the martensite and its tendency to form, and hence the weldability of the steel, are deduced from the carbon equivalent (CE) as defined by the International Institute of Welding formula:

$$CE = \%C + \%Mn/6 + \%Ni/15 + \%Cu/15 + \%Mo/5 + \%V/5$$

5.8.2 Deposition on Carbon and Low-Alloy Steels—PWHT Requirements

For steels with a low carbon equivalent (CE below about 0.50), it is usually possible to avoid martensite formation and HAZ cracking by preheating (typically to about 100–150°C) before welding, maintaining interpass temperatures during welding, and cooling slowly from the weld interpass temperature (e.g., by cooling in vermiculite).

A higher CE (about 0.50% and above) may mean that martensite formation is unavoidable. In this case, the best approach is to obtain the original tempering temperature of the steel as supplied before welding. This temperature can often be obtained from the steel certification and will usually be below 650°C (1200°F). Alternatively, a typical heat treatment temperature for the steel in question can be obtained

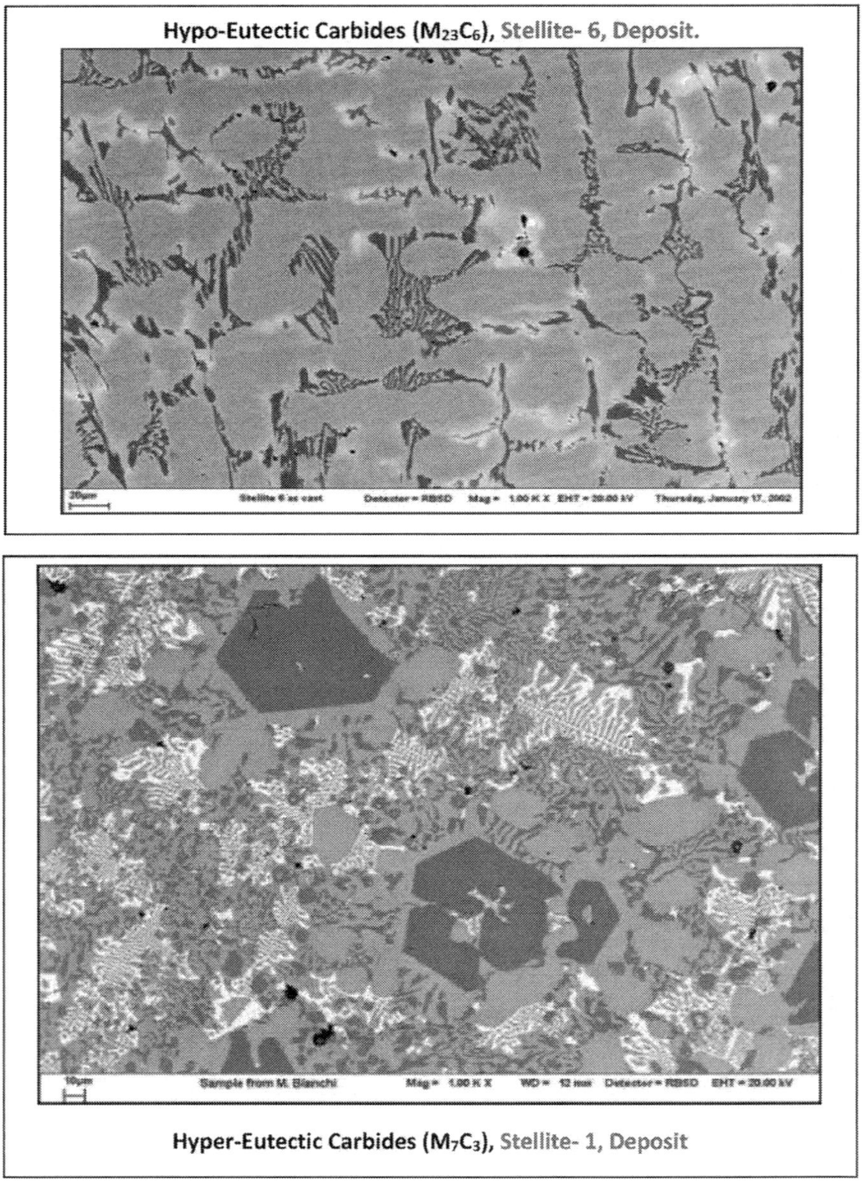

FIGURE 5.6 Depicting two types of stellite alloys. Carbide type—C > 0.8%, Strengthened by carbides of Cr, W, Mo.

from other sources. The optimum post-weld heat treatment (PWHT) temperature is about 50°C (100°F) below the usual tempering temperature. The aim is to temper the weld deposit to reduce hardness and brittleness in the HAZ sufficiently to prevent cracking, without over-tempering the rest of the component. Welded components are

TABLE 5.9
Hardness vs. Ductility, Various Stellite Alloys.

Alloy	Composition	Hardness
Stellite 21	27Cr, 0.2C, 5.5Mo	32 HRC
Stellite 6	28Cr, 1.2C, 5W	42 HRC
Stellite 12	29Cr, 1.85C, 9W	49 HRC
Stellite 1	33Cr, 2.45C, 13W	55 HRC
Stellite 20	22Cr, 2.45C, 17.5W	57 HRC

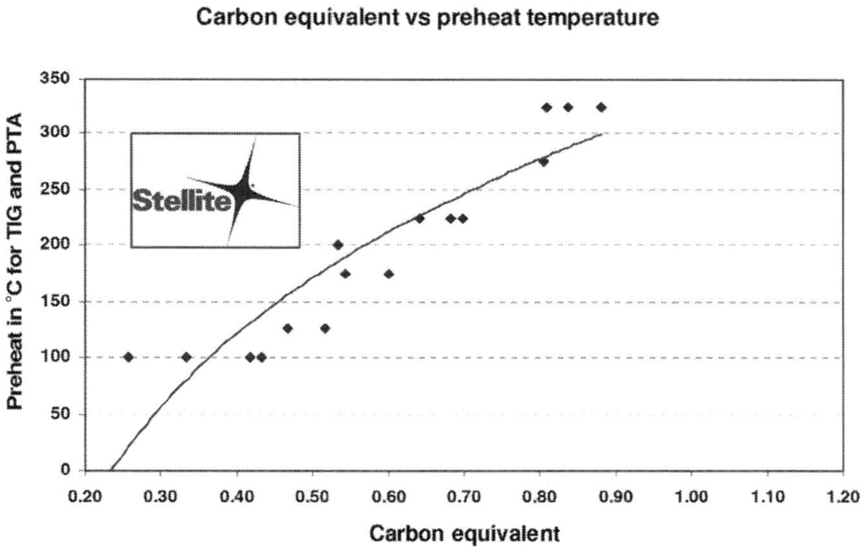

FIGURE 5.7 Typical preheat temperatures vs carbon equivalent for various steels.

usually placed in a furnace directly after welding, without loss of heat, and held at the PWHT for 1–2 hours followed by slow cooling to prevent martensite formation. It is not necessary to get the entire part thickness to reach this temperature, as it is only the surface layers which are at risk.

5.8.3 Medium-Alloy Steels, Martensitic Stainless Steels, Tool Steels, and Other Specialist Steels

The CE of these steels are often above 1.10, and they present a challenge for hard facing. These steels are generally heat-treatable. They are best handled on an individual basis after assessing their CE, hardenability, M_s and M_f temperatures, and typical tempering temperatures, as well as the individual component and hard facing requirements and customer specifications.

5.8.4 Austenitic Stainless Steels and Duplex Steels

From a hard facing point of view, stainless steels are handled quite differently to carbon steels, tool steels, and alloy steels. The CE is not important, rather the first thing to ascertain is whether the steel is austenitic, ferritic, martensitic, precipitation hardened, or a duplex steel.

5.8.5 Austenitic Stainless Steels

Unlike many other steels, austenitic steels do not form brittle martensite in the HAZ, because the high-temperature face-centered cubic (FCC) structure of the steel (austenite) is stabilized down to room temperature, mainly by the high nickel content and in some cases also by nitrogen, manganese, and/or copper.

In principle, weld hard facing onto these steels is not problematic when it comes to Stellite® deposit. They are usually welded using a low preheat temperature (50–150°C) followed by a slow cool under a blanket.

The difficulty with austenitic steels is that during hard facing, carbon can diffuse from the Stellite alloy into the HAZ, causing formation of carbides on the steel grain boundaries, and in the process, it depletes the Cr near the grain boundaries. The result is that the grain boundaries in the HAZ are less corrosion-resistant than the steel base metal overall, which can be a problem if the component is to be used in a corrosive environment (e.g., seawater). This process is known as "weld decay" or "sensitization" and is illustrated schematically in Figure 5.8. It is necessary to be

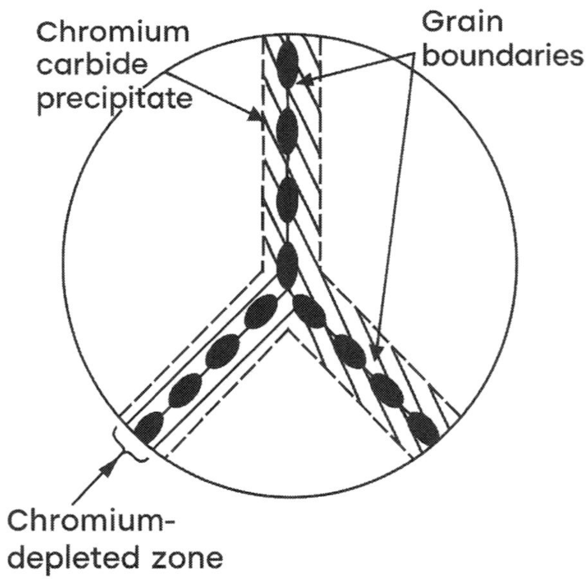

FIGURE 5.8 Schematic illustration of sensitization (weld decay).

aware of this and to evaluate each individual application as to the risks involved. Usually, additions of strong carbide-forming elements such as Nb or Ti help prevent sensitization, because the carbon then bonds preferentially with these elements rather than with the Cr. Such steels are called "stabilized," and the most common examples are AISI 321 and AISI 347 stainless steels. The alternative approach, namely lowering the steel's carbon content to reduce the risk of sensitization (e.g., the 316L, 304L etc. steels) does not apply when hard facing with Stellite alloys, as the overlay material provides the necessary carbon for sensitization to occur.

A commonly followed technique is the use of a nickel-based "barrier layer" as a first weld layer with little or no preheat and a very low interpass temperature. The nickel layer inhibits carbon diffusion, and the steel surface reaches lower temperatures during subsequent welding.

5.8.6 Duplex and Super Duplex Stainless Steels

Duplex steels consist of a mixed structure of austenite and retained ferrite (roughly equal proportions, with ferrite being the matrix). The duplex stainless steels alloying additions are either austenite or ferrite formers, extending the temperature range over which the phase is stable. Among the major alloying elements in duplex stainless steels, chromium and molybdenum are ferrite formers, whereas nickel, carbon, nitrogen, and copper are austenite formers. They are widely used in the oil and gas industry, typically where higher strength than that of standard austenitic steels is required. Also, they have better resistance to stress corrosion cracking (SCC) in chloride environments.

Typical duplex steel grades are as follows:

- Fe 23Cr 4Ni 0.1N (AISI 2304, UNS S32304)
- Fe 22Cr 5.5Ni 3Mo 0.15N (AISI 2205, UNS S32205)
- Fe 25Cr 5Ni 2.5Mo 0.17N Cu (AISI 2505)
- Fe 25Cr 7Ni 3.5Mo 0.25N W Cu ("Super Duplex" UNS S32760)

When hard-faced with Stellite alloys, these steels suffer not only from sensitization, as described previously for austenitic steels, but they are also susceptible to embrittlement due to the formation of secondary phases, i.e., intermetallics, carbides, and nitrides from extended exposure to elevated temperatures.

Typical sensitization issues on a stellite deposit on stainless steel.

Significant secondary phase precipitation may lead to a loss of corrosion resistance and loss of toughness.

From a weld hard facing point of view, their biggest drawback is the fact that when subjected to the typical thermal cycle of a welding process, they very easily and in a short time form a variety of brittle phases such as sigma (K), chi (L), and alpha-prime (M') phases. This occurs when they are subjected to temperatures of 300–1000°C (570–1800°F). This is virtually unavoidable because during weld hard facing, the surface undergoes melting, and so even if the bulk of the material remains below

300°C, there is always a zone in between which will experience temperatures in the embrittlement range, even if only for a brief period of time. This is why the minimum of preheat is recommended for duplex stainless steels. Depending on the component geometry and the type of Stellite alloy being deposited, the fastest possible cooling rate from the welding interpass temperature needs to be applied to minimize the time spent in the embrittlement zone. Normally, the interpass temperature is maintained at around 150–200°C to avoid precipitation temperatures.

Another approach is to apply a ductile "buffer layer" such as Nistelle® 625 (austenitic nickel–chromium-based super 625 alloys) alloys as a first weld layer with little or no preheat and a very low interpass temperature. During subsequent application of the Stellite overlay, the steel surface is shielded from the weld pool and reaches lower temperatures than it would otherwise have reached.

It is necessary for design engineers to be aware of these issues and to evaluate the potential effects in the intended application, for example, by considering the individual steel, the component geometry, and the speed at which brittle phases would be expected to form in each type of steels.

5.8.7 Dilution

Stellite alloys are cobalt-based. When they are applied to steels by hard facing, there is intermixing of the Stellite alloy and the steel, resulting in a higher iron content in the overlay, especially in the first layer. With respect to Stellite alloys, "dilution" has the following effects:

- It decreases the corrosion resistance of the Stellite alloy in extreme corrosion environments, because the Cr content drops to 20–25%Cr (but this is still better than many stainless steels).
- It decreases the hardness.
- It decreases the wear resistance by virtue of increase in the stacking fault energy of the Co-matrix, which affects mainly the galling properties.
- It decreases in the overall carbon content, which affects the general wear resistance and hardness.
- Dilution by Fe or Ni may be expected to reduce the brittleness by virtue of the fact that the content of brittle carbide phases is decreased.

5.9 CONCLUSION

Safe practices for surfacing are similar to those used for welding. In general, surfacing typically involves the use of consumables with higher alloy content than those used for joining. These consumables, when used in conjunction with open-arc processes such as flux-cored arc welding, may result in fume generation rates that are significantly higher than those experienced with the joining processes. Therefore, close attention should be paid to the total fume levels and the levels for specific elements such as Cr, Mn, Ni, and Co to stay within the exposure limits permitted. The consumable manufacturers' material safety data sheets (MSDS) and pertinent warning labels should be followed. Additional safety information can be found in the publications listed in References.

REFERENCES

1. Deloro Stellite, Danie De Wet, *Coatings in the Valve Industry*, Presentation, Valve Manufacturers Association Technical Seminar, 8–9 March 2012, Houston.
2. Jim Wu, *Advances in Wear Resistant Alloys and Products*, Presentation, Deloro Stellite Group, April, 2007.

6 Practical Aspects in Welding of Inconel/Inconel Alloys

Purvesh K. Nanavati

CONTENTS

6.1 INTRODUCTION

Solid-solution and precipitation-hardened Inconel alloys are renowned for their exceptional mechanical properties and corrosion resistance in high-temperature environments. Solid-solution alloys like Inconel 600 offer high-temperature strength and oxidation resistance, making them ideal for aerospace and chemical processing

DOI: 10.1201/9781003352419-6

applications. Precipitation-hardened alloys, such as Inconel 718, provide superior tensile strength and creep resistance due to the formation of intermetallic phases.

Welding these alloys presents significant challenges. Solid-solution alloys are prone to grain coarsening and phase segregation, leading to weakened weld areas. Precipitation-hardened alloys face issues like solidification cracking and strain-age cracking during post-weld heat treatment. Practical solutions include optimizing welding parameters to control heat input, using appropriate filler materials to match the alloy's composition, and employing advanced techniques like pulse current GTAW to minimize defects. Post-weld heat treatment is crucial for restoring the desired microstructure and ensuring the welded joint's mechanical integrity.

6.2 OVERVIEW OF INCONEL AND INCONEL ALLOYS

Inconel, a family of nickel–chromium super alloys, is a trademark of Special Metals Corporation, which pioneered advanced nickel-based alloys since its inception as International Nickel Company in 1902. It introduced alloys like INCONEL® and INCOLOY®, while another industry leader, Haynes International, developed corrosion-resistant, high-temperature HASTELLOY® and HAYNES® alloys since its foundation in 1912. Special Metals Corporation also debuted the first industrial alloy, MONEL metal.

Nickel has a face-centered cubic (FCC) "austenite" structure at all temperatures, and it can be alloyed with elements like copper (Cu), iron (Fe), chromium (Cr), molybdenum (Mo), and cobalt (Co) to create superior alloys, which act as solid solution strengtheners and form the basis for many of the binary (e.g., Ni–Cu and Ni–Mo alloys) and ternary (e.g., Ni–Fe–Cr and Ni–Cr–Mo alloys) types of Ni-based alloys. Chromium plays a vital role in forming a passive oxide film for corrosion resistance, and nickel's high solubility for chromium makes it an ideal basis for Ni-based alloys [1][2] Strength can be enhanced through the formation of precipitates of the Ni2(Mo,Cr) phase in high-temperature alloys. Aluminum and titanium promote the formation of the γ'Ni3(Ti,Al) precipitate which improves the high-temperature strength of the alloy [3][4]. However, these high-temperature alloys can be prone to strain-age cracking during post-weld heat treatment. Niobium (Nb) addition helps form the γ''Ni3Nb precipitate for strengthening, but it can also lead to solidification cracking. Boron and zirconium improve creep strength and ductility but adversely affect hot cracking resistance during welding [2].

Carbon content varies based on the intended use of the alloy. Corrosion-resistant alloys have low carbon content (below 0.01 wt%), while high-temperature alloys typically have carbon levels between 0.05 wt% and 0.1 wt% to form strengthening carbides. This is because lower carbon levels reduce carbide formation in the heat-affected zone (HAZ), which depletes the austenite matrix of important alloying elements like chromium and molybdenum, reducing corrosion resistance [5][6]. Despite low levels of carbon, impurities like sulfur, phosphorus, and lead can significantly impact the weldability of Ni-based alloys, necessitating the addition of magnesium and manganese to mitigate their effects [2].

The Ni-based alloys are used in a variety of applications in the engineering of systems that are exposed to extreme conditions, including highly corrosive and high-temperature environments. They are chosen as a material of construction for applications demanding aqueous corrosion resistance and heat resistance and high-temperature properties. For example, Inconel Alloy 625 with high strength and toughness up to 980°C coupled with low-temperature cryogenic properties. For example, Inconel alloy 718 can work from –253°C to 750°C.

Unlike other alloy systems, Ni-based alloys lack a comprehensive classification system and are usually identified by trade names or initial manufacturer-designated names/numbers. The use of a unified numbering system (UNS) helps cross-reference these alloys based on their chemical composition. Table 6.1 represents some commercial alloys and their UNS numbers with alloys system. Table 6.2 represents the commercial Inconel alloys with their nominal chemical composition.

Please note that the discussion regarding practical aspects of commercial Inconel 600 and 700 series alloy are under the scope of this chapter.

TABLE 6.1
Some Commercial Ni-based Alloys with Each Alloy System

Alloy Family	Alloy System		Commercial Alloy Examples
Corrosion-resistant	Solid-solution strengthened	Ni	Nickel 200
		Ni–Cu	400, K-500
		Ni–Mo	B-2, B-3®
		Ni–Cr–Mo	G-35®, 59
		Ni–Cr–Mo–W	C-276, C-22®, 686
		Ni–Cr–Mo–Cu	C-2000®
		Ni–Mo–Cr	HYBRID-BC1®
High temperature	Solid-solution strengthened	Ni–Fe–Cr	800H, RA330®, HR-120®
		Ni–Cr–Fe	600, 690
		Ni–Cr–Fe–Mo	HASTELLOY X
		Ni–Cr–Mo–Nb	625, 625SQ®
		Ni–Cr–Co–Mo	617
		Ni–Cr–W–Mo	230®
		Ni–Co–Cr–Si	HR-160®
	Precipitation-strengthened	Ni2(Mo,Cr)	242®, 244®
		γ'	Waspaloy, René 41, 282®
		γ''	718, 718 Plus®

Source: (with Permission) [7].

TABLE 6.2
Some Commercial Inconel Alloys with their Nominal Chemical Composition (in wt%)

Trade Name	Nickel (Ni)	Chromium (Cr)	Iron (Fe)	Molybdenum (Mo)	Others
Inconel 600	72 min	14–17	6–10	–	Cu: 0.5 max
Inconel 601	58.0–63.0	21–25	Bal.	–	Al: 1.0–1.7, Cu: 1 max
Inconel 625	58 min	20–23	5 max	8–10	Nb: 3.15–4.15, Al: 0.4 max
Inconel 718	50–55	17–21	Bal.	2.8–3.3	Nb: 4.75–5.5, Ti: 0.65–1.15, Al: 0.2–0.8
Inconel X-750	70 min	14–17	5–9	2.25–2.75	Nb: 0.7–1.2, Ti: 2.25–2.75, Al: 0.4–1.0

Source: Author-compiled information [8].

However, Table 6.1 is presented to provide the information of the solid-solution strengthening and precipitation-strengthening alloy systems exhibited by some of the major nickel-based commercial alloys.

Since their first use in the early 1900s, nickel-based alloys' composition has evolved to meet the increasing demands of diverse applications, particularly for corrosion resistance in aqueous environments and high-temperature service [9].

Inconel alloys are renowned for their exceptional heat and corrosion resistance, making them invaluable in a range of industrial applications. The aerospace industry is a significant consumer, where these alloys are utilized in the manufacture of aircraft and spacecraft components and rocket engines due to their ability to withstand extreme temperatures. In the energy sector, Inconel alloys are employed in gas turbine components and nuclear reactors owing to their high-temperature stability and resistance to neutron absorption. The chemical and petrochemical industries use these alloys in heat exchangers, reactors, and separators due to their superb corrosion resistance in harsh acidic and alkaline environments. Similarly, the marine industry uses Inconel for components exposed to seawater and in the manufacturing of submarines. In the automotive industry, they are used in exhaust systems and turbocharger rotors and seals. Moreover, they are also utilized in the medical field for making surgical equipment. Thus, Inconel alloys play a pivotal role across various industries, contributing to advances in technology and manufacturing.

Most system components in challenging applications require to be produced through welding. The welded joints' strength needs to match the base metal's strength under these conditions. Weldability, defined as a material's capability to be welded and operate effectively in the given environment, focuses on issues arising during welding and fabrication.

6.3 GENERAL WELDING CONSIDERATIONS

Nickel-based alloys are typically joined using gas tungsten arc welding (GTAW), gas metal arc welding (GMAW), shielded metal arc welding (SMAW), and submerged arc welding (SAW), with these processes detailed later in this book. In adhering to codes like EN/ISO, ASME, API, it is crucial to review welding procedures, materials, joint design, temperature control, and post-weld heat treatment (PWHT).

Ni-based alloys, which display lower molten metal fluidity and shallower penetration than steels, necessitate careful joint design and bead placement for adequate fusion. Wider grooves are required in the weld joint design of nickel-based alloys compared to carbon steel and stainless steel. A wider groove allows proper access to manipulate filler metal to obtain proper root and side wall fusion [10][11].

Figure 6.1 suggests suitable joint design for nickel-based alloys. Crater cracks are common, making start–stop grinding advisable. The joint area should be kept clean and the alloys solution annealed prior to welding to prevent cracking.

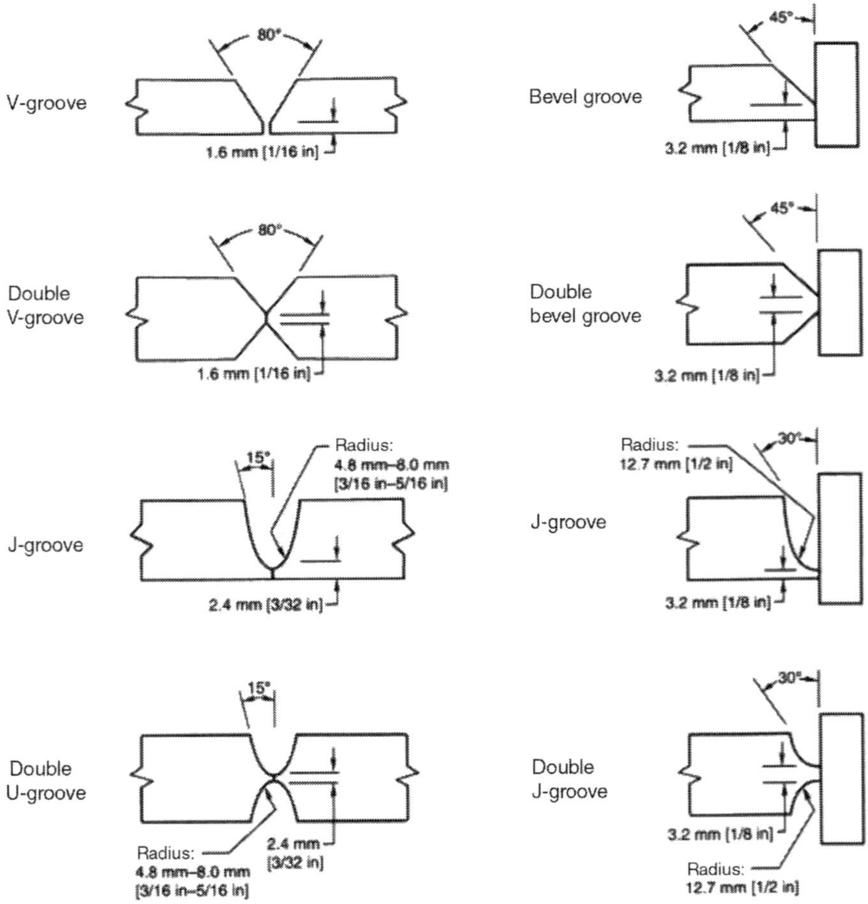

FIGURE 6.1 Weld joint designs for Ni-based alloys [12].

Joint preparation methods include plasma cutting, machining, grinding, and air carbon arc gouging, requiring bright metal conditioning before welding. The welding zone should be clean, and the surface polished using a solvent like acetone. Between weld passes, stainless steel wire brushing is effective for cleaning.

Nickel and its alloys risk embrittlement from elements like lead, sulfur, and phosphorus present in substances like grease, oil, and paint. Thorough cleaning, potentially with methylene chloride, alkaline cleaners, or specific compounds, is necessary before welding. If sodium carbonate-containing alkaline cleaners are used, they too must be cleaned off prior to welding.

Preheat is typically set at room temperature. The base material may need warming to bring it above freezing point or to stop moisture from condensing. An interpass temperature below 100°C–150°C is preferred.

Despite a lower thermal expansion coefficient reducing distortion and residual stress, Ni-based alloys still need controls like jigs, braces, bead placement, and weld sequences.

As-welded Ni-based alloys generally does not require PWHT. When needed, solution anneal or stress relief heat treatments at temperatures ranging 1050–1200°C can be utilized. However, temperatures around 600–650°C, while effective for stress relief, may lead to secondary phase precipitation, affecting corrosion resistance and mechanical properties [13].

PWHT can lead to the precipitation of secondary phases such as delta phase or gamma double prime (γ'') phase. These phases can significantly influence the mechanical properties of the alloy. For instance, the precipitation of the delta phase can lead to a decrease in tensile strength and an increase in ductility. In some cases, PWHT can also influence the corrosion resistance of Inconel 718 [2].

6.3.1 Gas Tungsten Arc Welding Guidelines

The gas tungsten arc welding (GTAW) process is highly adaptable and frequently used for Ni-based alloys due to its quality welds, minimal distortion, and spatter-free properties. It can function with or without filler metal, manually or automatically, and is ideal for tack and root pass welding. However, it lacks in productivity as deposition rates are low and demands higher skill levels.

Direct current electrode negative (DCEN) polarity is recommended with 2% thoriated tungsten electrodes (AWS A5.12 EWTh-2). The electrode should have a cone shape with a 1.6 mm flat point. Shielding gas should be 99.996% pure welding-grade argon, flowing at 12–15 liters per minute. A gas diffuser screen and a large gas cup ensure optimal weld puddle coverage. The weld metal should look bright-shiny after deposition. Sometimes, argon–helium or argon–hydrogen gases are used in high-speed, specialized welding systems. The root side of the weld joint requires a "back-purge" with argon at 2.5–5 liters per minute. Copper backing bars aid in achieving the desired bead shape [7].

The torch should be held perpendicular to the workpiece, using stringer bead techniques with enough current to melt the base and fuse the filler metal. The filler material tip should remain under the shielding gas to avoid oxidation.

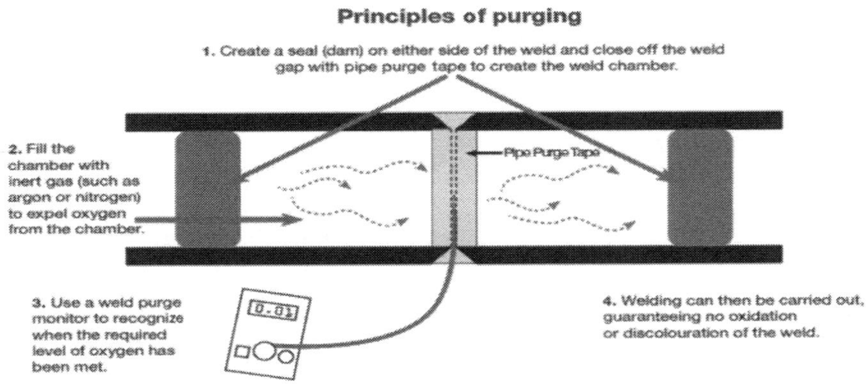

FIGURE 6.2 Schematic explaining the pipe purging process [14].

It goes without saying that gas purging of the root is essential when depositing a TIG root pass. Argon (Ar) or a mixture of the argon and hydrogen (H_2) is used as purging gas. Figure 6.2 shows the use of a seal (dam) and purge monitor device commonly used to weld nickel-based material used for pipe industries.

6.3.2 Gas Metal Arc Welding Guidelines

The gas metal arc welding (GMAW) process is more productive than GTAW, suitable for manual and automatic welding, with high deposition rates and simpler operation. Its four transfer modes are short-circuiting, globular, spray, and pulsed spray, chosen based on joint design, welding position, deposition rates, and welder skills.

Short-circuiting, suitable for all welding positions, provides excellent control but is prone to incomplete fusion. Globular mode is mainly used for cladding applications. Spray transfer, offering good fusion, productivity, and low spatter, is ideal for welding thick sections in flat positions but can induce secondary phase precipitation in the HAZ. Pulsed-spray, a variation of spray transfer, is used for thinner metals in all positions, lowering the overall heat input [7].

GMAW requires a direct current electrode positive (DCEP) and shielding gas flow rates of 15–20 liters per minute. Shielding gases can be pure Ar or Ar + He mixes, Ar + He + CO_2, and He + Ar + CO_2. CO_2-containing gases offer a stable arc and excellent welding characteristics but can increase the chance of incomplete fusion due to oxidation. Therefore, multi-pass welds with CO_2 gases should be lightly ground between passes. Back-purging, similar to GTAW, is necessary for the root side of the weld joint to avoid heavy oxidation [7].

6.3.3 Shielded Metal Arc Welding

When using shielded-metal arc welding (SMAW) to weld nickel-based alloys, the electrodes must be re-dried to 350°C for 2 hours before use.

To achieve maximum arc stability and control over the molten puddle in welding, a short arc length is crucial. The electrode should be angled backward toward the molten puddle (backhand welding) at a 20–40° drag angle. Stringer bead welding techniques are generally used, with some electrode manipulation to place the molten weld metal where required.

In vertical welding, a weave bead technique and larger diameter electrodes help maintain a flat bead profile. It is important to grind starts and stops to sound weld metal when welding nickel-based alloys. Initiating porosity can be minimized by using a start tab of the same alloy or grinding each start to sound weld metal. Small crater cracks at the stops can be reduced by using a back-stepping motion to fill the crater before breaking the arc [15][7].

6.3.4 Submerged Arc Welding

The submerged arc welding (SAW) process is often not recommended for nickel-based alloys due to its high heat input and slow weld cooling, which can increase weld restraint and promote cracking. Though it can be used successfully for solid-solution alloys like Alloys 400 and 600, it is mostly suitable for welding thick sections over 12.7 mm and not ideal for precipitation-strengthened alloys.

GMAW spool wire is utilized, with larger diameters used at higher weld heat inputs. SAW can operate with either DCEP or DCEN. DCEP is preferred for butt welding, as it produces a flatter weld bead with deeper penetration, while DCEN is favored for overlay applications due to its higher deposition rate and lower penetration [15].

The flux type is critical in successful welding and must be carefully selected for each alloy/wire. Welding can lead to loss of key alloying elements like chromium, and silicon pickup from flux can alter the weld deposit chemistry, increasing solidification cracking susceptibility [7].

6.4 COMMON ISSUES WITH WELDING OF NICKEL-BASED ALLOYS

Although it is crucial to clean the surface of nickel and its alloys right before welding, some imperfections may result due to a lack of inadequate cleaning process and/or welding procedures. Typically, the cleaning process involves degreasing the surface, eradicating all surface oxide through methods such as machining and grinding or using a stainless steel wire brush and then performing a final degreasing step.

6.4.1 Porosity

Porosity can be attributed to the presence of oxygen and nitrogen from air exposure and surface oxide or hydrogen from surface impurities. Mitigation methods include meticulous surface cleaning and using a deoxidant-containing filler material like aluminium and titanium. During the use of argon in TIG and MIG welding processes, the shielding efficiency of the weld pool, inclusive of a gas backing system, is significant. Argon–hydrogen gas blends typically result in more pristine welds during TIG welding. Utilizing argon with up to 10% H_2 as a backing gas can assist in the prevention of porosity, as hydrogen bubbles forming in the weld pool can accumulate diffusing hydrogen. However, an excess of hydrogen (>15%) in the shielding

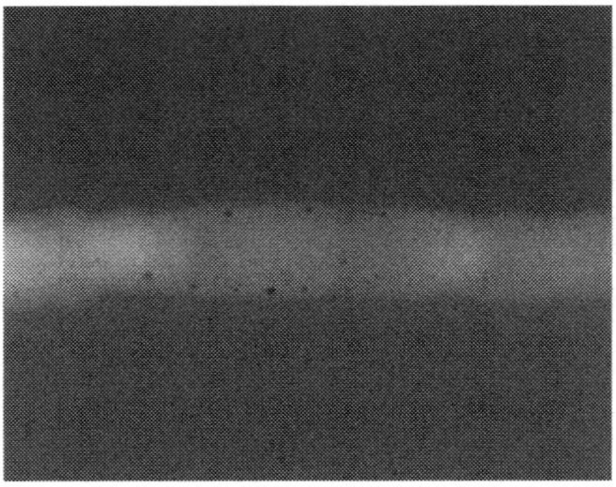

FIGURE 6.3 Porosity in a weld fill pass deposit as a tiny black spot in RT film [14].

gas could lead to hydrogen porosity. Figure 6.3 shows the porosity in a weld fill pass deposit as a tiny black spot in RT film.

6.4.2 Oxide Inclusions and Lack of Inter-run Fusion

The oxide that forms on nickel alloys has a higher melting point than the alloy itself and can stay solid during welding, leading to inclusions if it gets entrapped in the weld pool. Oxide or slag left on multi-run weld bead surfaces may cause fusion defects in subsequent runs as they do not melt away. Figure 6.4 macrograph shows the lack of inter-run fusion due to the presence of the oxide. High-temperature-formed surface oxide needs to be removed prior to welding through machining or abrasive grinding; wire brushing will not suffice as it polishes only the oxide. For multi-pass welding, it is imperative to remove any oxide and slag between each run.

6.4.3 Micro-fissuring

Much like austenitic stainless steel, nickel alloys can develop liquation cracks in reheated weld metal areas or the HAZ of the parent metal. It takes place in cases when the factors are beyond the welder's control, like grain size or impurity levels that influence this cracking. Certain alloys are more prone to this issue; for instance, welding some cast super alloys often results in liquation cracks. Figure 6.5 shows the micro-fissuring in welds developed in reheated weld metal area/HAZ.

6.4.4 Post-weld Heat Treatment Cracking

Also termed as strain-age or reheat cracking, this phenomenon often arises during the post-weld aging process of precipitation-hardening alloys. However, it can be mitigated by employing pre-weld heat treatment, like solution annealing, though over-aging offers the most robust resistance. Notably, Alloy 718 was designed explicitly to withstand such cracking.

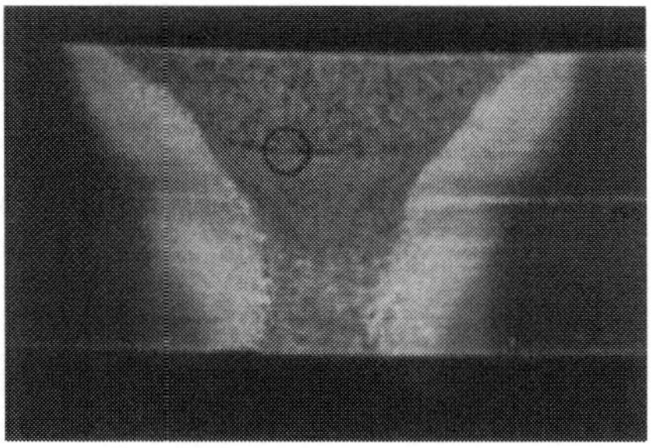

FIGURE 6.4 Macrograph showing the lack of inter-run fusion due to the presence of the oxide [14].

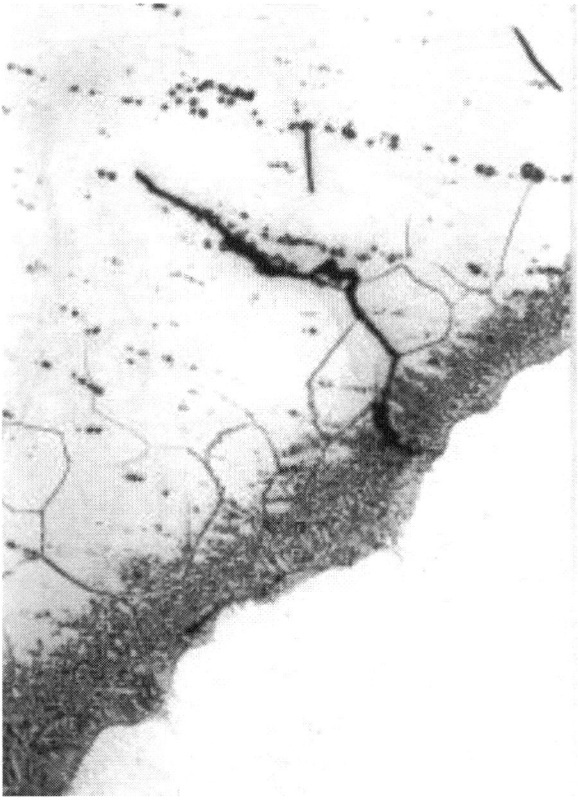

FIGURE 6.5 Image showing the micro-fissuring in reheated weld metal area/HAZ.

Source: (with permissions from TWI Global) [16].

6.4.5 Stress Corrosion Cracking

Welding typically does not render most nickel alloys prone to weld metal or HAZ corrosion. However, when Alloy 400 interacts with caustic soda, fluosilicates, or hydrofluoric acid, there is a risk of stress corrosion cracking. For such instances, post-weld thermal stress relief is performed. Additionally, high-temperature water can induce stress corrosion in Ni–Cr alloys. To combat this, a high-chromium filler metal has been formulated specifically for welds and overlays in such conditions.

6.5 SPECIFIC WELDING ISSUES WITH SOLID SOLUTION ALLOYS

Solid solution Inconel alloys are known for their high-temperature strength, excellent resistance to corrosion, and robust mechanical properties, making them ideal for various applications. However, when it comes to welding these alloys, some problems can occur, affecting their overall performance and reliability.

One of the primary issues with welding solution-strengthened Inconel alloys is the potential for microstructural changes to occur during the welding process. The high temperatures involved can lead to grain coarsening, phase segregation, and precipitation of secondary phases, which can all detrimentally impact the mechanical properties of the welded area [17].

Another significant problem during the welding of solution-strengthened Inconel alloys is solidification cracking. The solidification process of the alloy during welding can lead to hot cracking, especially in the presence of elements like sulfur and phosphorus. These impurities are not soluble in nickel and form low-melting eutectics, which can remain liquid and induce cracks in the weld solidification front [18][19].

Control of the heat input during welding is critical. Too much heat can cause excessive dilution and lead to a decline in corrosion resistance. Similarly, insufficient heat can lead to inadequate fusion and create weak joints [3].

Addressing these challenges requires careful optimization of welding parameters, judicious selection of filler materials, and where necessary, the application of post-weld heat treatment to restore the microstructure and mechanical properties [9].

6.6 SPECIFIC WELDING ISSUES WITH PRECIPITATION-HARDENED ALLOYS

Hot cracking, a serious issue encountered while welding precipitation-hardenable Inconel alloys, occurs when the material solidifies and contracts while still at high temperatures, leading to the formation of cracks. This is particularly prevalent in precipitation-hardenable Inconel alloys due to the formation of low-melting eutectics, or brittle intermetallic phases, which remain molten during solidification, leading to solidification cracking [9].

Several factors contribute to the likelihood of hot cracking during welding, such as the presence of impurities, alloy composition, filler metal choice, and welding procedure. Certain alloying elements, such as sulfur and phosphorus, which are generally kept at low levels, can lead to hot cracking by forming low-melting eutectics. Therefore, careful control of these elements is needed [20].

Additionally, the thermal cycle of the welding process plays a significant role. Rapid cooling rates, coupled with high tensile stresses during the welding process, may increase the susceptibility to hot cracking [19]. Addressing hot cracking in the welding of Inconel alloys requires a combination of careful alloy design, optimization of welding parameters, and post-weld heat treatment to minimize residual stresses and ensure homogeneous microstructure [21].

Second issue of concern for precipitation-strengthed Inconel alloy like Inconel 718 is strain-age cracking or PWHT cracking. During welding, the cooling process induces significant stress in the HAZ due to thermal gradients and external weld restraint. This residual stress gets relieved during the post-weld heat treatment (PWHT), which coincides with hardening phase precipitation. This reaction leads to higher strength within the grain compared to the grain boundaries, causing stress localization at the grain boundaries. This process can result in high stresses at grain boundary triple points, causing intergranular cracks. The schematic time–temperature regime in Figure 6.6 illustrates a welding operation followed by PWHT heating, where the dashed line indicates the precipitation curve [22].

The phenomenon of strain-age cracking is primarily influenced by the chemical composition of materials, with elements forming the hardening phases having the greatest impact. Several studies highlight this effect [23–25]. For instance, an elevated Al+Ti content tends to raise the cracking susceptibility, as depicted in Figure 6.7.

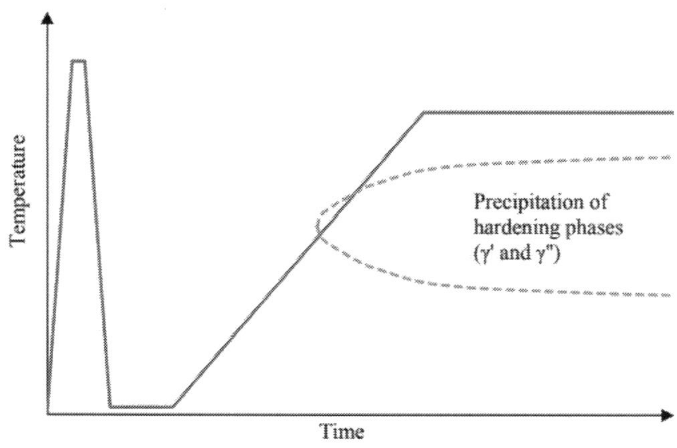

FIGURE 6.6 The time–temperature schematic represents a welding process followed by post-weld heat treatment (PWHT). The dashed line in the diagram denotes the precipitation curve in a time–temperature transformation (TTT) scenario.

Source: (with permissions) [22].

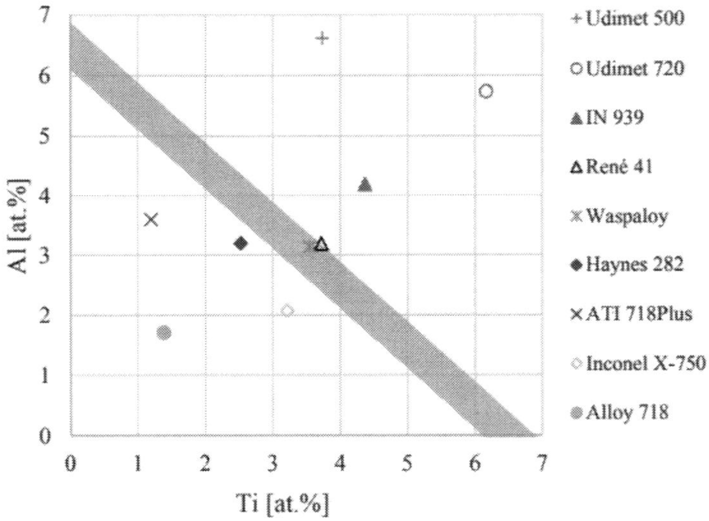

FIGURE 6.7 Figure depicts the concentrations of Al and Ti in various commercially available superalloys. Superalloys above the gray line are characterized by low weldability.

Source: (with permissions) [22].

6.7 GUIDELINES FOR THE SELECTION OF WELDING CONSUMABLES FOR INCONEL ALLOYS

Inconel 600, a solid-solution strengthened Ni–Cr–Fe alloy with notable immunity to chloride stress corrosion cracking due to high Ni content, offers resistance to oxidizing conditions at high temperatures and corrosive solutions due to chromium. However, even slight amounts of impurities like sulfur and phosphorous can compromise its weldability. While it can be autogenously welded in sheet-metal thickness, it is not recommended as a GMA filler for welding thicker sections due to poor weld metal cracking resistance. Matching filler metals like Inconel 62 and Inconel 82 containing Nb additions are usually used to weld this alloy, mitigating hot cracking issues. Covered electrodes, such as Inconel 182 and Inconel 132, can also be employed for welding Inconel 600. However, SMAW and SAW processes are generally avoided due to potential corrosion problems caused by residual slag.

Inconel 625, a solid-solution strengthened Ni-Cr-Fe-Nb alloy, shares similar weldability traits with Inconel 600. It has a matching filler metal (ERNiCrMo-3), frequently used to join Ni–Cr–Mo alloys [15].

Inconel 718, a nickel-based superalloy renowned for its excellent mechanical properties and corrosion resistance in high-temperature applications, requires a matching filler material for welding. ERNiFeCr-2, also known as "Inconel 718 filler metal," is often recommended. Although, Alloy 625 is an alternative consumable, if PWHT is not applied. This nickel–chromium–molybdenum alloy provides high strength and good corrosion resistance. For GTAW, FCAW, and GMAW processes involving

TABLE 6.3
Suggested Filler Metals with their Nominal Composition for Welding Inconel Alloys along with their AWS Classification Systems [15].

Inconel Alloy	GTAW	GMAW	SMAW	SAW	Nominal Composition
Inconel 600	ERNiCr-3 (AWS A5.14)	ERNiCr-3 (AWS A5.14)	ENiCrFe-3 (AWS A5.11)	ERNiCr-3 (AWS A5.14)	Ni: ~72%, Cr: 19%, Fe: ~balance
Inconel 625	ERNiCrMo-3 (AWS A5.14)	ERNiCrMo-3 (AWS A5.14)	ENiCrMo-3 (AWS A5.11)	ERNiCrMo-3 (AWS A5.14)	Ni: ~58%, Cr: 21.5%, Mo: 9%, Nb+Ta: 3.6%, Fe: ~balance
Inconel 718	ERNiFeCr-2 (AWS A5.14)	ERNiFeCr-2 (AWS A5.14)	ENiFeCr-2 (AWS A5.11)	ERNiFeCr-2 (AWS A5.14)	Ni: ~52%, Fe: ~balance, Cr: 19%, Nb: 5%, Mo: 3%
Inconel X-750	ERNiCr-2 (AWS A5.14)	ERNiCr-2 (AWS A5.14)	ENiCrFe-7 (AWS A5.11)	ERNiCr-2 (AWS A5.14)	Ni: ~70%, Cr: ~19%, Fe: ~balance
Inconel 800	ERNiCr-3 (AWS A5.14)	ERNiCr-3 (AWS A5.14)	ENiCrFe-3 (AWS A5.11)	ERNiCr-3 (AWS A5.14)	Ni: ~72%, Cr: 19%, Fe: ~balance
Inconel 825	ERNiFeCr-1 (AWS A5.14)	ERNiFeCr-1 (AWS A5.14)	ENiFeCr-1 (AWS A5.11)	ERNiFeCr-1 (AWS A5.14)	Ni: 58%, Fe: 22%, Cr: 20%, Mo: 3%, Cu: 2.2%

Inconel 718, using ERNiFeCr-2 as bare or flux-cored wire typically yields the best results. Table 6.3 represents the suggested filler metals with their nominal composition for welding Inconel alloys along with their AWS classification systems. [26]

Note: The readers may please be noted that actual compositions may vary slightly due to manufacturing tolerances and specific product formulations. Always refer to the product specifications or manufacturer's data sheets for the most accurate information.

To summarize, in addition to selecting the appropriate filler material, the welding procedure and PWHT should be carefully planned and controlled to prevent the formation of secondary phases and maintain the alloy's desirable properties.

As always, the exact selection of welding consumables and procedures should be determined by a welding engineer or a similar professional based on the specific requirements of the job and the standards and codes that the welded product needs to adhere to.

6.8 FUTURE WELDING TRENDS WITH NICKEL-BASED ALLOYS

Based on my existing knowledge, some potential future trends in welding technology of nickel-based alloys are with additive manufacturing (AM), which is becoming a popular method for producing complex parts from nickel-based alloys. This could improve the speed and efficiency of welding processes for nickel alloys. However, in wire and arc additive manufacturing (WAAM) of nickel-based alloys, several challenges arise. The rapid heating and cooling cycles associated with the process can create high residual stress levels, potentially causing distortion or cracks. The complex phase transformations of alloys like Inconel 718 make it difficult to control microstructure during WAAM. The process can also result in defects such as a lack of fusion or solidification cracking, which affect product quality. Moreover, controlling the interpass temperature is crucial to prevent undesirable microstructures and defects. WAAM tends to create a rough surface finish that demands further post-processing. Lastly, optimizing the process parameters to achieve specific mechanical properties and geometric precision is a complex task due to the numerous variables involved.

6.9 CASE STUDIES

The author of this chapter is an academician with a keen professional engagement in industrial matters. He maintains a technical blog: https://weldingknowledge.co.in/, where he addresses industrial issues into feasible solutions. He leverages his deep expertise in the field to tackle topics such as filler metal selection for similar and dissimilar metal welding and welding procedure qualifications, offering valuable insights to his readers. Despite his academic roots, his work is firmly grounded in real-world industry applications and challenges.

Following are three case studies related to Hastelloy C-22 grade welding practical guidelines. The author presents them as case studies in this chapter with an intention to create interest among readers in welding behavior of Haste-alloy besides Inconel alloys, which has been the scope of this chapter.

6.9.1 Case Study 1: Welding Procedure Guidelines for Hastelloy C-276

Question:

Does Hastelloy C-276 material require any special care during welding?

Does it need guidance on procedure qualification for tube-to-tube welding application?

Answer:

Hastelloy C-276 is a corrosion-resistant nickel-based superalloy. The main alloying elements are Cr, Mo, Fe, and W. The alloy design is based on solid solution strengthening and no precipitation hardening mechanism; so the alloy C-276 is not hardenable by martensitic structure thereby eliminating any chance of martensite formation in HAZ due to entirely austenitic structure up to the end of solidification. So, unlike C–Mn or low alloy steel, you need not worry about controlling preheat, but you are suggested to maintain the interpass temperature below 150°C during multi-pass welding.

Alloy C-276 is generally supplied in the solution-annealed condition. Solution annealing would dissolve all alloying elements in the austenitic matrix. Standard solution annealing treatment comprises soaking at 1120°C followed by rapid quenching to room temperature to prevent any re-precipitation of carbides and/or embrittling phases.

Alloy C-276 is weldable, but there are two alloying elements that are of main concern, Mo and W.

The welding metallurgy of alloy C-276 becomes upset by the development of some deleterious phases, known as μ and P phases. Both μ and P phases belong to the group of topologically close-packed (TCP) phases. These TCP phases in the fusion zone during welding has adverse effects on tensile strength, impact toughness, and corrosion resistance. Besides this, at the end of solidification, there is also a risk of hot cracking due to segregation tendencies of elements like W and Mo present in the weld metal composition. Both the TCP phases P and μ are rich in Mo and W, i.e., these elements preferentially partition into P and μ phases, depleting the nickel-based gamma matrix of these elements. Many researchers [27–31] have reported the phenomenon of segregation tendencies of Mo and/or W differentially in dendrite core region and dendrite arm spacing with optical microscopy techniques and using SEM-EDAX techniques in subgrain boundary and subgrain interior. The higher the level at which Mo and W are present in the Ni–Cr–Mo-based alloy, the higher is the possibility for the formation of these brittle phases. Unfortunately, the alloy C-276 contains approximately, 15 wt% Mo in base metal composition.

So you are welding autogenously (without filler metal) and using filler metal; in both cases, pulse GTAW technique is highly recommended. The pulse current has a better advantage of low heat input coupled with a faster cooling rate that helps better the distribution of the elements into the matrix. At the same time, it helps suppress the microsegregation and avoid the formation of TCP phases. Pulse current also helps refine the microstructure, so you will get better mechanical properties, particularly toughness and corrosion resistance in the welded joint.

If you are joining it with filler metal, then your choice for filler metal becomes very important. Many works of the literature (see References) suggest using higher Mo-content filler metal than base metal composition. Among them, ERNiCrMo-4 (with 17 %Wt Mo approximately) and ERNiCrMo-10 (with 13 %Wt Mo approximately), covered under A5.14/A5.14M:2018 Specification for Nickel and Nickel-Alloy Bare Welding Electrodes and Rods, are the best choices. This is because high Mo from the filler metal will supplement any deficiency of less Mo in dendrite arm spacing than in dendrite cores. So the weld metal will not face any corrosion issues during application in seawater (naval/marine application).

Although, ERNiCrMo-3, under matching (with 9%wt approximately), is also a tempting choice, but there are two problems using this filler metal. One is less Mo, and the second is the presence of Nb (3.5%Wt approximately) in filler metal composition. Nb has a tendency for segregation and thereby causes hot cracking in the weld joints.

Refer to the literature related to the investigation given in References [30–31]. In their experimental work, they have successfully welded with the following pulse current parameters and obtained better weld metal mechanical and chemical properties as compared to normal GTAW. They have worked with 4 mm thick Hastelloy C-276 material, welded with 2.4 mm tungsten electrode using Pulse GTAW technique, shielded by 15 liter/min pure argon as shielding gas.

TABLE 6.4
Suggested Welding Parameters from the Literature Source [30–31].

Pulse Current (A)	Background Current (A)	% on Time	Pulse Frequency (Hz)
165	77	60	5

Depending on your job thickness, you may optimize and establish suitable welding process parameters. Second, for your Tube-to-Tube welding process, you need to back purge to protect the root side of the weld joint from oxidation during welding with an inert shielding gas (usually welding- grade argon). Refer to References [31–33] for the procedure.

For procedure qualification run, in addition to the mechanical test, I advise you to conduct pitting or crevice corrosion test (ASTM G48 Method A or B) or as required by your client.

I hope you will find this information useful to you!

6.9.2 Case Study 2: Corrosion Resistance Overlay of Inconel 625 over SS 316L—Dissimilar Metal Welding

Question: What are the potential challenges or considerations involved in applying a corrosion-resistant Inconel 625 overlay on SS 316L base material for seawater applications?

Answer: Inconel 625, a Ni-based solid solution-strengthened alloy, well known for its excellent strength and good corrosion resistance at extreme environments, used in thermal plants, boiler tubes, petrochemical industry, and power plant.

Nickel Alloy 625 is used for both its high strength and outstanding aqueous corrosion resistance, having outstanding resistance to chloride pitting and crevice corrosion cracking and immunity to chloride ion stress corrosion cracking (Ch-SCC).

The presence of Cr content (approximately 20%wt) along with Mo-rich Nb and Fe makes this Inconel 625 (Ni–Cr–Mo–Nb) austenitic alloy achieve excellent corrosion resistance property.

SS 316 L is a low-carbon version of 316, a standard Mo-containing grade, next in importance to SS 304 amongst the austenitic stainless steels and the most immune from sensitization (grain boundary carbide precipitation).

In SS 316 L, the Mo gives 316 better overall corrosion resistant properties than grade 304, particularly higher resistance to pitting and crevice corrosion in chloride-containing environments. So the selection of the base metal SS 316L and Inconel 625 filler metal is most appropriate with respect to the service application in seawater; but here the prime concern is about maintaining the Inconel 625 chemistry on the top overlay surface, i.e., functional surface.

During the overlay welding process, a dilution of the weld overlay filler metal with the base metal will take place. Generally, to save the cost of precious filler metal like nickel-based alloys, the fabricators would try to keep the overlay layer as thin as possible. While doing so, the fabricator tends to apply higher heat input to complete the process in a single pass, which may result in a higher dilution of contents from base material in the chemistry of the deposited weld metal. As a result, the changed chemical composition of the overlay material will also change the corrosion resistance. Dilution affects the PREN number, which is an index of the relative corrosion resistance of the material as calculated by the following formula:

$$\text{PREN} = \%\text{Cr} + 3.3\%\ \text{Mo} + 16\%\ \text{N}$$

So the purpose of overlay is not fulfilled for the given application.

Many literatures [33–36] do suggest that the process of Inconel 625 overlay on CS, LAS, and austenitic SS using conventional arc welding processes, like GTAW (Hot/Cold Wire), GMAW, FCAW, and SAW, including electro-strip cladding process, results in higher dilution in the first two layers so their corrosion performance is often compromised with "more than 5% Fe content" diluted from the base metal into the Inconel 625 weld overlay chemistry.

The reason why Fe content badly affects the corrosion resistance of Ni-based alloy is rather complex and requires extensive research study. Probably, one of the reasons is that Fe increases susceptibility to chloride ion attack on corrosion-resistant overlay of alloy 625; so as a result, the fabricated product suffers from issues like severe pitting and stress corrosion cracking.

To keep the iron (Fe) content within the specified limit (less than or equal to 5% as reported by many literatures), there are some useful guidelines obtained through various research outcomes:

- At least, two layers deposition (or three layers, until you get the pure Inconel 625 chemistry on the top layer using a controlled or low heat input arc welding process—such as cold metal transfer (CMT), a method patented by Fronius International) offering a droplet deposition mechanism, which controls the short circuit by the retraction motion of the wire from short-circuiting action. There are electronically (digital wave form technology-enabled) controlled short-circuit MIG welding process variants as well as high-deposition TIG welding variants giving the benefits of high-deposition rate and good control on dilution, reduced heat input, stable arc, uniform weld profile, and a spatter-free welding.
- The study conducted by Evangeline A. et al., [35] on Inconel 625 overlay on SS 316L using CMT reported successful weld overlay with optimum corrosion properties: Current = 180 amps, Voltage = 17 volts, Torch angle = 80°, and Welding speed = 150 mm/min, attains less dilution (9.3%), and more clad bead width (8.956 mm).
- Use of solid (GMAW preferably) Inconel 625 weld consumables, preferably, already low in Fe content in undiluted weld metal chemistry. You, your client, and your filler metal supplier have to sit together to figure out the permissible limit of Fe% in top undiluted layer being tolerated without affecting the performance of the weld overlay components. Probably, you can sit with filler wire manufacturer/technical representative to work out any Fe% deviation in undiluted weld deposits with consumable vs the filler metal specifications, i.e., AWS A5.14 (GTAW/GMAW) and AWS Sec II, Part-C.

- Procedure qualification tests including simulated corrosion tests, such as ASTM G48 Practice A, B, are highly recommended to assess the pitting and crevice corrosion resistance performance of test coupons under simulated conditions.
- Spectroscopy (i.e., chemical analysis) of the weld overlay top layer in addition to simulated corrosion tests will give an idea about the relationship between Fe% and corrosion resistance.

One can also theoretically check the weld metal dilution by using the following formula:

$$D(\%) = (CFZ - CFM)/(Cs - CFM) \times 100$$

D = % weld dilution
CFZ = elemental composition of the fusion zone (weld)
CFM = elemental composition of the filler metal (wire/strip)
Cs = elemental composition of the substrate (base metal)

As a rule of thumb, the higher the dilution the lower will be the corrosion resistance.

Hope this information will be helpful to you for the successful procedure qualification of the fabrication involving Inconel 625 overlay to SS 316L components for seawater application.

6.9.3 Case Study 3: Use of Nickel-Based Filler Metal for Dissimilar Metal Welding

Question: Which is a suitable filler metal for joining SS Type 310 and Creep resisting 2.25Cr-1Mo steel for high-temperature furnace application?

Answer: Generally, for joining CS/LAS to 300 type stainless steel, austenitic filler materials like E/ER 308, 309, 316, and 347 are used, but these filler metals are used when the service application is below 425°C. For higher temperature service applications, I would suggest using high-nickel filler metals ERNiCrFe-2 and ENiCrFe-2 and ENiCrFe–3.

From a metallurgical point of view, another advantage of Ni-base weld metal is that it can effectively reduce the carbon migration from the ferritic steels to the weld metal. It is a well-known fact that carbon migration into stainless steel weld metal beyond a certain level causes solidification cracking due to less or nil presence of ferrite in the solidifying weld metal [2][7].

In the petrochemical industries, the problem of cracking of joints between creep resisting ferritic steel 2.25Cr–1Mo steel and austenitic stainless steels, 304H, SS 310, are common. Often, cracking is observed in HAZs of the ferritic steel just after a few millimeters away from the fusion boundary towards the ferritic steel side. This phenomenon occurs when the welded joint is in service application at an elevated temperature of around 565°C. Most of the time, this occurs due to stress concentrations and/or thermal fatigue generated owing to differences in the coefficient of thermal expansion (CTE) values between either sides of the base metals and weld metal. So the choice of filler metal and resultant weld metal CTE value should be matching with the CTE values of the base metal(s) being welded.

Another aspect for selection of filler metal is its tolerance for dilution from the base metal. Generally, filler metal choice for joining austenitic stainless steels (Type 304L, 316L, 321, 347, etc.) with creep resisting ferritic steel 2.25Cr–1Mo, our obvious choice is for a filler metal which can tolerate dilution by both base metals without the formation of defects such as solidification/liquation cracking, precipitation of embrittling phases, and so on in the joint and at the same time, providing optimum mechanical strength and corrosion resistance is the first priority.

A higher-alloy filler metal like type 309 with a Ferrite Number 8–10 approximately, or Type 312 with FN 20–25 approximately or ferritic-austenitic type Type 2209 with Ferrite number 35–65 approximately seem to be an attractive choice for their high capacity to tolerate dilutions from base metals. But the potential threat with using these high-ferrite filler metal is embrittlement of welds (and severe loss of ductility) due to transformation of delta ferrite in to brittle sigma phase when the weldment is heat treated or exposed to service environment at temperatures between 704°C and 593°C. So if the weldment requires PWHT in this range, as much as possible low ferrite in weld metal composition is preferred.

Another approach is to first apply a buttering (weld overlay) layer of 13 mm or more using SS 309 on the 2.25 Cr-1Mo side and subsequently filling pass to complete the groove using E/ER 308 filler metal. This should improve the weldability, but again, this technique is not suitable for higher-temperature application requirement.

Because the coefficient of thermal expansion (CTE) of high-nickel alloys is approximately in between the ferritic steel and austenitic stainless steels. Since one of the referred literatures suggests that mean CTE in the temperature range 0–600°C for the E/ERNiCrFe-2, E/ERNiCrFe-3 filler metal 15 μm m-1K-1 is in between the 14.0 μm m-1K-1 for ferritic steels 2.25Cr-1Mo and 18 μm m-1K-1 for SS of types 310 [15].

So, in summary, the present weld joint between austenitic stainless steel type 310 and 2.25Cr–1Mo for high-temperature furnace parts applications can be successfully produced with ERNiCrFe-2 or ERNiCrFe-3 as per AWS A5.14/A5.14M:2018, Specification for Nickel and Nickel-Alloy Bare Welding Electrodes and Rods. And for fill pass, ENiCrFe-2, ENiCrFe-3, as per AWS A 5.11 Specification for Nickel and Nickel-Alloy Welding Electrodes for Shielded Metal Arc Welding to obtain optimum mechanical strength and corrosion resistance.

However, the weld metal is likely to solidify in 100% austenitic mode, which may pose problem of weld solidification cracking, so a weld procedure qualification test is highly recommended.

REFERENCES

1. Babu, S. S.; David, S. A.; Park, J. W.; Vitek, J. M. (2004). Joining of nickel base superalloy single crystals. *Science and Technology of Welding and Joining*, 9–14.
2. DuPont, J. N.; Lippold, J. C.; Kiser, S. D. (2009). *Welding Metallurgy and Weldability of Nickel-Base Alloys*. Hoboken, NJ: John Wiley & Sons.

3. Reed, R. C. (2006). *The Superalloys: Fundamentals and Applications*. Cambridge: Cambridge University Press.
4. Donachie, M. J. (2002). *Superalloys: A Technical Guide* (2nd ed.). Materials Park, OH: ASM International.
5. Tien, J. K.; Caulfield, T. (1981). *High-Temperature Alloys: Their Exploitable Potential*. New York: Elsevier Science.
6. Zakharov, V. V. (1999). Development of nickel alloys for medical implants. *Metal Science and Heat Treatment*, 41, 496–498.
7. Sowards, J.; Caron, J. (2014). *Weldability of Nickel-Base Alloys*. Oxford: Elsevier. https://doi.org/10.1016/B978-0-08-096532-1.00615-4.
8. https://www.specialmetals.com/documents/technical-bulletins/inconel.... Access Date 01–07–2023 IST 16:00.
9. Hodge, F. Galen. (2006). The History of Solid-Solution-Strengthened Ni Alloys for Aqueous Corrosion Service. JOM: *The Journal of The Minerals, Metals & Materials Society*, 58(9), 28–31.
10. American Welding Society (AWS). (2011). *Welding Handbook, Vol. 4: Materials and Applications, Part 1* (9th ed., Chapter 4: Nickel and Cobalt Alloys, pp. 275–349). Miami, FL: American Welding Society.
11. ASM International. (2011) *ASM Handbook, Vol. 6A: Welding Fundamentals and Processes* (ed. T. L. Anderson, pp. 847–862). Materials Park, OH: ASM International.
12. American Welding Society (AWS). (2012). *Guide for the Evaluation of Hot Cracking Test Data for Weld Metal*. AWS G2.1M/G2.1:2012. Doral, FL: American Welding Society.
13. Sravan Shashank, S.; Rajakumar, S.; Karthikeyan, R.; Nagaraju, D. S. (2020). Weldability, Mechanical Properties and Microstructure of Nickel Based Super Alloys: A Review. *E3SICMED 2020 Web of Conferences*, 184, 01040. https://doi.org/10.1051/e3sconf/202018401040.
14. Andersen, L.; Andersen, T. E. *Consulting Welding of Nickel & Nickel Alloys-Technical Update*. https://www.huntingdonfusion.com/index.php/en/technical-support/white-papers-40017/3173-weld-purging-of-nickel-alloys. Access Date 05–07–2023 13:00 IST.
15. Richard E. Guidelines (2018). *For the Welded Fabrication of Nickel Alloys for Corrosion-Resistant Service, a Practical Guide* (2nd ed., pp. 11–27). The Nickel Institute.
16. https://www.twi-global.com/technical-knowledge/job-knowledge/weldability-of-materials-nickel-and-nickel-alloys-022. Access Date 05–07–2023 13:00 IST.
17. Radhakrishna, C.; Vasudevan, M.; Ganesan, V.; Parameswaran, P.; Laha, K.; Bhaduri, A. K. (2016). Microstructure and mechanical properties of Inconel 625 superalloy. *Journal of Nuclear Materials*, 469, 114–125.
18. Lippold, J. C.; Kiser, S. D. (2005). *Welding Metallurgy and Weldability of Nickel-Base Alloys*. John Wiley & Sons, Inc.
19. Lippold, J. C.; Kiser, S. D.; DuPont, J. N. (2019). *Welding Metallurgy and Weldability*. John Wiley & Sons, Inc.
20. Savage, W. F. (1989). *Hot Cracking Phenomena in Welds*. Springer-Verlag.
21. Tzelepis, D.; Lippold, J.; Ramirez, A.; Mills, M. (2018). Weldability of nickel-base alloys. In S. Lampman (Ed.), *Weldability of Materials*. Materials Park, OH: ASM International.
22. Hanning, F.; Andersson, J. (2016). *A Review of Strain Age Cracking in Nickel Based Superalloys*. Conference Proceedings of the 7th International Swedish Production Symposium, Lund, SE.
23. Hughes, W. P.; Berry, T. F. (1967). A study of the strain-age cracking characteristics in welded René 41—Phase I. *Welding Journal*, 46, 361s–370s.
24. Thamburaj, R.; Goldak, J. A.; Wallace, W. (1979). The Influence of Chemical Composition in PostWeld Heat Treatment Cracking in René 41. *SAMPE Quarterly*, 4, 6–12.

25. Carlton, J. B.; Prager, M. (1970). *Variables Influencing the Strain–Age Cracking and Mechanical Properties of René 41 and Related Alloys. Welding Research Council Bulletin 150* (pp. 13–23). Welding Research Council.
26. Tharappel, J. T.; Babu, J. (2018). Welding processes for Inconel 718 – A brief review. *IOP Conference Series: Materials Science and Engineering* (Vol. 330, Article 012082). IOP Publishing.
27. Dupont, J. N.; Banovic, S. W.; Marder, A. R. (2003). Microstructural evolution and weldability of dissimilar welds between super austenitic steel and nickel-based alloys, *Welding Journal*, 82, 125s–135s.
28. Cieslak, M. J.; Headley, T. J.; Romig, A. D. (1986). The welding metallurgy of Hastelloy alloys C4, C22, C276. *Journal of Metallurgical Transactions A*, 17A, 2035–2047.
29. Manikandan, M. et al. (2015). *Welding Metallurgy of Corrosion-Resistant Superalloy C-276*. IntechOpen. http://dx.doi.org/10.5772/61104
30. Manikandan, M. et al. (2015, December). Microsegregation studies on pulsed current gas tungsten arc welding of alloy C-276. *International Journal of Scientific & Engineering Research*, 6(12), 33. ISSN 2229-5518.
31. Sharma, S. et al. (2018). Influence of filler material composition on the microstructural evolution, mechanical properties and corrosion behaviour of nickel-based superalloy C-276. *Materials Research Express*. https://doi.org/10.1088/2053-1591/aaf9c4.
32. http://www.haynesintl.com/es/tech-briefs/general-haynes-international-information/back-purging-and-welding-requirements-for-fabrication-of-hastelloy-alloy-pipe-systems.
33. Gittos, M. F.; Gooch, T. G. (1996). Effect of iron dilution on corrosion resistance of Ni-Cr-Mo alloy cladding. *British Corrosion Journal*, 31(4), 309–314.
34. Elango, P.; Balaguru, S. (2015). Welding parameters for Inconel 625 overlay on carbon steel using GMAW. *Indian Journal of Science & Technology*, 8(31).
35. Evangeline, A. et al. (2019). Cold metal arc transfer (CMT) metal deposition of Inconel 625 super alloy on 316L austenitic stainless steel: microstructural evaluation, corrosion and wear resistance properties. *Materials Research Express*, in press. https://doi.org/10.1088/2053-1591/ab0a10.
36. Zhou, H. et al. (2020). Effect of iron ion on corrosion behavior of Inconel 625 in high-temperature water. *Scanning*, Article ID 9130362. https://doi.org/10.1155/2020/9130362.

7 Practical Aspects in Welding of Aluminium Alloys

Aishwary Mishra, Agilan M, and Chakravarthy P.

CONTENTS

7.1 INTRODUCTION

A Roman scientist "Pliny the elder" mentioned an interesting story in his book *Historia Naturalis* about a goldsmith who presented a dinner plate (very light and shiny) to his emperor which was very light and bright as silver indicating the accidental discovery of aluminium some 2,000 years ago. Unfortunately, the emperor feared the loss in value of his treasures and ordered the beheading of the goldsmith. However, it was reintroduced in 1808 by Davy and Wohler producing aluminium powders between 1827 and 1845 through chemical synthesis [1]. However, the production of aluminium was limited and its usage, a luxury and laboratory curiosity until the conception of the Hall-Héroult thermoelectric process for the extraction of aluminium from its ore bauxite, in the year 1886, which paved the way for the easy production of aluminium from its ore [2].

 DOI: 10.1201/9781003352419-7

In the modern world, aluminium and its alloys are widely used in many strategic industries owing to its high specific strength, high specific modulus, tailorable properties through heat treatment, ease of fabrication, and high corrosion resistance. Aluminium, when alloyed with various elements, provides a range of mechanical and physical properties, and based on the alloying elements, aluminium alloys are classified under eight different series as per the Aluminium Association (AA) and are tabulated in Table 7.1.

The first digit in the series represents the major alloying element, while the second digit designates the mill control or alloy modifications, and the last two digits do not carry any significance. For example, with the alloy AA5352, the second digit (3) indicates that the alloy has been modified three times but has a similar composition to earlier versions AA5052, AA5152, and AA5252.

Aluminium and its alloys are amenable to a variety of fabrication processes starting from conventional ingot metallurgy route to the recent additive manufacturing processes. Though components of near net shape are manufactured, it becomes indispensable to resort to many joining processes for the purpose of an assembly or manufacture of a component of larger geometry. Welding of metallic materials has been recognised as a versatile process to fabricate components or products as integral units, and its development dates back to bronze age, around 3000 BC, where archaeologists have unearthed jewellery made by hard soldering and swords made by hammering. Till date, many welding processes have been developed to suit a variety of materials based on the requirements. However, establishing a joint between two faying edges not only leads to the physical and chemical coalescence of interfaces but also results in significant changes in the

TABLE 7.1
Various Series of Wrought Aluminium Alloys with Major Alloying Elements, Strengthening Mechanism, and Weldability

Series of Aluminium Alloy	Major Alloying Element	Strengthening Mechanism	Weldability
1XXX (commerciallypure)	Nil	Cold working	Weldable
2XXX	Copper (Cu)	Age hardening	Selectively weldable
3XXX	Manganese (Mn)	Cold working	Weldable
4XXX	Silicon (Si)	Cold working	Weldable
5XXX	Magnesium (Mg)	Cold working	Weldable
6XXX	Si+Mg	Age hardening	Weldable
7XXX	Mg+Zn	Age hardening	Selectively weldable
8XXX	Other elements	Age hardening	Selectively weldable

microstructure and the attendant properties thereon. Hence, the choice of welding processes for a particular application is very crucial in determining the overall properties and, hence, the life of the component.

7.1.1 Temper Conditions of Aluminium Alloys

Aluminium alloys, employed in the strategic sectors, are available in a variety of thermally and mechanically treated conditions, a few of which are tabulated in Table 7.2.

The type of strain hardening and thermo-mechanical processing carried out on the aluminium alloy is represented with the letter "H," viz. H1, H2, H3, and H4. For example, **xxxx-H3** represents that the aluminium alloy has undergone strain hardening and stabilised by low-temperature thermal treatment.

Similarly, the temper condition of the aluminium alloy is represented by the letter "T," viz. T1, T2, T3, T4, T5, T6, T7, T8, T9, and T10, while T6, T7, and T8 are the most commonly used temper conditions for aluminium alloys.

Based on the type of stress relieving carried out, these temper designations are sometimes further elaborated. For example, **xxxx-T6** designates that the alloy is subjected to solution heat treatment followed by water quenching and artificial ageing. Temper T651 is a subset of T6 and represents that the alloy is subjected to a stress-relieving operation by tensile forces immediately after the water quenching process. Based on the form of the raw materials or component, the stress-relieving process can be by applying either tensile forces or compressive forces. In general, the sheets/plates are subjected to tensile forces, while bulk materials are subjected to compressive forces. The temper treatment, T652, represents that the alloy has undergone solution heat treatment followed by water quenching, stress relieving by compression, and artificial ageing. Another notable example is **xxxx-T8** in which the aluminium alloy undergoes solution heat treatment followed by water quenching, cold working, and various degrees of artificial ageing.

The designations described here can help one decipher the history of thermo-mechanical processing the aluminium alloy has undergone and hence get an idea about its strength and microstructure before the alloy is taken up for any kind of material processing. For example, **AA6061-T6** represents Al–Si–Mg alloy in solution heat treated and artificially aged conditioned.

TABLE 7.2
Various Temper Conditions of Aluminium Alloys and their Corresponding Designations

xxxx-**F**	As fabricated, no special control
xxxx-**W**	Solution heat-treated (used only on alloys that naturally age harden)
xxxx-**O**	Annealed (wrought alloys only)
xxxx-**H**	Strain hardened (cold worked to increase strength), wrought alloys only
xxxx-**T**	Thermally treated to produce effects other than F, O, or H

7.2 PHYSICAL METALLURGY OF ALUMINIUM ALLOYS

Aluminium crystallises in a face-centred cubic (FCC) structure with aluminium (or alloying element) atoms occupying the corners as well as the face-centred positions of a cubic lattice. This makes the aluminium and its alloys processed through forming route available as wrought products for industrial use. Owing to their crystal structure, these alloys are used in applications where sub-zero temperatures prevail. To cite an example, it is used in the cryogenic propellant tanks in launch vehicles. Aluminium alloys used for structural applications undergo any or a combination of the following strengthening mechanisms.

- Solid solution strengthening
- Precipitation strengthening or age hardening
- Cold working or strain hardening
- Grain size refinement

Not all these four strengthening mechanisms are undertaken in all aluminium alloys, and in most of the alloys, employed in either the commercial sector or aerospace sector, it is their combinations which provide the required physical and mechanical properties to the alloy. One must also note that the quotient of strengthening is also unequally shared between the four strengthening mechanisms mentioned earlier, with precipitation strengthening contributing the highest and grain size refinement contributing the least to the overall strength of aluminium alloys.

- *Solid Solution Strengthening*: Based on the nature of the system, most of the alloying elements substitute the aluminium lattice by either forming solid solutions (without changing the FCC crystal structure) or resulting in the formation of intermetallic compounds with specific stoichiometry. The alloying elements when substituted with aluminium in its lattice without changing the crystal structure are depicted in Figure 7.1(a), resulting in a strengthening

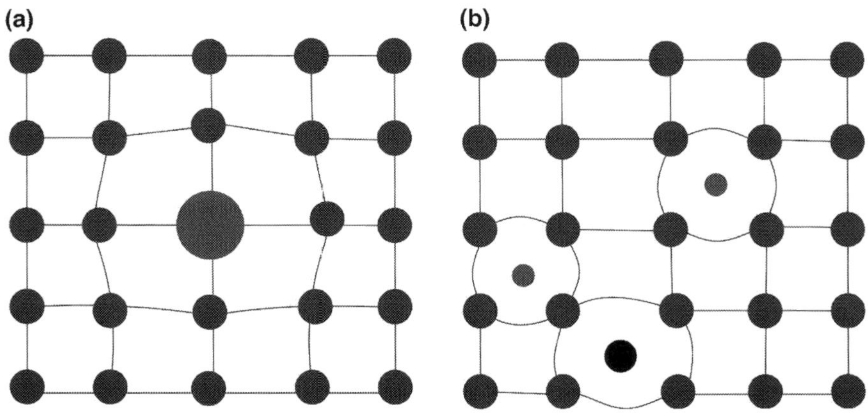

FIGURE 7.1 Schematic of (a) substitutional solid solution and (b) interstitial solid solution.

called *substitutional solid solution strengthening.* Though impurities are not desirable, their presence in very small quantities is unavoidable and based on their size, they may occupy either substitutional positions or interstitial positions [Figure 7.1(b)]. Occupancy of a smaller atom in the lattice site will result in a localised tensile field around the solute (alloying element) atom, whereas the presence of a larger atom will result in a localised compressive field. The alloying elements do not have the similar atomic size as compared to aluminium, and when present in the lattice positions of aluminium, they tend to distort the lattice. The strengthening is derived from the obstructions offered to dislocation motion, and the dislocations are not free to wander in the distorted lattice. On the other hand, these solute atoms, based on their size comparable to the aluminium atom, occupy a favourable position to partially cancel the stress fields associated with the dislocations, resulting in anchoring/pinning of dislocations and requiring much higher levels of stress to free the dislocations from the solute atmospheres.

- *Precipitation Strengthening*: The requirements for any system to exhibit strengthening by precipitation hardening are that the system should exhibit a decrease in the solubility of the alloying elements with a decrease in the temperature, and the precipitates that form in the matrix must be coherent in nature. This phenomenon is analogous to increase in the solubility of sugar in water as the water temperature is raised. As the water is cooled, the sugar solubility reduces, and sugar precipitates down. Similarly, in solid solutions, as the temperature is lowered (below the solvus temperature), phases that are thermodynamically more stable at lower temperatures precipitate out of the host matrix which can settle both within the grain and at the grain boundaries. The presence of these coherent precipitates distorts the lattice and hinders the movement of dislocations, resulting in strengthening. The fine precipitates also anchor the movement of dislocations restricting their free movement. The dislocations getting anchored due to the applied stress is shown in Figure 7.2.

  Depending on the size of the precipitates, the strengthening mechanism in the alloy can be due to either Orowan strengthening or chemical hardening. The smaller precipitates offer strengthening by forming dislocation loops around them, whereas the bigger-sized precipitates offer resistance to dislocations by chemical hardening. Precipitation strengthening is observed to have the highest effect on the strength of aluminium alloys. Both the size of the precipitates and the precipitate spacing affect the strength and are usually determined by the temperature and duration of the ageing treatment carried out for the alloy.
- *Strain Hardening/Cold Working*: Cold working or plastic deformation results in the multiplication of dislocations, and the interaction/intersection between the dislocations results in strengthening. Like dislocations repel each other, and a larger applied stress is required to move the trailing dislocation closer to the leading dislocation. Intersections of dislocations also lead to the generation of sessile dislocations, resulting in strengthening.
- *Grain Size Refinement:* Smaller the grain size, higher will be the stress required to initiate plastic deformation as per the Hall-Petch relation.

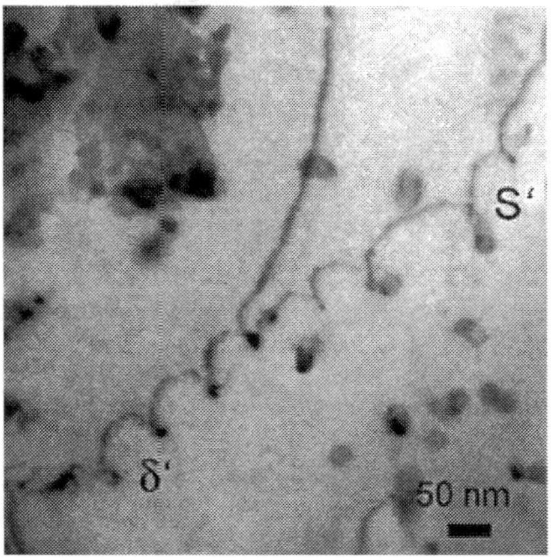

FIGURE 7.2 Orowan loops around precipitates in AlCuMgLiAgZrSc alloy in T6 condition [3].

All these strengthening mechanisms increase strength at the cost of loss in ductility except for grain size refinement, which increases strength without significant loss in ductility.

7.3 CHALLENGES IN WELDING OF ALUMINIUM ALLOYS

Welding of aluminium and its alloys are difficult due to the innate characteristics it possesses such as high thermal and electrical conductivity, high thermal expansion coefficient, and the tenacious oxide layer on the surface. In addition to the mentioned properties, these alloys also have high affinity towards atmospheric gases, in particular hydrogen, at higher temperature solidification range. These attributes of the alloy make them sensitive to the formation of various defects during the solidification process when subjected to fusion welding. The mechanical properties of aluminium at higher temperature degrade to a considerable extent that makes the material susceptible to cracking and distortion due to hot shortness. The process parameters for aluminium alloys that respond to heat treatment and especially the precipitation hardening alloys must be carefully chosen, as the heat input may result in altering the microstructure to a larger extent and thus requiring the need for post-weld heat treatment. Apart from the judicious selection of process parameters for obtaining efficient joints, the performance of the welded joints of aluminium is determined by its temper condition.

Most of the problems associated with welding of aluminium can be tackled by the selection of the right joining process (shielded metal arc welding processes, high-energy density processes, friction-based solid-state joining processes), filler wire, process parameters, joint configuration and process sequence, and so on. Typical problems associated with welding of aluminium alloys are detailed here.

- *Gas Porosity*: Most of the aluminium alloys have a remarkably high gas solubility (especially hydrogen) at elevated temperatures, which reduces by

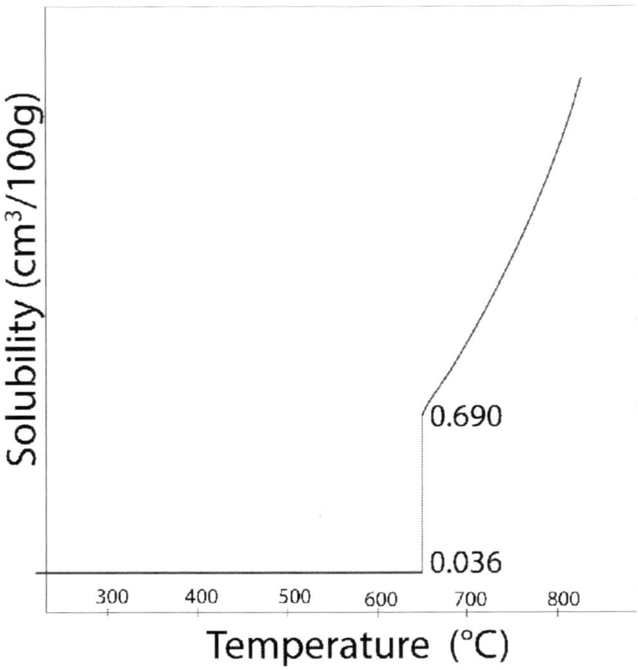

FIGURE 7.3 Solubility of hydrogen in aluminium.

a factor of 20 or more as the alloy starts to solidify (Figure 7.3). This makes it near impossible to obtain a porosity-free weld in aluminium alloys. These porosities can be as large as 3–5 mm in size.

Non-autogenous welds usually have higher porosities because of additional contamination from the filler wire. Additionally, higher currents result in higher welding temperatures resulting in additional gas solubility and henceforth a higher porosity level. On the contrary, in flat welds, a higher welding current results in a higher gas evolution rate compared to absorption, thereby reducing the porosity levels.

- *Oxide Layer Removal:* Aluminium forms a tenacious and rapidly forming aluminium oxide layer that melts above 2050°C, which is more than three times the melting temperature of aluminium (660°C). It is this passive layer that provides excellent corrosion resistance; however, this oxide film is more of a problem when it comes to welding. Having such a high melting temperature, it is very unlikely for the oxide layer to completely melt during welding, and hence it gets entrapped in the weld pool. To avoid this, gas-shielded arc welding processes usually employ direct current electrode positive (DCEP) in which the ion bombardment ruptures the oxide layer and provides a cleaning effect. However, much heat is generated at the electrode, resulting in the melting of the tungsten electrode. Hence, tungsten inert gas

(TIG) welding processes usually use alternating polarity to avoid detrimental tungsten contamination in the welds.

- *Hot Cracking:* Known by many other names such as solidification cracking or centreline cracking, hot cracking is a common phenomenon linked to any fusion welding process and not specifically with aluminium alloys. It is associated with solidification process and not commonly observed in pure metals and eutectic alloys.

  During the process of solidification, the weld metal is subjected to tensile stresses due to solidification shrinkage. The two-phase mushy region is weak during solidification, and once the tensile stresses exceed their critical value, the solidification cracking occurs along the centreline as shown in Figure 7.4(a). Solidification of the molten metal also results in the segregation of low-melting point eutectic liquids in the grain boundaries. The presence of this liquid film in the grain boundaries of the solidified grains makes the solid very weak, and hence any strain arising from solid/liquid contraction or geometric constraints results in the failure of the alloy as shown in Figure 7.4(b). The cracks propagate along the solidifying thin liquid film present at the grain boundaries, and these cracks are intergranular in nature. Alloying composition hence plays a major role in avoiding the tendency towards hot cracking. A longer solidification duration makes an alloy more susceptible to hot cracking, and hence there are reasons why only specific alloys in various series of aluminium alloys are weldable.

  The 1xxx series aluminium alloys do not have any alloying elements and exhibit a sharp melting point due to the absence of any other significant phases and hence do not undergo any eutectic segregation, unlike the other alloyed series. This makes the 1xxx series immune to hot cracking and hence showcases excellent weldability. However, their applications are limited to places where the joint requires a high thermo-electrical conductivity, since these alloys are soft and lack structural strength due to the absence of strengthening mechanisms viz. precipitation strengthening and/or solid solution strengthening.

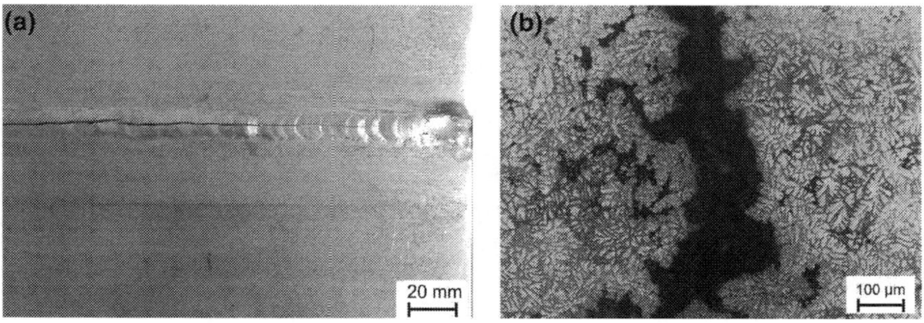

FIGURE 7.4 (a) Photographs of hot cracking in AA7075 alloy (b) Micrographs of hot cracking in AA2195 alloy.

Another reason that certain alloys in various other series of aluminium alloys, apart from the 1xxx series, are less susceptible to hot cracking is the solute content (alloying element concentration). The susceptibility towards hot cracking is indicated in Figure 7.5. The volume fraction of eutectic liquid segregation is directly proportional to the solute content in the alloy/filler material for lower concentrations of solute. A low solute content implies a lesser volume fraction of segregated liquid and hence a lesser hot cracking susceptibility. On the contrary, a higher solute content and hence a higher eutectic liquid volume in the grain boundaries tends to fill up cracks, if any, and the cracks get healed resulting in a reduced tendency towards hot cracking.

- *Thermal Conductivity:* Most of the aluminium alloys possess a high thermal conductivity, which means that the heat from any external source, e.g., an arc is readily dissipated to the other parts of the base metal resulting in a poor fusion towards the faying edges. To compensate this, a higher welding current resulting in a higher heat input is usually required. However, a higher welding current has its own disadvantages, and hence coarse grains in the weldment are usually observed in fusion welding of aluminium alloys. Optimisation of welding parameters for an arc welding process becomes an inevitable task for the joining of aluminium alloys. Sometimes, high-energy density welding processes like electron beam welding or laser beam welding can help overcome this concern; however, these processes come at their own costs and sophisticated set-ups. The thermal conductivity of aluminium alloys depends not only on the alloying elements but also on

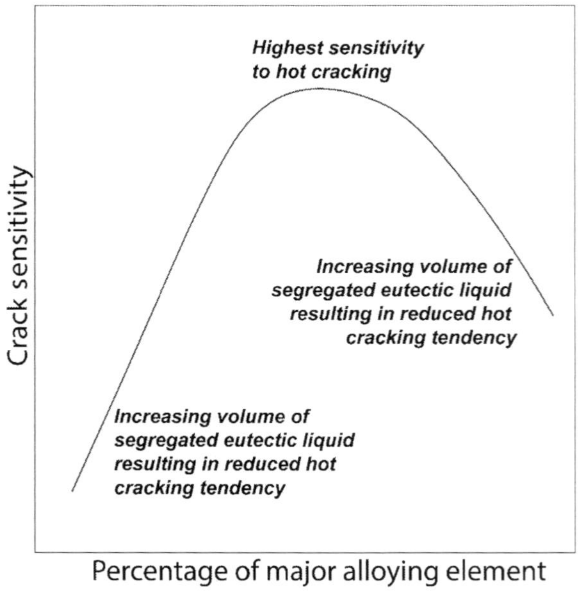

FIGURE 7.5 Crack sensitivity with respect to solute content in aluminium alloys.

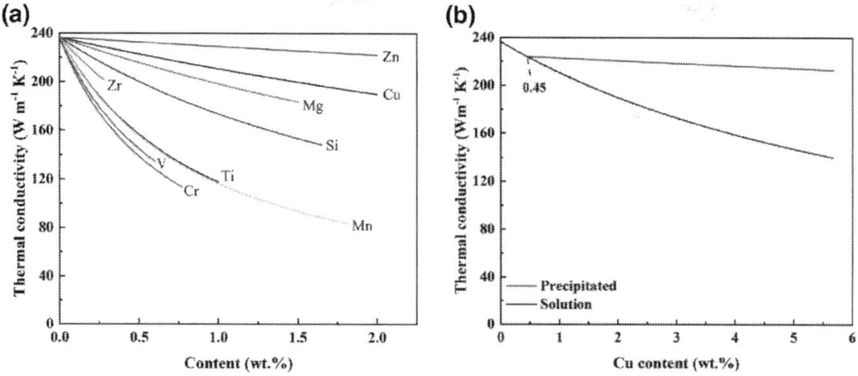

FIGURE 7.6 Variation of thermal conductivity against the alloying content for (a) various alloying elements (b) different strengthening mechanisms [4].

the mode of strengthening, i.e., solid solution strengthening or precipitation strengthening. A trend in the thermal conductivity of aluminium alloys based on the alloying element percentage is shown in Figure 7.6(a). It has been observed that for all the alloying elements, the thermal conductivity reduces with the increase in alloying.

It is also observed that for the same alloying element concentration, a solid solution-strengthened aluminium alloy shows a lower thermal conductivity than a precipitation-hardened alloy as can be observed from Figure 7.6(b).

In general, the highest thermal conductivity is observed for the 1xxx series aluminium alloys because of the absence of any second phases or precipitates. They are followed by the 6xxx, 3xxx, 2xxx, 4xxx, 5xxx, and the 7xxx series in the decreasing order of thermal conductivities. One must note that this sequence may change not only for specific alloys in a particular series but also with the heat treatment conditions. For eg., aged aluminium alloys will have a higher thermal conductivity when compared to alloys that are solution treated and quenched. One frequent practice to avoid issues related to higher thermal conductivity of aluminium alloys is that they are usually preheated to suitable temperatures (e.g., 150–200°C), so that temperature difference between the weldment and the base metal is reduced thereby reducing the thermal current.

7.4 CHOICE OF FILLER WIRES

Choosing the apt filler wire often requires good knowledge of the variety of filler wires available and their chemistry and compatibility with the base metal to produce a sound and defect-free weld in aluminium alloys. A wrong choice of filler wires leads to higher susceptibility towards eutectic segregation and can also promote coarse grain structures in the weldment, making the weld more prone to cracking at elevated temperatures. The choice of the filler material employed for performing the joining operation in case of fusion welding of aluminium alloys plays an equally significant role as that of the correct choice of welding parameters.

One common rule of thumb for welding of engineering alloys is to choose the same filler wire as that of the base metal to maintain the chemical composition and to avoid residual stresses in and around the weld pool, which may arise due to differential thermal expansion between the base metal and the filler metal used. However, the choice of filler wire to be used for joining of aluminium alloys becomes more of a complicated metallurgical problem to solve than to just consider thermal expansion behaviour between the filler and the base metal. Aluminium has an exceptionally low melting point (660°C) compared to steels, copper alloys, and superalloys. Fillers chosen in most of the applications (except for the 1xxx series) are usually different from that of the base metal to comprehend the deteriorating effects of low-temperature eutectic liquid segregation along the grain boundaries during solidification. The most applied engineering criteria (but not all) considered for welding of the different series of aluminium alloys are discussed here.

- *1xxx:* Usually, the best industrial practice for welding of the 1xxx series is to use the same filler wire as that of the parent metal to avoid any degrading effects due to the alloying elements. Eg., AA1100 is usually welded with AA1100 filler itself. For such welds, usually the HAZ forms the weakest link due to high thermal gradients and hence the most common area for failure under tensile loads. However, the 1xxx series filler wires are usually not used in welding of other aluminium series alloys or even dissimilar combinations involving the 1xxx series as one of the base metals, since the weld will draw in alloying additions from the base alloys being welded, resulting in an increased susceptibility towards cracking.
- *2xxx:* This is an Al–Cu alloy system and one of the most widely used aluminium alloys in the aerospace sector due to their active response towards ageing treatments and hence achieving tailored properties for various functional applications. Despite a large number of 2xxx series aluminium alloys, such asAA2011, AA2018, AA2024, AA2218, AA2219, AA2319, AA2195, AA2050, AA2081, AA2519, and so on, being developed and studied, AA2219 and AA2519 are the only two alloys in the 2xxx series that possess good weldability. These alloys are welded with the only available filler materials for the 2xxx series viz. AA2319, AA4043, and AA4195. This unique base metal filler combination makes AA2219 the most widely used aluminium alloy in the aerospace sector, providing a combination of high specific strength and good weldability with AA2319 and AA4043 filler metals. AA2219 welded with AA2319 filler responds to post-weld heat treatment resulting in an increased strength of the joint. On the contrary, AA4043 does not respond to heat treatment, and hence AA2319 becomes a major choice in structural applications. AA2319 contains copper (Cu), manganese (Mn), and titanium (Ti); moreover, the presence of titanium enables the weldment to achieve finer grains in the fusion zone. These finer grains will have a higher grain boundary area, and hence any eutectic liquids segregated along the grain boundaries during solidification will get dispersed uniformly in

a less concentrated manner, making the solidifying weld less susceptible to hot cracking. Apart from being used as a filler metal in the welding of AA2219, AA2319 does not find any industrial applications. AA4043 is a binary Al–Si alloy with approximately 5% Si. This filler is characterised by excellent fluidity and particularly good resistance to weld cracking at the cost of much reduced mechanical strength. AA4043 is more suited for welding of dissimilar series of Aluminium alloys, i.e., joints which will not undergo any post-weld strengthening heat treatment. Another commercially available filler material used for welding the 2xxx series aluminium alloys is AA4195. It contains around 5% Si and 4% Cu and is mostly used in the welding of aluminium alloys, in particular the 2xxx and 6xxx series. When used in aluminium systems which do not have Cu, weld dilution happens and results in a much-reduced Cu content in the weldment (~2% Cu), making the weldment highly prone to hot cracking. Sometimes, advanced Al–Cu alloys with lithium addition such as AA2195 and AA2050 show susceptibility to hot cracking. The reason is due to the nature, number, and amount of eutectic constituents formed during the solidification of welds. Though numerous studies have reported the usage of AA2319 and AA4043 as filler wires for these alloys, fusion welding is always challenging due to multiple repairs. Hence, it is recommended to weld these materials using solid-state joining processes such as friction stir welding.

- *3xxx:* This series of aluminium alloys exhibits excellent weldability along with formability. However, due to their poor strength, they are seldom used in aerospace and other structural applications, and their usage is limited to domestic, automobile, and electronics sectors. They are usually welded with the 4xxx series fillers because of the excellent wettability of the 4xxx series alloys due to the presence of silicon. Another common industrial practice is to use the 5xxx series filler viz. AA5356 for welding of the 3xxx series alloys. AA5356 is an Al–Mg alloy that offers superior strength compared to AA4043 however at the expense of making the weld more susceptible to solidification cracking. The 5xxx fillers should not be used to weld the A3xx casting alloys, such as A356 and A319, because these materials contain a lot of Si, which combines with Mg from the 5xxx filler to produce an extremely brittle intermetallic of magnesium silicide, which drastically lowers the ductility of the weld.
- *4xxx:* Having silicon as the prime alloying element, they are mostly used as fillers for welding and brazing of other series of aluminium alloys. However, they are themselves weldable as well with fillers from their own series viz. AA4043 and AA4047. The Al–Si filler metals are characterised by excellent fluidity and have good resistance to weld cracking. In addition, they are less prone to develop porosity in the weld than the 5xxx fillers. Some of the 4xxx fillers, such as AA4643 and AA4010, have a small amount of Mg, which helps to make the weld heat-treatable to match the base metal properties more closely. This is especially important in applications where the welded assembly will be completely reheat-treated, i.e., solution heat-treated, quenched, and aged after welding.

- *5xxx:* The most common industrial practice to weld the 5xxx series aluminium alloys is to use the AA5356 as the filler material. Additionally, some 5xxx series alloys are also weldable with AA4043. However, as a general engineering practice, it is not recommended that welders use the 4xxx fillers to weld Al–Mg 5xxx alloys, such as AA5083. The high Si content of the 4xxx alloys combines with the high Mg of the 5xxx base material, producing the intermetallic magnesium silicide. Consequently, the weld tends to have poor ductility and toughness. The one exception is welding AA5052 using AA4043, and this combination is possible due to AA5052 having a low Mg content. The industrial practice is to use AA5554 or AA5754 to weld alloys with low Mg content like AA5052, AA5154, and AA5454, and AA5356 is used to weld intermediate-strength alloys such as AA5086 and AA5083. Sometimes, AA5183 is preferred for its better mechanical properties. AA5183, AA5556, and AA5087 are used to weld the highest-strength 5xxx alloys.
- *6xxx:* These are moderate strength alloys which are pretty tricky to weld due to their high hot cracking susceptibility. The 4xxx fillers are widely used to weld 6xxx extrusions and sheets/plates. They are also used to weld castings to 6xxx alloys and to themselves. In fact, during the welding of 6xxx alloys, which are prone to cracking, the Al–Si chemistry of the 4xxx filler, when diluted with the 6xxx base alloy, provides a weld chemistry that is highly crack-resistant. Additionally, in equally important applications, the 5xxx series

TABLE 7.3
Suggested Filler Wires to be used for Welding for Most Commonly used Aluminium Alloys

Parent Metal (Aluminium Alloy)	1050 1080 1200	2219	3103 3105	5005 5083 5251 5454	6061 6063 6082	7005 7019 7020 7039	8090
8090				5556			5556
7039	5556		5556	5356	5556	5556	
7019	5356		5356		5356	5356	
7005						5039	
6061	5356		4043	5356	5556		
5083	5356		5356				
5005	5356						
3103	5356	2319	5356	5356	5356	5556	5556
3105			4043		5056	5356	
	4043	4043					
2219	2319	2319					
	4043						
1050	4043	2319					
1080	1050	4043					

fillers, specifically AA5356, often are used to weld the 6xxx base alloys such as AA6061 and AA6063 when superior mechanical properties are desired.
- *7xxx:* This too is a family of high-strength aerospace alloys. Like the 2xxx alloys, most of them are not weldable using Gas Tungsten Arc Welding (GTAW) or Gas Metal Arc Welding (GMAW) because of their susceptibility towards hot cracking and stress-corrosion cracking. The exceptions are AA7003 and AA7005 extrusion alloys and AA7039 plate alloys. All three of these alloys are readily weldable using AA5356 filler. AA7075, even though an extensively used aerospace alloy, is usually never welded non-autogenously. However, recently, AA7075 alloy was made weldable by the proper selection of filler wire and its constituents. To eliminate hot cracking, 7075 filler wire is added with TiC nanoparticles, which produces a fine-grain microstructure and modified secondary phases.
- *8xxx:* These are newly developed Al–Li alloys which are unweldable by the currently available fusion welding techniques and filler wires.

Table 7.3 lists the commonly used filler wires for welding of the various series of aluminium alloys.

7.5 CHOICE OF SHIELDING GASES

The two most commonly used shielding gases for arc welding of aluminium alloys are helium (He) and argon (Ar). They are used in the following composition, and their properties are depicted in Table 7.4.

- Pure helium (i.e., 99.99% or higher purity)
- Pure argon (i.e., 99.99% or higher purity)
- Mixture of argon and helium (e.g., 80% Ar and 20% He)

Helium is a pretty expensive gas (due to market shortage) and much lighter compared to argon and hence requires much higher flow rates compared to argon for effective shielding, which adds up to its cost. Helium moroever has higher thermal

TABLE 7.4
Property Comparison between He and Ar as Shielding Gases

	Argon	Helium
Ionisation potential, eV	15.8	24.6
Arc initiation	Good	Poor
Arc stability	Good	Poor
Thermal conductivity (at 300K) mW/mK	17.9	156.7
Density (relative to air)	1.38	0.137
Cleaning action	Good	Poor

conductivity and ionisation potential compared to argon. This results in a higher heat generation in the arc, resulting in a wider bead when helium is used. On the other hand, argon is preferred due to easy arc initiation, arc stability, and effective shielding along with the easy availability and cheaper cost. However, many engineering applications use a mixture of these shielding gases to produce combined effects. For example, a mixture containing 75% Ar and 25% He is often used in many GTAW and GMAW applications. While argon helps in a higher depth-to-width ratio, the higher heat input due to the presence of helium helps in higher feeding rates and hence faster welding operations. The other difference seen in the appearance of welds using helium or argon as shielding gases is that while argon produces a much shinier and brighter weld, helium or mixtures of helium and argon produce dull weld surfaces which usually require post-weld cleaning operations. The main reason behind this is the arc stabilising action of argon. It smoothly envelops and stabilises the weld current of the torch, preventing the exposed areas from getting oxidised. Furthermore, it assists in a cleaner finish of weldment imparting a shinier appearance. Despite multiple advantages of argon over helium, most of the welding operations in the aerospace sector utilise a combination of these two shielding gases. The main reason behind this is the high thermal conductivity of aluminium, leading to a lack of fusion towards the penetration side as a common weld discontinuity. This is compensated by the high thermal capacity of the He–Ar gas mixture.

It is also to be noted that AC (alternating current) is the most common polarity setting used for arc welding of aluminium alloys in the industry, and argon serves as the best choice of shielding gas when compared to helium or He–Ar mixtures, since argon not only helps in an easier arc initiation but also sustains a much more stable arc compared to the other two alternatives. Pure helium or high percentages of helium (He-90%, Ar-10%) shielding gas are used primarily for GTAW with direct current electrode positive (DCEP) polarity setting. Often designed as seam welders, the combination of GTAW–DCEP and the high heat input from the helium gas used can provide faster welding speeds and outstanding penetration. This configuration is often used to produce full penetration butt welds, welded from only one side.

For instance, a 3.4-mm-thick AA2219 adaptor, welded to a tankage structure used for storing liquid oxygen, employs a single-pass GTAW with DCEP polarity with a 99.998% helium as shielding for welding a square butt joint with Ø1.6 mm ER2319 filler with a welding current of 90A. The presence of helium gas allows a single pass to suffice to fuse the weld root and a welding speed as high as 200 mm/min. A similar 4.5-mm-thick AA2219 adaptor is gas tungsten arc welded in two passes to another tankage structure with 80% Ar–20% He gas mixture as shielding gases with alternating polarity (AC) for welding a vee-groove joint with Ø2.4 mm ER2319 filler with peak current welding current of 185A. The maximum welding speed as per the weld procedure qualification record is only 80 mm/min.

7.6 PROCESSES USED FOR WELDING AND JOINING OF ALUMINIUM ALLOYS

Because of reasons stated earlier, there are only a few series of aluminium alloys which are weldable by fusion or solid-state welding processes. The weldable series are joined by a variety of processes which are described subsequently in this section. Recently, there is an increase in the development and employment of high-energy density processes (like electron or laser beam welding processes) and solid-state welding processes (like friction stir welding) compared to conventional fusion welding processes (like GTAW/GMAW) in both aerospace and other industrial sectors. The process being employed depends primarily on the following factors:

- Weldability of an alloy
- Availability of
  - welding equipment/facilities
  - qualified welders or operators (for person- or process- (respectively) oriented joining processes
  - testing techniques (destructive and non-destructive)
  - heat treatment facilities
- Geometric complexity of the joint
- Required physical and mechanical properties (and hence microstructure)
- Production quantity
- Cost of joining

Aerospace components being joined will have more stringent quality requirements compared to a component being used on the ground.

7.6.1 Gas Tungsten Arc Welding (GTAW)

Gas tungsten arc welding (GTAW) was conceptualised in 1890 by C.L. Coffin to weld in a non-oxidising gas atmosphere, and the concept was further refined in the late 1920s by using helium and argon for shielding. This process was ideal for welding magnesium and also for welding stainless steels and aluminium. The GTAW process has become one of the most important fusion welding processes adopted for aluminium and its alloys. Known by many other names, GTAW employs a tungsten electrode, filler wire (if necessary), and shielding gas/es. A typical GTAW set-up is demonstrated in Figure 7.7.

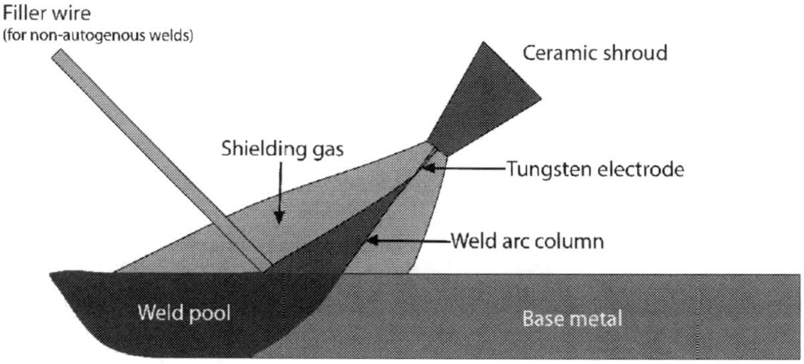

FIGURE 7.7 Schematic of the GTAW process.

The GTAW set-up described here is usually common for all types of fusion welding processes involving a tungsten electrode. Based on the polarity of the electrode and the workpiece, GTAW processes are divided into the following types.

- *Direct Current Electrode Negative (DCEN):* The electrode is connected to the negative terminal, and the workpiece is connected to the positive terminal. More heat is generated at the workpiece resulting in a higher depth of penetration. Oxide layer inclusions (described in Section 7.2) are very prominent for this polarity setting.
- *Direct Current Electrode Positive (DCEP)*: Electrode is connected to the positive terminal, and the workpiece is connected to the negative terminal. This reversed polarity helps in removing the oxide layer due to bombardment of ions on the workpiece; however, since the majority of the heat is generated at the electrode, i.e., tungsten, chances of tungsten getting melted and being a part of the weld bead in the form of inclusions is significantly high, making the weld brittle. The electrode tip is accordingly designed to account for least inclusions when DCEP is adopted.
- *Alternating Current (AC):* This is the most used polarity setting for aluminium alloy welding. Because of the continuous reversal of polarity, it achieves intermediate effects of DCEN and DCEP, i.e., a combination of oxide removal and moderate penetration.

GTAW is a commonly used process in the industry for joining aluminium alloys, considering its easy and quick set-up and less sophisticated and portable equipment. Many industries have introduced automated set-ups for carrying out long and circumferential seams to avoid human errors. Otherwise, movement of the tungsten electrode along with proper feeding of the filler wire is a welder's skill to produce a sound and defect-free weld. Thicker sections usually require edges to be prepared accordingly to deposit the filler wire to the root side, failing to do so may result in improper fusion.

7.6.1.1 Microstructure

The strengthening in aluminium alloys is primarily due to the presence of fine and uniformly distributed precipitates (e.g., $CuAl_2$ in the 2xxx series aluminium alloys) which hinder the movement of dislocations, resulting in strengthening. In GTAW processes employed for the welding of aluminium alloys, usually the temperature of the arc is significantly high to dissolve these precipitates. A high cooling rate post-welding further does not give sufficient time for the diffusion of atoms within or across the grains, which inhibits the formation of such precipitates. As a result, the fine and uniform distribution of the precipitates, which is present in the parent metal, will not be observed in the weldment. Since the size and distribution of the strengthening precipitates are small, optical microscopy may not be helpful to capture the microstructure of aluminium alloys, and hence transmission electron microscopy is usually used. Optical micrographs can however easily reveal the elongated structure of grains normal to the weld seam direction, which is due to the directional solidification post-welding. Figure 7.8(a) shows elongated grains in the base metal (AA2219)

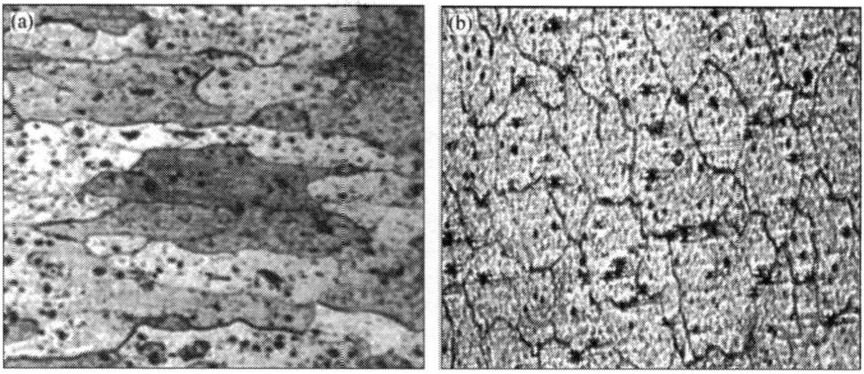

FIGURE 7.8 Microstructure of (a) thebase metal, AA2219 and (b) GTAW weld fusion zone

Source: (with permissions from Elsevier publications) [5].

TABLE 7.5
GTAW Parameters Used for Welding of 7-mm- [6] and 5-mm-Thick AA2219 [5]

Plate thickness, mm	7	5
Current, A	350	150
Voltage, V	15	30
Speed, mm/min	200	180
Polarity	AC	AC
Shielding gas	99.99% Ar	99.99% Ar
Shielding gas flow rate, L/min	16	14

in the rolling direction and Figure 7.8(b) shows elongated grains perpendicular to the weld seam in the GTAW joint's weldment. The parameters used for the welding are indicated in Table 7.5. Figure 7.9 depicts the TEM images of the base metal and GTAW weldment, respectively. One can clearly observe the partial dissolution of the fine precipitates and other precipitates coarsening out in the GTAW weldment as per the reasons explained earlier. Energy dispersive spectroscopic (EDS) analysis shows that the undissolved precipitates in the GTAW weldment are 5.2% Cu and 94.4% Al by weight percentage. [5] EDS results stand as a proof towards the non-employment of any filler wire for the GTAW process, and hence the chances of formation of any other precipitates are eliminated.

The dissolution of these strengthening precipitates forms precipitate-free zones (PFZs) in the fusion zone and offers no resistance to any macroscopic shear stresses, resulting in a reduction in strength of the weldment, and hence the weld becomes the weakest link and the most common location for failure under tensile loads. This is evident from the results of the tensile test as described in Table 7.6. It is also to be noted that the fusion zone exhibits minimum Vickers hardness as compared to the heat-affected zone and the base metal.

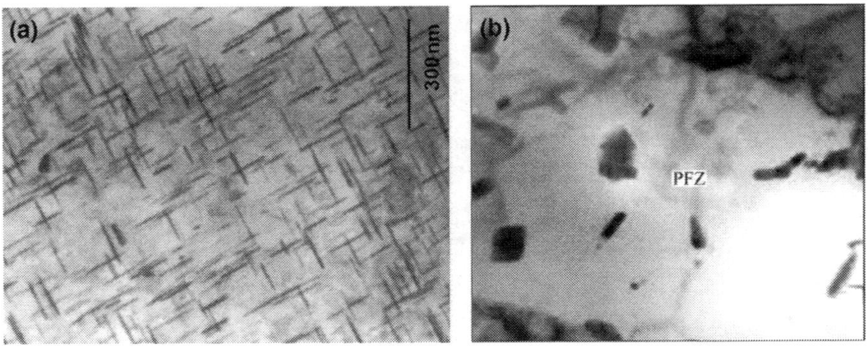

FIGURE 7.9 TEM image of (a) base metal, AA2219 [6] and (b) fusion zone of AA2219.

Source: (with permissions from Elsevier publications) [5].

TABLE 7.6
Mechanical Property Comparison between AA2219 Base Metal and GTAW Specimen

	AA2219 Base Metal 5 mm	AA2219 GTAW Specimen 7 mm [6]	AA2219 GTAW Specimen 5 mm [5]
Yield strength (MPa)	390	163	220
Ultimate tensile strength (MPa)	470	264	242
% elongation	15	6.1	8.8
% reduction in area	10.5	–	6.2

FIGURE 7.10 TIG welding of NASA's SLS's liquid hydrogen propellant tank.

Source: (Courtesy of NASA).

A large variety of space sector components that include propellant tanks for earth storable engines and cryogenic engines and other structural elements have gas tungsten arc welded aluminium alloys in them. Many of these welding processes have been automated especially for large-sized components. GTAW is also employed as a repair welding process, used to eliminate several types of defects in automated gas metal arc welding, electron beam welding, and friction stir welding. Figure 7.10 shows a welder welding structures inside NASA's SLS hydrogen propellant tank.

7.6.2 Gas Metal Arc Welding (GMAW)

Also known as metal inert gas (MIG) welding, GMAW is a fusion welding process and employs a consumable electrode, which serves as a filler wire as well. It employs a continuously fed wire which serves as both the filler and the electrode unlike GTAW, which uses a separate non-consumable tungsten electrode. A typical GMAW set-up is depicted in Figure 7.11.

GMAW offers many advantages over GTAW, such as:

- Higher deposition rates resulting in lesser welding time even for thicker sections;
- Narrower heat-affected zones compared to GTAW;
- Good oxide film removal during welding in DCEP mode;
- Easier handling since the electrode and filler are same; hence the welder need not adjust movement of both hands (in case of manual GMAW).

These advantages make GMAW the most-used fusion welding process for joining of aluminium and its alloys.

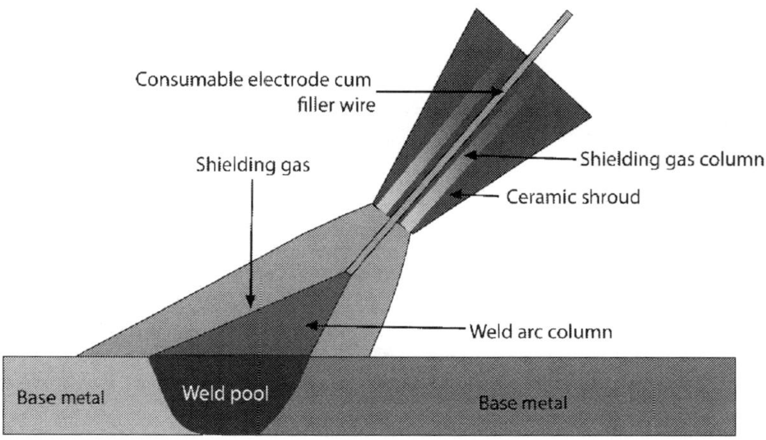

FIGURE 7.11 Schematic of GMAW process.

Like GTAW, GMAW uses DCEN, DCEP, and AC polarities of current. However, based on the current intensity, the following two types of material transfers happen in GMAW.

- *Dip Transfer:* Welding at low amperages results in incomplete melting of the continuously fed wire. As a result of this, the unmelted wire touches the melt pool and short circuits. The heat due to the short circuit melts the wire, and molten wire gets drawn to the weld pool by means of surface tension forces. The arc re-establishes, and the cycle repeats. Dip transfer mode of metal transfer usually results in excessive spatter if the welding parameters are not optimised appropriately.
- *Spray Transfer:* Welding at high welding currents causes complete melting of the fed wire before it can short circuit with the melt pool. In this case, the molten metal comes out in the form of small droplets which form a spray resulting in little-to-no spatter and high penetration.
- *Pulsed Transfer:* A relatively modern technique which uses a low background current superimposed with a high-intensity pulsed current resulting in a low overall average current value. The intensity of the background current is not sufficient to result in the melting of the filler wire, however, the superimposed pulsed wave causes melting and results in a controlled stream of droplets at low average current values resulting in a very efficient metal transfer. The background and pulsed currents can be adjusted to achieve the required metal deposition rate.
- *Cold Metal Transfer*: It is a modification to the dip transfer method, where the short circuit is detected, and the power supply is interrupted. Subsequently, the feed wire is retracted to enable the transfer of the molten metal at the tip of the feed wire without the aid of any electromagnetic forces. This results in less spatter and a low heat input to the weld metal. After the wire is retracted to sufficient length, the power supply resumes to re-establish the arc.

Considering the low melting point of aluminium, most of the GMAW processes employed in the aerospace and industrial sector use a spray or pulsed mode of metal transfer for joining of aluminium alloys. Dip transfer finds applications in the

TABLE 7.7
Suggested Wire Diameters for Various Metal Transfer Modes in GMAW

Transfer Mode		Wire Diameter (in mm)
Dip		0.8
Pulsed		1.2 and 1.6
Spray	Low current	1.2 and 1.6
	High current	1.6 and 2.4

welding of thin sections where high welding currents may result in a burn-through. Table 7.7 suggests the recommended wire diameters for the three modes of metal transfers mentioned earlier.

7.6.2.1 Power Sources for GMAW

One important characteristic of a GMAW process is the power source dynamics. Since the rate at which the wire is fused at the tip must match with the rate at which the wire is fed, a suitable welding current–voltage relationship needs to be established.

- *Constant Voltage Characteristic:* In most of the industrial applications which work on manual GMAW, a self-adjusting arc set-up is employed which uses a flat or *constant voltage characteristic*. It becomes difficult for a welder to manually maintain a constant arc length throughout the seam. As a result of such fluctuations in arc length, the arc voltage will vary. With a flat voltage set up in the power source, the source voltage gets marginally changed based on the arc length to match the wire fusion rate with the wire feed rate. For instance, an increase in the arc length due to unsteady hands of the welder will trigger an increase in the arc voltage accompanied by a large drop in the weld current. This reduced weld current results in a reduced wire burn-off rate, thereby reducing the arc length and arc voltage and increasing the current, and this cycle goes on. As a result, the arc length remains almost constant throughout the arc welding process. Figure 7.12 depicts a typical constant voltage characteristic power source.
- *Constant Current Characteristics:* Also known as drooping characteristics, the current sensitivity with respect to voltage fluctuations is reduced, as a result of which a substantial change in arc voltage triggers only a minor

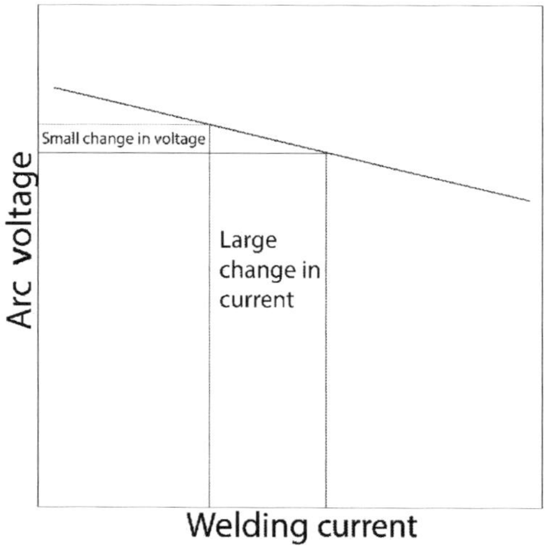

FIGURE 7.12 V-I characteristics of a self-adjusting GMAW arc.

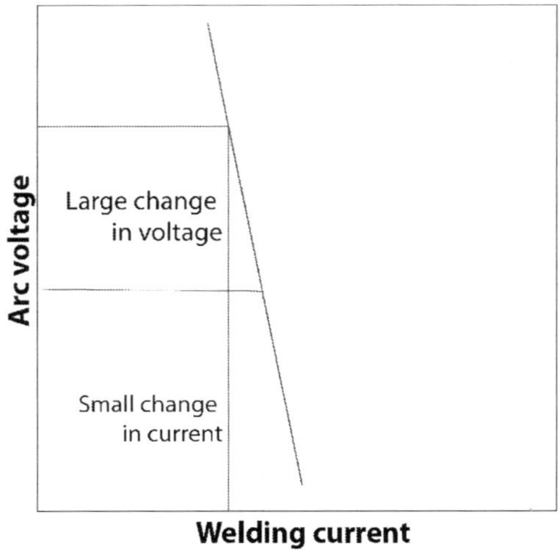

FIGURE 7.13 V-I characteristics of a drooping power source GMAW arc.

change in the welding current. This results in almost the same amount of heat generation unlike the constant voltage characteristic giving a more consistent penetration. Figure 7.13 depicts the typical drooping characteristics of a GMAW power supply.

One major disadvantage of a drooping voltage power source is the dependence of arc stability on the skill of the welder. Since in drooping characteristic, the arc length gets changed based on the motion of the hand of the welder, which on the other hand is compensated by adjustment in arc length in case of flat voltage characteristic, and hence arc instability becomes a common issue in case of constant current characteristic. This arc instability often leads to excessive weld spatter unless the welder is skilled enough to understand the cause behind it and adjust the wire feeding accordingly to overcome the arc instability.

Despite arc stability issues, the constant current characteristic finds application in the welding of thick aluminium sections which require substantial penetration depths. The reason flat voltage characteristic power is not used in the welding of such thick sections is the possibility of picking up the arc voltage from the side wall rather than the root of the weld. This results in uneven penetration and weld spatter. Similar issues can be expected while welding a fillet with flat voltage characteristics.

Arunkumar et al. have drawn a comparison between microstructure and mechanical properties of 25-mm-thick AA2219 aluminium alloy welded by GMAW and GTAW using ER2319 filler wire. The welding parameters used for welding processes are tabulated in Table 7.8.

The microstructures of the base and metal fusion zones in case of GMAW and GTAW observed by them are depicted in Figure 7.14.

TABLE 7.8
Welding Parameters used for GMAW and GTAW of 25-mm-thick AA2219 with ER2319 Filler Wire [7].

	GMAW	GTAW
Electrode	2% Zirconated	–
Joint configuration	Double V groove (AWS D1.2)	
Filler wire diameter (mm)	1.2	2.5
Voltage (V)	28–30	10.5–11.8
Current (A)	185–195	125–138
Welding speed (cm/min)	25–28	10–11
Shielding gas	Argon (99.99%)	Argon (99.99%)
Shielding gas flow rate (L/min)	15	15
Polarity setting	DCEP	AC

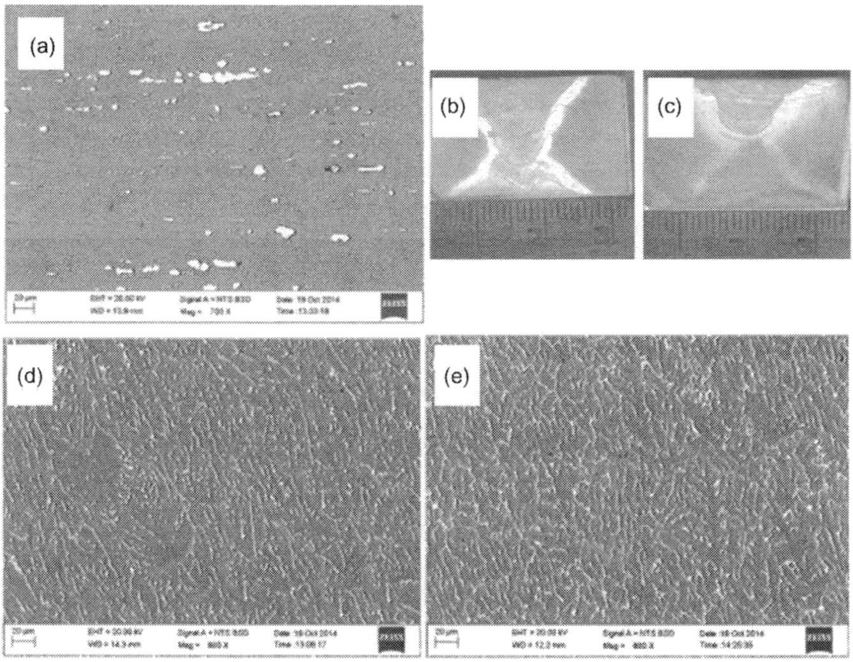

FIGURE 7.14 (a) Optical micrographs of AA2219 base metal, (b) macrostructure of GTAW (c) macrostructure of GMAW weld,(d) fusion zone microstructure of GTAW, and (e) fusion zone microstructure of GMAW weld.

Source: (with permissions from Taylor and Francis [7].

TABLE 7.9
Suggested List of Parameters and Shielding Gases for GMAW of Aluminium Alloys

Thickness (mm)	Current (A)	Voltage (V)	Filler Wire Diameter (mm)	Shielding Gas	Weld Speed (mm/min)
1.5 and below	60–90	16–18	0.6	Argon	1,000
1.6–2.4	90–110	18–20	0.6	Argon	1,000
2.4–3.2	110–140	20–24	0.6/1.2	Argon	800–750
3.2–6.4	140–200	24–28	1.2/1.6	Argon	750
6.4–10	200–280	28–30	1.6	Ar/Ar-He	750–900
10–16	280–300	24–28	1.6	Ar/Ar-He	750–900
16–25	280–350	26–30	1.6/2.4	Ar/Ar-He	750–1,000
50–75	550–650	30–32	4.8/5.6	Helium	250

The optical micrographs (Figure 7.14) of both GMAW and GTAW weld fusion zones show dendritic grains which are attributed to rapid cooling in both the arc welding processes. The dendritic arm spacing however is found to be wider for the GMAWed specimen and narrower for the GTAWed specimen. This spacing of the dendritic arms significantly influences the tensile mechanical properties, where the GMAWed specimens have shown a marginally lower ultimate tensile strength and yield strength compared to the GTAWed specimens. The wider spacing in the dendritic arms of the GMAWed specimen can be attributed to the higher heat input for the GMAW process compared to the GTAW process since the GMAW is done with a direct current reversed polarity (DCEP) setting as it is usually carried out for aluminium alloys which develops around two-thirds of the heat at the tip of the filler wire while on the other hand, GTAW is carried out with an alternating polarity setting which reduces the overall welding current substantially and hence reduces the net heat delivery to the weld pool resulting in less coarse grains compared to GMAW process and hence better room temperature tensile properties.

The logic behind the choice of shielding gases is similar to that of GTAW and has been discussed previously in Section 7.5. A rule of thumb is to select helium or He–Ar mixtures for thicker sections and Argon alone for thinner sections. Table 7.9 summarises some parameters which can be considered for welding of various thicknesses of aluminium alloys. Other welding parameters like number of passes, joint configuration, and so on can be optimised based on the specific alloy, wire feeding rate and weld speed, comfort and skill of the welder (in case of manual welding), and the presence of weld defects like porosity, lack of fusion, and so on.

7.6.3 Electron Beam Welding (EBW)

EBW is one of the high-energy density autogenous fusion welding processes used to join aluminium alloys. The high-energy density beam is developed by accelerating electrons by heating up a tungsten filament to a high temperature. This high-energy beam can be as wide as 2.5 mm and is focused onto the weld line with the help of electromagnetic coils, which form part of the EBW machine itself. A typical EBW set-up is depicted in Figure 7.15.

EBW is extensively used in the welding of thick aluminium sections since they can cause penetration as deep as 400 mm for aluminium alloys. Electron beam-welded joints usually have a very narrow heat-affected zone and fusion zone and display minimum weld shrinkage and distortion due to their smaller fusion zone compared to arc welding processes. The major process parameters include:

- Accelerating voltage (usually in kV)
- Current (usually in mA)
- Travel speed

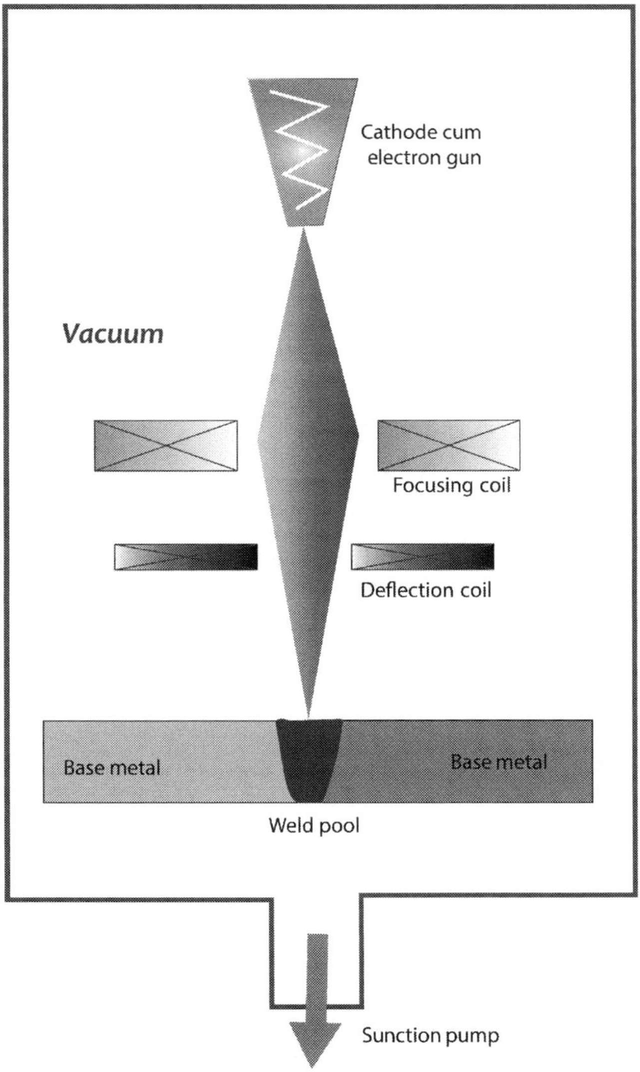

FIGURE 7.15 A typical EBW set-up.

The EBW machines usually have one of the three capabilities:

- Stagnant beam and moving workpiece
- Moving beam and stagnant workpiece
- Moving beam and moving workpiece

Since the incident high-energy beam is a narrow channel of accelerating electrons which possess charge on them, they are highly prone to getting deviated due to any disturbance in the form of contamination, air, or presence of any other charged species. To avoid this, EBW is always carried out in a sealed chamber with a vacuum level of 10^{-3}Pa or better to avoid any deflections in the electron beam. This mandatory requirement of vacuum for any EBW process makes it an expensive process since the creation of elevated levels of vacuum will require specialised diffusion pumps. Higher the size of the hardware to be welded, larger will be the requirement of the vacuum chamber demanding more powerful pumps to be installed and hence costlier will be the process. It is not uncommon for pumps to take several hours to create a vacuum level which produces a defect-free weld. Because of the cost factor and sophisticated set-up, EBW is not the most widely used process in the industry to join aluminium alloys; however, it is the most preferred one when finished components are to be welded, and distortion becomes an area of non-conformance. EBWed joints show superior strength as compared to GMAWed or GTAWed joints due to the higher energy density that melts a very narrow region in a short span of time with an overall lower heat input to the base metal. This results in a narrower fusion zone as well as heat affected zone which are the most common areas of failure for aluminium alloys. Another common trait of EBWed joints is their higher weld efficiencies when compared to GTAW and GMAW.

EBW can be carried out for almost all the weldable aluminium alloys which show lower potential towards hot cracking. Over the course of time, aluminium alloys are welded with mainly two current settings viz:

- Constant current (CC)
- Pulsed current (PC)

The microstructure evolution and the associated mechanical properties are found to be entirely different for these two current settings. While GTAW and GMAW welds are usually characterised by dendritic network of grains in the fusion zone, EBWed fusion zones show cellular grain structures which are finer towards the bulk and coarser towards the faying surfaces (Figure 7.16(a) and (b)).

Biju et. al. have shown that pulsed EBW results in a higher cooling rate when compared to the conventional constant current EBW [8]. This can be attributed to the pattern of pulsed power supply, in which high power pulses, followed by periods of low (background) power, results in melting of small packets of material, which solidifies quickly in the bulk neighbourhood of the unmelted material which acts as a massive heat sink. This higher cooling rate in pulsed EBW, can be attributed for finer grains in the fusion zone, higher internal residual stresses in the weld joint, and

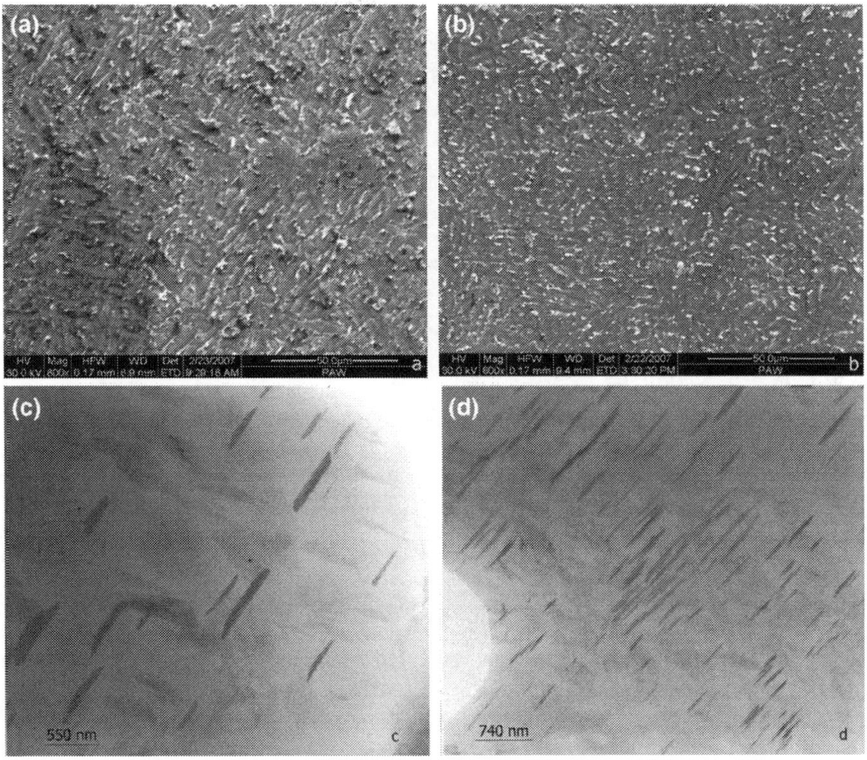

FIGURE 7.16 Scanning electron micrographs of the fusion zone of EBWed joints corresponding to (a) constant current and (b) pulsed current. Transmission electron micrographs of the fusion zone of EBWed joints corresponding to (c) constant current and (d) pulsed current.

Source: (with permissions from Taylor and Francis) [8].

a higher dislocation density compared to constant current EBWed joints. This goes in line with the observations recorded through transmission electron micrographs (Figure 7.16(c) and (d)) in which precipitate dissolution is observed in constant current EBW due to higher temperatures associated with the process, resulting in a lower strength for CC-EBW compared to PC-EBW, where precipitate dissolution phenomena are less significant and impart a higher strength. Table 7.10 describes a list of welding parameters used for welding of 7.2-mm-thick AA2219-T87 plates by EBW in constant current and pulsed current modes and the properties obtained are provided in Table 7.11 [8].

For both these processes, the focus current was kept near to 540 mA. For the pulsed EBW, both the pulse on and pulse off were kept as 7.5 ms with an amplified frequency of 66 Hz.

PC-EBWed joints exhibited superior mechanical properties in uniaxial tension tests. The minimum values are described in Table 7.11.

TABLE 7.10
Parameters used for Carrying out EBW with Constant Current and Pulsed Current Modes [8]

	Accelerating Voltage (kV)	Current (mA)	Velocity (mm/sec)	Working Distance (mm)
EBW-CC	60	90	30	260
EBW-PC	60	150 (Peak)	23	260

TABLE 7.11
Mechanical Properties in Constant Current and Pulsed Current EBW of 7.2-mm-Thick AA2219-T87 Alloy

	Ultimate Tensile Strength (UTS) (MPa)	0.2% Yield Strength (YS) (MPa)	%Elongation (Gauge length = 25 mm)
EBW-CC	301	196	4.6
EBW-PC	317	211	4.8

7.6.4 Friction Stir Welding (FSW)

Invented in the early 1990s by TWI, UK, FSW has emerged as one of the most widely used processes for the joining of aluminium alloys. It is a solid-state welding process which joins two substrates by means of plastic deformation and diffusion without melting either of the substrates. The establishment of a joint without melting in FSW is highly advantageous in the following ways:

- *Higher Weld Efficiency and Joint Strength Compared to any Fusion Welding Process:* Since temperatures involved are far benign compared to temperatures in any fusion welding process, be it GTAW, GMAW, EBW or LBW, grains don't tend to get coarsened, and what is usually expected in an FSW weldment are fine grains providing superior mechanical strength to the joint.
- *No Hot Cracking Post Welding due to Absence of any Melting:* This is one of the major reasons for the non-weldability of a majority of aluminium alloys.
- *Flexibility in Joining Dissimilar Alloys*: Dissimilar alloys which cannot be fusion welded due to the high susceptibility of welding defects like gas porosities and solidification cracking can be easily welded by FSW.
- Other advantages:
  - *No consumables* like shielding gases, filler wires, and so on.
  - *No requirement of weld edge preparation:* The faying surfaces usually are in square butt configuration, sticking closely to one another.
  - *Lesser weld distortion* compared to fusion welding processes.

- *Moderate immunity towards presence of oxide layers*: During the stirring and ramming action of the tool, the oxide layers usually tend to break and get dislodged from the joint surfaces resulting in a lesser oxide inclusion in the weldment.

A typical friction stir welding set-up is depicted in Figure 7.17. It uses a wear-resistant tool which, in most industrial applications, is made of tool steels like H13, AISI 440C, and so on. The rotating tool is plunged into the edges to be joined and traversed to the entire length, and the stirring action of the tool establishes the joint along the edges.

The FSW machine is responsible for providing:

- The required amount of force in the depth direction (called the z-force or plunge force).
- The required amount of force in the longitudinal direction (called the x-force or axial force).
- The plunge depth is the depth from the surface of the joint that the tool shoulder is made to plunge (i.e., the interference depth).
- The required rotational speed, feed rate, and direction of motion of the tool.

The following are some of the essential parameters of a FSW process:

- Plunge force
- Axial force
- Tool rotation speed
- Tool feed rate
- Plunge depth
- Tool's pin morphology (shape, diameter etc.)
- Tool's shoulder diameter

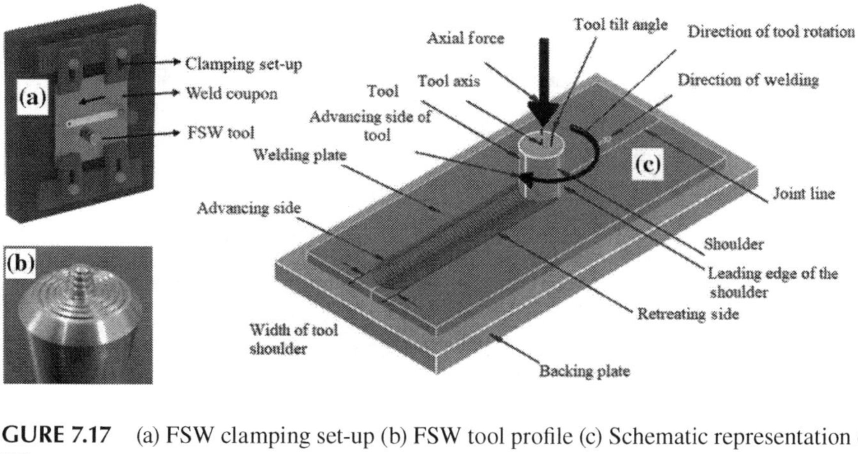

FIGURE 7.17 (a) FSW clamping set-up (b) FSW tool profile (c) Schematic representation of FSW process.

Parameters like tool angle and x-traverse distance are other parameters which one needs to define in the machine before taking up the welding process.

Figure 7.18 shows some advanced friction stir welding and processing tools by MegaStir exhibited at Fabtech Show held on November 11–13, 2014, Georgia World Congress Center, Atlanta, GA USA.

A major disadvantage of the FSW process is the presence of an"end-hole" (also called an exit-hole) towards the end of the joint, i.e., where the tool is retracted. This is because of the non-availability of base material to be plastically deformed and fed into the weld joint as the tool starts to emerge out of the joint end (i.e., the x-traverse distance). As a result of this, the end of the joint will form a void in the shape of a cone. This is one major reason FSW mostly finds applications in longitudinal seam welding and not circumferential seam welding. The presence of a conical end-hole towards the end of the x-traverse distance is compensated by using a longer plate for welding and then subsequently eliminating the portion with the end-hole by appropriate machining methods like wire-cutting or conventional machining. Figure 7.19 depicts the defects such as flash and end-hole during FSW of AA6061-T6 aluminium alloy. The current developed technique with retracting tool technology paves way for welding circumferential joints without any end-hole when the tool exits the workpiece.

Because of their lower melting points, FSW finds widespread applications in joining aluminium, magnesium, and zinc alloys, followed by copper, mild steel, and titanium alloys.

Plate thicknesses as high as 80 mm can be joined by FSW. Usually for higher plate thicknesses, the process is carried out from both the sides, i.e., the top and the bottom. Figure 7.20 shows a macro-section of 75-mm-thick AA6082 plates double-sided friction stir welded.

FIGURE 7.18 Some friction stir welding tools with different profiles.

Source: (courtesy of NearEMPTiness) [9].

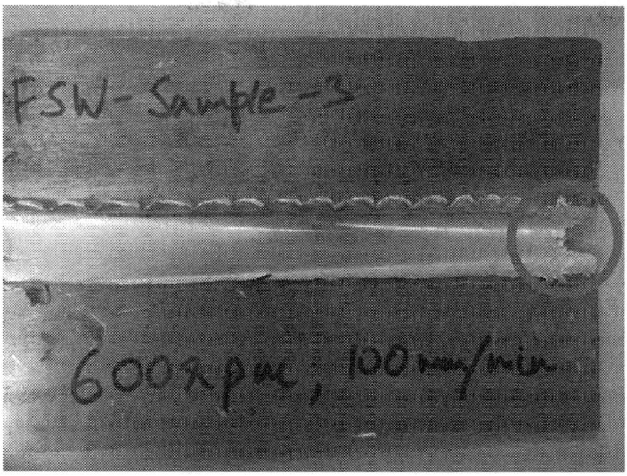

FIGURE 7.19 Presence of flash and end-hole defect in a friction stir welded AA6061 alloy.

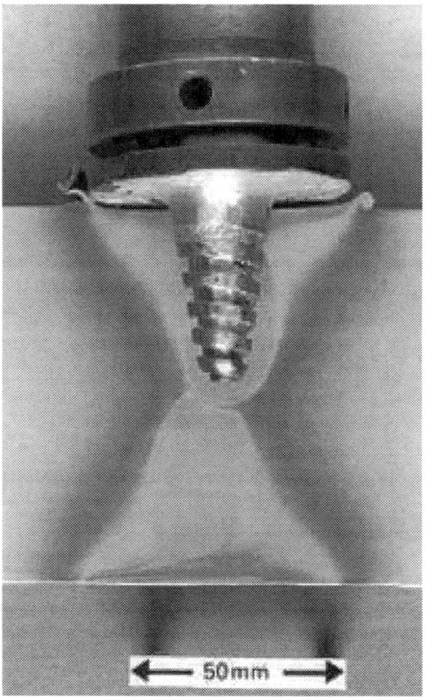

FIGURE 7.20 Double-sided FSW of 75-mm-thick AA6082 alloy illustrating a Whorl™ tooltip.

Source: (Courtesy of TWI Ltd.) [10].

Another emerging area in the field of FSW is self-reacting friction stir welding (SRFSW) also called bobbin tool friction stir welding (BTFSW). It uses a twin shoulder stepped tool which plunges from both sides of the weld, i.e., the crown side and the root side. Simultaneous plunging from the crown and root sides of the weld results in a drastic reduction of the component of axial force experienced by the tool during its x-traverse. This reduces the overall loads on the machine, enhances the life of the tool, and eliminates the requirement of extensively heavy machinery and clamping which is usually the case with conventional FSW, which uses a pin-shaped tool. Another advantage of using this tool is that it prevents bending of sheets/plates due to the applied axial force and avoids the requirement of rigid supports or additional fixtures. Another application of SRFSW comes in the welding of thick aluminium sections which usually require welding from both the top and the bottom sides; however, with bobbin-type tools, thicknesses as high as 30 mm can be welded by SRFSW in a single pass, thereby reducing the time and effort spent in re-clamping and resetting while welding the opposite side in case of a pin-type tool. Figure 7.21 shows a bobbin-type tool used for the welding of 6-mm-thick AA6061-T6 plates along with the experimental set-up. The tool is made with H13 tool steel. Table 7.12 shows some of the welding and tool geometry parameters.

FSWed joints show superior mechanical properties compared to fusion welding processes such as GTAW. The prime reason being a solid-state joining process with no melting and re-solidification. Strengthening is due to the homogeneous microstructure within the weld region. FSW on 5-mm-thick AA2219-T87 sheets were carried out using the following parameters mentioned in Table 7.13.

Table 7.14 Mechanical property comparison between GTAW and FSW of 5-mm-thick AA2219-T87 alloy.

FIGURE 7.21 (a) Bobbin type tool and (b) Set-up used for self-reacting friction stir welding of 6-mm-thick AA6061 alloy.

TABLE 7.12
Parameters used for Bobbin Tool Friction Stir Welding of 6-mm-Thick AA6061-T6 Alloy

Welding machine	ETA Tech., India
Tool's top shoulder diameter, mm	20
Tool's bottom shoulder diameter, mm	19
Tool's shoulder-to-shoulder distance, mm	5.6
Tool rotational speed, RPM	600
Tool feed rate, mm/min	100
Tool z direction plunge-top, mm	0.3
Tool z direction plunge-bottom, mm	0.1

TABLE 7.13
Parameters used for FSW of 5-mm-thick AA2219 sheets [11]

Welding machine	ETA Tech., India
Tool pin profile	Taper threaded pin with the scrolled shoulder
Tool shoulder diameter, mm	20
Tool pin diameter, mm	6 to 4
Tool pin length, mm	4.8
Tool rotational speed, RPM	800
Axial force, kN	25
Tool feed rate, mm/min	300

TABLE 7.14
Comparison between the achieved mechanical properties of FSWed joint compared with GTAW joint for 5-mm-thick AA2219-T87 sheets [11]

	UTS (MPa)	0.2%YS (MPa)	%Elongation (Gauge length=25 mm)
Base metal (AA2219-T87)	477	384	12.8
GTAW	267	164	6.2
FSW	**348**	**181**	**12.6**

FSW does not employ any filler material, and there is no melting of the base metals. As a result, no dilution of alloying elements can be expected in the weldment. Due to the extensive stirring in the pin region of the rotating tool, the stir zone is subjected to severe plastic deformation, which causes the coarse elongated grains in the base metal as shown in Figure 7.22(a) to get fragmented to fine and equiaxed grains as shown in Figure 7.22(b) through the process of recrystallisation and thus

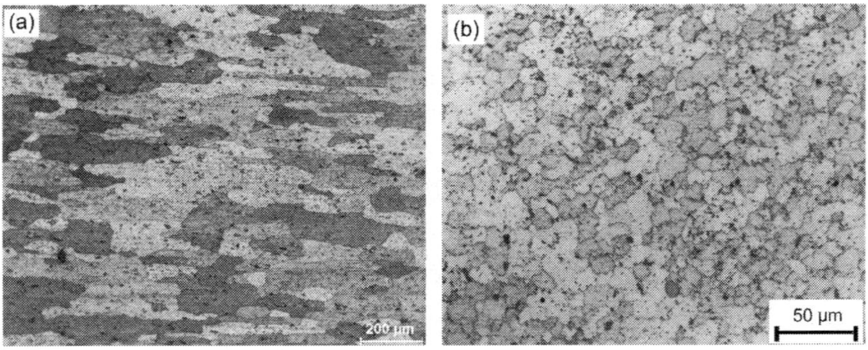

FIGURE 7.22 Optical micrographs of AA2219-T87 (a) base metal and (b) stir zone [11].

providing a higher strength to the joint. Another reason for higher strength can be expected due to the breaking up of the coarse strengthening precipitates to fine uniformly distributed precipitates in the severely deformed zone of the FSWed joints.

Applications of FSW are plenty, right from welding of huge tankage systems for storing propellants for liquid rocket engines to welding of human space capsules carrying astronauts to space. Figure 7.23 depicts the bulkhead to nose-cone friction stir welding of Orion spacecraft, welding of the Orion space capsule using FSW at NASA's Michoud Assembly Facility and propellant tanks and FSW of propellant tank's dome.

Defects in Friction Stir Welding: Although friction stir welding overcomes a lot of defects usually associated with fusion welding processes, the huge stirring and forging forces felt by the alloy due to the rotary motion of the tool in FSW leads to some commonly observed defects linked with this solid-state welding process. Most of these welding defects are due to the improper pairing of the tool's rotational speed with the tool feed rate. Both the aforementioned parameters need to be optimised in accordance with the choice of the tool size and geometry. Some of the most observed defects are listed here.

- *Wormhole Porosity:* This type of defect occurs usually in high ratios of tool feed to tool rotational speeds. The temperatures developed aren't high enough to cause adequate plastic deformation and stirring, resulting in a void which usually extends throughout the weld seam, i.e., the x-traverse.
- *Overheat Porosity/Surface Porosity:* This type of defect occurs at low ratios of tool feed to tool rotational speeds. Elevated temperatures result in superficial melting and ejection of the material away from the tool, resulting in a porosity that is usually very near the surface. This porosity can be either localised or observed throughout the length of the weld.
- *Cross Section Reduction*: Due to the excessive plunge forces experienced by the substrate from the tool shoulder, a reduction in cross sectional area along the weldment is usually expected. This results in higher and non-uniform stresses in the weldment.

FIGURE 7.23 (a) Bulkhead to nose-cone welding of Orion spacecraft [12]. *Source*: (courtesy of NASA) (b) Welding of the Orion space capsule using friction stir welding at NASA's Michoud Assembly Facility [13]. *Source*: (courtesy of NASA) (c) Friction stir welding of propellant tank domes [14].

- *Burrs/Flashes*: Due to the centrifugal stirring forces experienced by the base metal, the material is thrown out axially, resulting in the formation of sharp burrs or flashes that need to be removed completely. The main culprit behind excessive flash is high plunging force or high plunging of the tool shoulder into the joint.

7.6.5 Laser Beam Welding (LBW)

Like EBW, LBW is a high-energy density fusion welding process used to join a variety of alloys. Here, unlike EBW, a high-energy beam of photons is used as the energy source in place of a beam of accelerating electrons. One major reason for limited application of LBW for welding of aluminium alloys is the high reflectivity of aluminium resulting in a poor coupling (the fraction of incident laser beam absorbed by the substrate) between the laser and the aluminium alloy. It is experimentally seen that for aluminium alloys, the coupling factor reduces with the increase in wavelength of incident laser. Hence, shorter wavelength lasers like solid-state lasers are usually used for LBW of aluminium alloys. The shorter wavelengths can also be transmitted through optic fibres resulting in an altogether higher flexibility in the motion of the laser source.

Yoon et al. carried out LBW for a variety of aluminium alloys with different laser sources. The parameters used for LBW of 1.2-mm to 2-mm-thick aluminium alloy sections and achieved ultimate tensile strength and joint efficiencies in the welded condition are listed in Table 7.15. The alloys were welded in T6 temper condition.

The advantages of LBW are similar to those of EBW; however, LBW comes at an additional advantage of simple and less expensive set-up due to non-requirement of any vacuum chamber for carrying out the welding process as in EBW. The 6xxx and 7xxx series alloys which cannot be autogenously welded can be welded by LBW using wires (e.g., AA4047 or AA4032) that can help prevent the solidification cracking. Shielding gases like helium, argon, nitrogen, or their combinations are usually fed coaxially along with the laser beam to improve the surface appearance of the welded joint. Applications of LBW are usually in the automobile industry. LBW is industrially carried in the following three modes.

- *Pulsed Laser Welding:* In this, the laser beam is turned on and off at a very high frequency (e.g., between 200 Hz and 1,000 Hz). The pulsing laser supplies energy to the weld joint in packets creating small overlapping pockets of molten material. These small molten pockets cool amazingly fast supplying minimum heat to the surrounding metal resulting in reduced distortion and hot cracking tendency. Considering the high thermal conductivity of aluminium alloys, this is the most preferred mode of laser beam melting.

TABLE 7.15
Joint Efficiencies and Achieved Ultimate Tensile Strength of Aluminium Alloys Welded by Gas and Solid-State Laser

Alloy	Autogenous 5 kW CO_2 Gas Laser		Pulsed 2 kW Nd: YAG Solid-State Laser	
	UTS (MPa)	Joint efficiency (%)	UTS (MPa)	Joint efficiency (%)
AA5083	283	90	264	84
AA6061	189	58	250	68
AA7475	233	49	126	26
AA8090	242	84	236	82
AA2219	–	–	263	58

Source: (Courtesy of TWI Ltd.) [15].

- *Continuous Laser Welding:* In this, a steady non-pulsing laser beam is directed towards the weld joint producing deeper penetration compared to pulsed laser welding. The material towards the leading edge of the laser melts, while the material towards the trailing edge solidifies. Care should be taken to optimise the feed rate to avoid overheating of the molten metal. Unlike pulsed laser welding, which involves multiple melting and solidification process, continuous laser welding is used to weld thick aluminium alloy sections that are more prone to solidification cracking.
- *Laser Stir Welding:* Here, a continuous beam of laser is directed towards the weld joint, and the beam is made to oscillate at a high frequency which results in stirring action in the molten weld metal. The stirring action results in crack-free joints and allows more control over the weld pool geometry, i.e., the oscillation frequency of the laser beam can be adjusted to weld varying, geometrically complex, and asymmetrical cross-section thicknesses with the additional advantage of higher feeding rates. Usually, laser stir welding is performed without filler wires, thereby lowering the cost of consumables.

7.6.6 Variable Polarity Plasma Arc Welding (VPPAW)

Variable polarity plasma arc welding (VPPAW) is a direct current plasma arc welding process operated with change in polarity to produce high-quality welds with higher penetration. The basic difference between alternate current (AC) and variable polarity is the waveform shape. In AC, the waveform is sinusoidal, and polarities (positive and negative) are varied in a balanced way. In variable polarity, the ratio of polarities can be varied independently of current and duration. Normally, the waveform is rectangular in nature for variable polarity.

Aluminium welded with direct current electrode positive (DCEP) or direct current electrode negative (DCEN) was unsuccessful due to weld metal porosity, electrode degradation, and incomplete fusion. To overcome the weldability issues due to the formation of aluminium oxide layer, an alternating current (AC) power supply is used. The success in AC power supply was limited due to electrode deterioration, lack of weld pool control, and the need for the superimposed high-frequency and decreased amperage capability of the welding torch. Variable polarity welding techniques give better solutions to overcome these problems. This process takes the advantages of both DCEP and DCEN processes. Figure 7.24 shows the optical microstructure of VPPA-welded AA2219 alloy having as-solidified dendritic microstructure without porosities. The following advantages made VPPAW process attractive for welding of aluminium alloys.

- Reduction in the number of weld passes and 100% joint penetration (single pass up to 12.7 mm for aluminium)
- Less required joint preparation (square butt joints)
- High-quality welds than GTAW and minimum number of repairs
- Consistent X-ray quality
- Minimal distortion

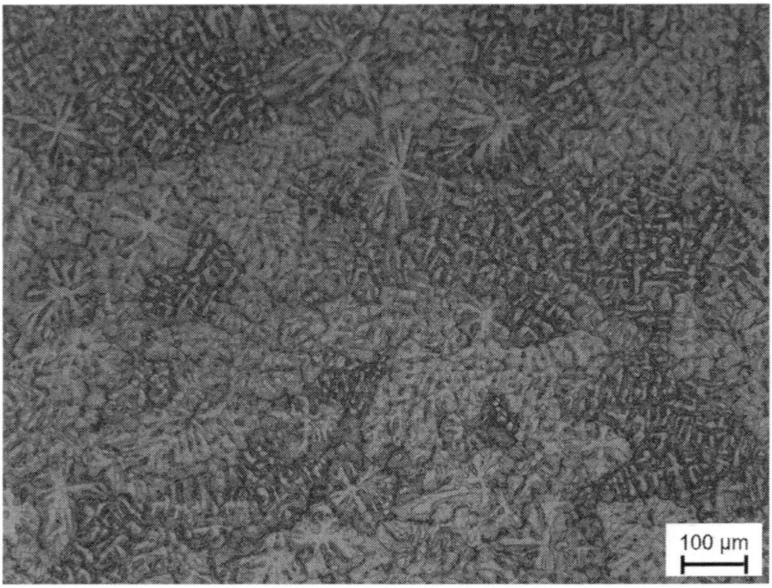

FIGURE 7.24 Microstructure of AA2219 material welded by VPPAW using AA 2319 filler material.

The electrode in VPPAW alternates between positive and negative polarity. While in the electrode negative polarity (DCSP), penetration is more, and electrode deterioration is minimal. When the electrode is connected to the positive terminal, the positive ions clean the workpiece surface (cathodic cleaning), and the electrode deterioration is more. Electrode negative operation (straight polarity) is necessary for plasma arc. The system spends most of its time in the more efficient straight polarity mode for penetration and part of its time in reverse polarity for the cleaning benefit. A flat-tipped electrode and minimum reverse cycle time is required to keep electrode erosion low. The process provides many benefits for welding of aluminium applications.

NASA has utilised the VPPAW technology for manufacturing of propellant tanks in space shuttles.

7.6.7 Brazing of Aluminium

Brazing is a joining process by which a permanent joint is achieved by not melting the faying edges of the base metal but melting the filler metal alone. The brazing temperature is above the liquidus of the braze filler and below the solidus temperature of the material to be joined. During brazing, the filler metal melts, spreads, and wets the closely fitted joint surface by capillary action. Generally, Al–Si- (typically 4047 alloy) based braze fillers are used for similar and dissimilar welding of aluminium alloys.

The demand for brazing of aluminium alloys has steadily increased due to the characteristics such as lightweight, good thermal conductivity, good formability, and high corrosion resistance. Brazed aluminium heat exchangers became prevalent in radiators, condensers, evaporators, air-conditioning equipment, and so on.

The presence of an inherent aluminium oxide layer over the surface makes the aluminium brazing challenging. The thin, impervious oxide layer has a melting point of around 2038°C and it prevents the wetting of braze filler with the parent metal and results in a poor joint. Flux-based brazing and filler metal with magnesium content are developed to overcome this problem. During brazing, 1.0–2.0% Mg content in the braze filler diffuses on the surface, vapourises, and disrupts the oxide layer and facilitates brazing. Magnesium acts as a reducing agent and when combined with the surface layer of aluminium (aluminium oxide), reduces aluminium oxide to aluminium and promotes wettability for brazing.

Brazing of aluminium is preferred during the following instances:

- Joining of parts having thin sections.
- Very large joint area and many junctions per unit area.
- Dissimilar metal joining.
- Elevated temperature operation is not preferred.
- Less distortion of the parts is allowed.
- Production cost is low.

For certain insulation applications, aluminium-to-alumina metal–ceramic joints are used. Mo–Mn metallisation is applied over ceramic surfaces for better wetting characteristics. Figure 7.25 shows the typical metal–ceramic joints made using AA4047 braze filler. On the one side, the braze alloy AA4047 wets and flows over the metallised (Mo–Mn) alumina layer. On the other side, the wettability of AA3003 alloy is improved by the presence of magnesium chips inside the furnace chamber, which acts as a catalyst for the wetting of aluminium base metal.

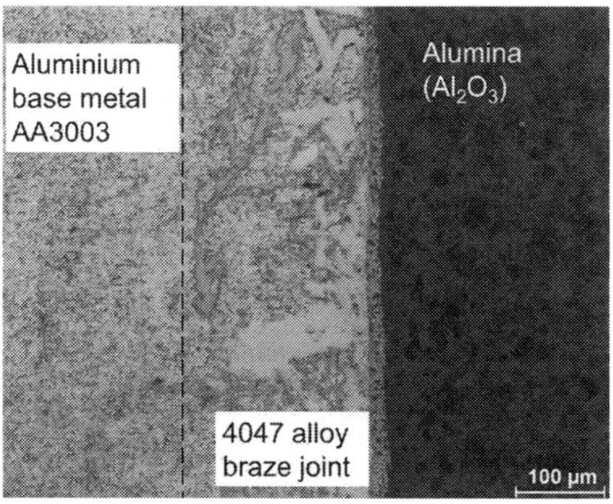

FIGURE 7.25 Typical braze joint between aluminium andalumina ceramic using 4047 braze filler.

REFERENCES

1. D.G. Eskin, *Physical metallurgy of direct chill casting*, CRC Press, Taylor & Francis Group, 2008.
2. https://aluminiumleader.com/history/timeline/#1886 (Accessed on 04.07.2023)
3. Marcello Cabibbo and Chiara Paoletti, High-temperature equal-channel angular pressing of a T6-Al-Cu-Li-Mg-Ag-Zr-Sc alloy, *Journal of Manufacturing and Materials Processing, MDPI*, 2021, 5, 6. https://doi.org/10.3390/jmmp5010006
4. Ailing Zhang and Yanxiang Li, Thermal conductivity of aluminium alloys—a review, *Materials*, 2023, 16, 2972. https://doi.org/10.3390/ma16082972
5. S. Malarvizhi, V. Balasubramanian, Effect of welding processes on AA2219 aluminium alloy joint properties, *Transactions of Nonferrous Metals Society of China*, 2011, 21, 962–973.
6. Ganji Venkata Narayana, *Fracture behaviour of aluminium alloy AA2219-T87 welded plates at room and cryo temperatures* (PhD Thesis), Department of Metallurgical Engineering and Materials Science, IIT Bombay, 2006.
7. S. Arunkumar, P. Rangarajan, K. Devakumaran, P. Sathya, Comparative study on transverse shrinkage, mechanical and metallurgical properties of AA2219 aluminium weld joints prepared by gas tungsten arc and gas metal arc welding processes, *Defence Technology*, 2015, 11, 262–268. http://dx.doi.org/10.1016/j.dt.2015.05.008
8. Biju S. Nair, G. Phanikumar, K. Prasad Rao, P. P. Sinha, Improvement of mechanical properties of gas tungsten arc and electron beam welded AA2219 (Al–6 wt.-%Cu) alloy, *Science and Technology of Welding and Joining*, 2007, 579–585.https://doi.org/10.1179/174329307X227210
9. https://commons.wikimedia.org/wiki/File:Advanced_friction_stir_welding_and_processing_tools_by_MegaStir.JPG (Accessed on 04.07.2023)
10. C.J. Dawes, W.M. Thomas, *Development of improved tool designs for friction stir welding of aluminium*, TWI, Granta Park, Great Abington, Cambridge, CB1 6AL.
11. M. Agilan, *Welding and weldability studies on AA2195, Al-Cu-Li alloy* (PhD Thesis), Department of Metallurgical and Materials Engineering, IIT Madras, 2022.
12. https://en.m.wikipedia.org/wiki/File:Bulkhead_%26_Nosecone_Of_Orion_Spacecraft_Joined_By_Friction_Stir_Welding.jpg (Accessed on 01.07.2023)
13. https://www.nasa.gov/sites/default/files/thumbnails/image/artemis_iii_orion_weld_maf.jpg (Accessed on 01.07.2023)
14. Guoqing Wang, Yanhua Zhao, Yunfei Hao, Friction stir welding of high-strength aerospace aluminium alloy and application in rocket tank manufacturing, *Journal of Materials Science and Technology*, 2018, 34(1), 73–91. https://doi.org/10.1016/j.jmst.2017.11.041
15. https://www.twi-global.com/technical-knowledge/published-papers/laser-welding-of-aluminium-alloys-using-different-laser-sources-june-1998 (Accessed on 25.06.2023)

8 Practical Aspects of Corrosion-Resistant Weld Overlays

Raghavendra Darji, Divij Vaishnav, and Gaurang Joshi

CONTENTS

8.1 INTRODUCTION

Corrosion is a natural process that gradually deteriorates metals and alloys due to various environmental factors. It is a ubiquitous phenomenon with significant economic and safety implications. Understanding the causes, mechanisms, and prevention methods of corrosion is crucial for industries and researchers alike. This chapter

DOI: 10.1201/9781003352419-8

aims to explore the corrosive process, and potential preventive measure, i.e., corrosion-resistance weld overlay to mitigate its effects. Corrosion can be defined as "Corrosion is the chemical or electrochemical reaction between a metal and its surrounding environment, resulting in the deterioration of the metal's properties" or "Corrosion is when a refined metal is naturally converted to a more stable form such as its oxide, hydroxide or sulphide state this leads to deterioration of the material" (Silman 1972). However, corrosion occurs in between the metal/alloy and its surroundings itself, but it also occurs within the metal/alloy. One such example of this type of corrosion is "Intergranular Corrosion." Corrosion is a diverse phenomenon, and there are several types of corrosion that can occur depending on the specific conditions and environmental factors (Silman 1972). Here are some common types of corrosion.

8.1.1 Uniform Corrosion

Uniform corrosion, also known as general corrosion, is the most common type of corrosion. It occurs when the entire surface of a metal corrodes uniformly at a relatively constant rate. This type of corrosion is often caused by exposure to moisture, oxygen, and other corrosive elements in the environment.

8.1.2 Galvanic Corrosion

Galvanic corrosion, also referred to as bimetallic corrosion, occurs when two dissimilar metals are in contact with each other in the presence of an electrolyte. The more active (less noble) metal experiences accelerated corrosion, while the more noble metal remains relatively unaffected. The driving force behind galvanic corrosion is the difference in electrochemical potentials between the two metals.

8.1.3 Pitting Corrosion

Pitting corrosion is a localized form of corrosion characterized by the formation of small pits or cavities on the metal surface. It typically occurs in areas where the protective oxide layer has been compromised, allowing aggressive ions to attack the metal. Pitting corrosion can be highly destructive and can lead to rapid penetration of the metal, resulting in structural failure.

8.1.4 Crevice Corrosion

Crevice corrosion occurs in narrow crevices, gaps, or confined spaces where stagnant or trapped electrolytes exist. It is often observed in joints, seams, gaskets, or areas with poor ventilation. The restricted access to oxygen or the accumulation of corrosive agents within the crevice leads to localized corrosion and can cause severe damage.

8.1.5 Intergranular Corrosion

Intergranular corrosion is a form of corrosion that specifically affects the grain boundaries of a metal. It occurs due to the different chemical composition or segregation of impurities along the grain boundaries. Intergranular corrosion can lead to grain boundary weakening, resulting in the loss of structural integrity.

8.1.6 Stress Corrosion Cracking (SCC)

Stress corrosion cracking is a combination of corrosion and mechanical stress. It occurs when a metal is exposed to a corrosive environment while under tensile stress. The corrosive environment acts as a catalyst, accelerating crack initiation and propagation. SCC can lead to catastrophic failures even under relatively low applied stresses.

8.1.7 Erosion Corrosion

Erosion corrosion results from the combined effects of corrosion and mechanical erosion. It occurs when a metal surface is exposed to a high-velocity fluid or abrasive particles. The fluid flow enhances the corrosion process by removing the protective oxide layer, leading to accelerated metal loss.

8.1.8 Microbiologically Influenced Corrosion (MIC)

Microbiologically influenced corrosion is caused by the presence and activities of microorganisms, such as bacteria, fungi, or algae. These microorganisms can create corrosive by-products or promote the formation of localized environments on the metal surface, accelerating corrosion.

Understanding the various types of corrosion is essential for implementing effective preventive measures and selecting appropriate materials for specific applications. It enables industries to develop strategies to mitigate corrosion-related risks and ensure the longevity and reliability of metal structures and components (Shekari et al. 2017). Corrosion poses a significant economic burden on various industries in India. While it is challenging to provide an exact figure for the overall loss due to corrosion in the country, several studies and reports indicate substantial financial implications (Feng et al. 2019; *No Title*; Shekari et al. 2017; *No Title* 2023; Valdez et al. 2016). Here are some key areas where corrosion-related losses are observed in India and other countries:

1. *Infrastructure:* Corrosion affects critical infrastructure, including bridges, highways, pipelines, and buildings. The corrosion-induced deterioration of these structures leads to increased maintenance, repair, and replacement costs. A study by the National Association of Corrosion Engineers (NACE) estimated that the annual cost of corrosion for infrastructure in India could be around 3–4% of the country's GDP.
2. *Oil and Gas Industry:* India has a significant presence in the oil and gas sector, and corrosion poses a substantial challenge in this industry. Corrosion in pipelines, storage tanks, refineries, and offshore installations results in leaks, production downtime, and environmental risks. It leads to financial losses due to repairs, production disruptions, and the need for corrosion control measures.

3. *Transportation:* Corrosion affects various modes of transportation, including railways, automobiles, and aviation. Rusting of railway tracks, corrosion-related failures in vehicles, and deterioration of aircraft structures increase maintenance and replacement costs. The Indian Railways, for example, incurs significant expenses in combatting corrosion-related issues in its infrastructure.
4. *Marine and Coastal Infrastructure:* India has a vast coastline and numerous marine structures, such as ports, harbours, and offshore platforms. The corrosive effects of saltwater and marine environments require extensive corrosion prevention and control measures. Failure to address corrosion in these structures can lead to costly repairs, decreased operational efficiency, and safety concerns.
5. *Manufacturing and Industrial Processes:* Corrosion impacts various manufacturing industries, such as steel, chemicals, pharmaceuticals, and automotive. The loss of equipment, machinery, and process systems due to corrosion leads to production disruptions, increased downtime, and decreased productivity. Corrosion also affects storage tanks, boilers, and other critical components, adding to the financial burden.
6. *Electrical and Electronics:* Corrosion in electrical and electronic systems, including power transmission infrastructure, electrical panels, and circuitry, can lead to operational failures, safety risks, and increased maintenance and replacement costs. Corrosion-induced failures in electrical systems can have severe consequences, particularly in industries that rely heavily on uninterrupted power supply.

It is important to note that these losses are estimates and can vary over time due to factors such as infrastructure development, industry growth, and advancements in corrosion prevention and control practices. Moreover, they highlight the substantial economic impact that corrosion has on various sectors in India. However, it is not right to consider only India; all countries face the challenges due to corrosion in various sectors as mentioned earlier. Efforts to raise awareness, implement preventive measures, and invest in corrosion-resistant materials and technologies are crucial to mitigating these losses and ensuring sustainable growth. The losses may depend upon the geography and weather conditions since the corrosion is a process governed by the environment. Apart from that, corrosion is influenced by various factors, both environmental and metallurgical, that can accelerate or mitigate the corrosion process. Understanding these factors is crucial for implementing effective corrosion prevention and control measures. Here are some key factors that influence corrosion (Silman 1972; Peabody 2001).

8.1.8.1 Environmental Factors

8.1.8.1.1 Moisture

The presence of moisture, such as humidity, rain, or water, is a primary factor in initiating and sustaining corrosion. Moisture provides the electrolyte necessary for electrochemical reactions to occur.

8.1.8.1.2 Oxygen

Oxygen plays a vital role in the corrosion process. It combines with metal atoms to form metal oxides, which can further react and deteriorate the metal surface. Higher oxygen levels generally lead to increased corrosion rates.

8.1.8.1.3 Acidity or Alkalinity (pH)

The pH of the surrounding environment can significantly influence corrosion. Extreme pH conditions, either highly acidic or highly alkaline, can accelerate corrosion rates. The optimal pH range for minimizing corrosion varies depending on the metal and the specific environment.

8.1.8.1.4 Salinity

Salts, particularly chloride ions, can intensify the corrosion process. Coastal and marine environments with high salt content are more corrosive, as chloride ions can penetrate the passive oxide layer on metals and accelerate corrosion reactions.

8.1.8.1.5 Temperature

Higher temperatures generally increase the rate of corrosion. Elevated temperatures can enhance the chemical reactivity of corrosive agents and promote faster electrochemical reactions on the metal surface.

8.1.8.1.6 Humidity and Climate

Humid environments and climates with high levels of rainfall contribute to increased moisture content and create favourable conditions for corrosion. Additionally, atmospheric pollutants, such as sulphur dioxide and nitrogen oxides, can react with moisture and accelerate corrosion processes.

8.1.8.2 Metallurgical Factors

8.1.8.2.1 Alloy Composition

The composition of the metal or alloy plays a crucial role in its corrosion resistance. Certain alloying elements, such as chromium, nickel, and molybdenum, can enhance the formation of a protective oxide layer, thereby reducing corrosion susceptibility.

8.1.8.2.2 Microstructure

The microstructure of a metal, including grain boundaries, phases, and inclusions, can affect its corrosion resistance. Irregularities or impurities in the microstructure can create sites for corrosion initiation and propagation.

8.1.8.2.3 Surface Finish

The surface finish of a metal can influence its corrosion resistance. A smooth and clean surface is generally more resistant to corrosion than a rough or contaminated surface. Surface coatings, such as paints or protective films, can provide an additional barrier against corrosive agents.

8.1.8.2.4 Galvanic Effects

When two dissimilar metals or alloys are in contact in the presence of an electrolyte, galvanic corrosion can occur. The difference in electrochemical potentials between the two metals leads to accelerated corrosion of the less noble (more active) metal.

8.1.8.2.5 Mechanical Stress

Mechanical stress, including tensile, compressive, or cyclic loading, can influence corrosion. Stress can accelerate corrosion processes by altering the metal's protective oxide layer, creating preferential sites for corrosion initiation.

Understanding these factors and their interactions is essential for assessing and managing corrosion risks. By considering these influences, industries and researchers can implement appropriate preventive measures, select corrosion-resistant materials, and design structures and systems that withstand the corrosive environment effectively.

8.1.8.3 Corrosion Prevention and Control (Peabody 2001; Silman 1972)

8.1.8.3.1 Protective Coatings

Applying protective coatings, such as paints, enamels, or electroplating, can create a barrier between the metal surface and the corrosive environment.

8.1.8.3.2 Cathodic Protection

Cathodic protection involves connecting a sacrificial anode or an impressed current system to the metal structure to shift its potential and reduce corrosion.

8.1.8.3.3 Alloy Modification

Altering the alloy composition or introducing corrosion-resistant alloy systems can enhance a metal's resistance to corrosion.

8.1.8.3.4 Environmental Modification

Controlling the environment through dehumidification, maintaining proper pH levels, or reducing exposure to corrosive elements can help minimize corrosion.

8.1.8.3.5 Regular Maintenance and Inspection

Implementing proactive maintenance practices, including cleaning, inspection, and timely repair, can prevent corrosion-related failures.

8.2 REQUIREMENT AND ADVANTAGES OF THE CRO

Based on the previous discussion, it is possible to infer that corrosion is a complicated and ubiquitous process that affects metals and alloys in a variety of sectors. Understanding the many forms, causes, and preventative strategies is critical for limiting its harmful consequences. Industries may prevent financial losses, maintain safety, and extend the life of their metallic assets by adopting effective corrosion management measures. Ongoing materials science research and technology breakthroughs provide potential options for generating more corrosion-resistant materials and new preventative strategies. Given the importance of corrosion in economic and technical advancement, it is desirable to investigate corrosion prevention measures in the various industries indicated previously. Oil and gas, pharmaceutical, refining, and food processing sectors all need massive vessels, reactors, and heat exchangers. These machines are constantly in touch with corrosive environments at high temperatures and pressures. Given the activities performed in these types of equipment, the size of the equipment is enormous, such as the dimensions of a "Blast Furnace." It is apparent that welding is one of the key procedures used to create them. When it comes to their working environment, three variables have a direct impact on their performance: 1) pressure, 2) temperature, and 3) environment (which is responsible for corrosion). Finally, while developing the component, especially during the material of construction choosing all these parameters must be considered. Assuming the selected material has excellent mechanical properties but resistance to corrosion and other damage mechanisms, such as creep resistance, stress corrosion cracking, degradation in surface properties, and so on, and is less or insufficient for the service requirement, the selected material must be changed if it lacks all the required properties. Any single superior characteristic of any metal/alloy will be of no use; rather, the metal/alloy should have a mix of traits that may be used with certain design and manufacturing modifications. Metals/alloys such as SA 516 Gr 60/70 and all grades of SA 387 (low-alloy steels) have strong mechanical qualities, creep resistance, and exceptional toughness; nevertheless, all these metals/alloys contain less than 10% Cr, Mo, and V in the chemical composition, implying that roughly 90% is Fe. Because Fe is the most abundant element in these alloys, it is more likely to corrode under service conditions when the environment is corrosive. Regardless, the pressure and temperature requirements must be such that the designer cannot avoid utilizing this material. Under such conditions, particularly in corrosive environments, the corrosion resistance overlay (CRO) offers the best option by combining all the needed qualities in a single component.

8.3 PROCESSES USED FOR THE CRO

CRO is essentially the cladding process on the surface of the base metal, the other alloy/metal with greater corrosion resistance is to be cladded by various methods such as electro slag strip cladding (ESSC) (Sandes et al. 2020), flux-cored arc welding (FCAW) (Alvarães et al. 2019), gas metal arc welding (GMAW) using metal-cored wires (Liu, Long, et al. 2022), shielded metal arc welding (SMAW) (Sandes et al. 2020), gas tungsten arc welding (GTAW) (Guo, Xiao, Wang, He, et al. 2021), and the most recent method used for CRO, the hot wire gas tungsten arc welding (HW-GTAW) (Arivazhagan et al. 2020). In general, any procedure has benefits and disadvantages. ESSC can provide an excellent deposition rate that is unrivalled by any other process, but when CRO is required inside small manholes, nozzles, flanges, and so on, the process is difficult to select because the equipment used in the process is large. In such cases, GMAW and FCAW are better processes because they provide a higher deposition rate. SMAW is used to repair the CRO, which is discussed later in this chapter.

The goal of the CRO is to offer corrosion resistance to the base metal while avoiding direct contact with the base material (which has adequate mechanical qualities but no corrosion resistance), allowing the equipment to perform well for the specified life duration. As shown in Table 8.1, the metals/alloys indicated for seawater have variable corrosion resistance in various conditions. If the environment changes, the metals/alloys' position in this series may change. This leads to the general conclusion that the CRO metal/alloy should be chosen based on the environment in which the CRO will be in contact. The overlay procedure may be chosen according on the CRO metal/alloy. Using the ESSC over the wide surface area of the base metal, an overlay of SS304L over SA 516 Gr 60/70 carbon steel may be readily done. However, if titanium must be clad over the same carbon steel, no fusion welding procedure may be used since metallurgical compatibility is critical when depositing. However, methods such as explosive bonding and friction surfacing are now available for incompatible metals/alloys. Taking into account all of the preceding topics, the chapter addresses the applications of CRO, its obstacles, and the outcomes given by other studies, as shown in Table 8.2.

TABLE 8.1
Galvanic Series in Seawater at 25°C

Sr No.	Metals/alloys starting from most noble to least noble (cathodic to anodic/more corrosion resistance to least corrosion resistance)
1.	Platinum
2.	Gold
3.	Graphite
4.	Titanium
5.	Silver
6.	SS 316 (passive)
7.	SS 304 (passive)
8.	SS 410 (passive)
9.	Nickel 200 (passive)
10.	Copper alloys (yellow brass, Al bronze, Si bronze)
11.	Hastelloy B
12.	Nickel 200 (active)
13.	Copper alloys (Muntz metal, Mn bronze, naval brass)
14.	Tin
15.	Lead
16.	SS 316 (active)
17.	SS 304 (active)
18.	SS 410 (active)
19.	Cast iron
20.	Low-carbon steel
21.	Aluminum alloys (5052, 3004, 3003)
22.	Galvanized steel
23.	Zinc
24.	Magnesium alloys
25.	Magnesium

Source: (Amani et al. 2015).

TABLE 8.2
Review of the Literature on the Corrosion Resistance Overlay

Title	Remarkable Outcome	Reference
Microstructure and Corrosion Properties of Single Layer Inconel 625 Weld Cladding Obtained by the Electroslag Welding Process	At 1 mm from the fusion line, excellent results were achieved at low dilution. The findings show that using ESW instead of traditional overlay procedures may drastically lower the cost of overlay.	(Mendes and Farneze 2020)
Microstructure and Corrosion Resistance of Inconel 625 Overlay Welded by Pulsed TIG Process	A single layer of Inconel 625 deteriorated, but a two-layer overlay showed to be superior for corrosion resistance, even when pitting.	(Guo, Xiao, Wang, He, et al. 2021)
Effect of Heat Treatment Temperatures on Microstructure and Corrosion Properties of Inconel 625 Weld Overlay Deposited by PTIG	Because of the precipitate phases generated during the post-weld heat treatment, higher post-weld heat treatment temperatures were detrimental to corrosion resistance.	(Guo, Zheng, et al. 2016)
Comparison of Corrosion Resistance and Corroded Surfaces of Welding Metal in Overlay-Welded Inconel 600 and Inconel 625 by Gas Metal Arc Welding	Inconel 625 had better corrosion resistance than Inconel 600.	(Jung and Lee 2016)
Corrosion Resistance of Inconel 625 Overlay Welded Inside Pipes as a Function of Heat Treatment Temperature	After post-weld heat treatment, no test specimen showed sensitization for all four cases.	(Ban et al. 2016)
High Temperature Corrosion Resistance of Candidate Nickel-Based Weld Overlay Alloys in a Low NOx Environment	Microsegregation of the molybdenum showed a detrimental effect on the corrosion resistance.	(Deacon et al. 2007)
Performance of the INCONEL 625 Alloy Weld Overlay Obtained by FCAW Process	Productivity was improved along with the optimum corrosion resistance despite of second phase generation and dilution.	(Alvarães et al. 2019)
Wear and Corrosion Behaviours of Inconel 625-Cladded Layer on P460N Steel in Different Environments	Corrosion rate reduced by three times with the overlay of Inconel 625 on P460N steel.	(Zafari et al.)
Improved Mechanical Properties, Wear and Corrosion Resistance of 316L Steel by Homogeneous Chromium Nitride Layer Synthesis Using Plasma Nitriding	Plasma nitriding over the austenitic phase improves corrosion resistance, wear resistance, and mechanical characteristics.	(Naeem 2020)

(Continued)

TABLE 8.2 (*Continued*)
Review of the Literature on the Corrosion Resistance Overlay

Title	Remarkable Outcome	Reference
Microstructural Characterization of Inconel 625 Nickel-Based Alloy Weld Cladding Obtained by Electroslag Welding Process	ESW proved to be a better process for the overlay because only 4% dilution was observed at 1 mm height from the fusion line; however, corrosion tests were not performed; however, considering the chemical composition and austenitic microstructure, it can be concluded that ESW was a better choice; also, it reduces the cost of overlay.	(Sandes et al. 2020)
Microstructure, Hardness and Corrosion Behavior of Gas Tungsten Arc Welding Clad Inconel 625 Super Alloy over A517 Carbon Steel Using ERNiCrMo3 Filler Metal	Overlay welding was done using welding currents of 80 A, 100 A, and 120 A, with the 100 A current producing the best results (0.143 mpy corrosion rate). The obtained corrosion rate was minimal when compared to the base metal corrosion rate of 24.2 mpy.	(Ostovan et al. 2020)
Effect of Titanium Addition on the Neutral Corrosion Resistance of Welding Overlay	Titanium was added in different quantities to carbon steel flux-cored self-shielded filler wire, and it was found that the inclusion of titanium decreases corrosion resistance.	(Liu, Wu, et al. 2022)
Effect of Inert and Active Shielding Gases in the Corrosion Resistance of IN625 Weld Overlays	Two to three layers of deposition in an inert gas environment during the GMAW weld overlay generated the corrosion resistance of Inconel 625 itself.	(Silva and Osorio)
Influence of Nb Additive on Alkaline Corrosion Behavior of Slag-Free Self-shielded Metal-Cord Welding Overlay	The quantity of carbides generated during the overlay process influences corrosion behaviour.	(Liu, Long, et al. 2022)
A Comprehensive Review of Corrosion Resistance of Thermally Sprayed and Thermally Diffused Protective Coatings on Steel Structures	The use of Si and rare elements enhances corrosion resistance by acting as a seal. The corrosion resistance of Ni-based alloys is related to passive oxide films formed by passive components such as titanium oxide, chromium oxide, and other oxide-producing passive elements.	(Galedari et al. 2019)
Dissimilar Cladding of Ni–Cr–Mo Superalloy over 316L Austenitic Stainless Steel: Morphologies and Mechanical Properties	Inconel 625 protects the substrate from the 3.5% NaCl environment.	(Sathiya and Evangeline 2021)

TABLE 8.2 (*Continued*)
Review of the Literature on the Corrosion Resistance Overlay

Title	Remarkable Outcome	Reference
Improved Hardness, Wear and Corrosion Resistance of Inconel 718 Treated by Hot Wall Plasma Nitriding	The surface qualities were enhanced by hot wall plasma nitriding, and the corrosion resistance was increased as well.	(Maniee et al. 2020)
Microstructure and Phase Analysis of Multilayer Ni–Cr–Mo Clad for Corrosion Protection	When the Fe level is less than 2%, a passive layer of Mo, Nb, Ni, Cr, and oxygen is generated on the surface, protecting the substrate from corrosion.	(Moshkbar et al. 2021)
Laves Phase Formation and Segregation of Nb in Ni–Cr–Mo Superalloy over 316L by Hot Wire (HW) TIG Cladding Process	In contrast to the findings achieved utilizing the hot wire GTAW, the corrosion resistance of the pure base metal was lowered.	(Arivazhagan et al. 2020)
Microstructural Development and Corrosion Behavior of 13Cr–4Ni-1Mo Martensitic Stainless Steel Clad	The sensitization citations were raised after a post-weld heat treatment of a four-layer overlay of 13 Cr–4Ni–1Mo on AISI 4140.	(Farzadi 2020)
Effect of Electric Potentials on Microstructure, Corrosion and Wear Characteristic of the Nitride Layer Prepared on 2Cr13 Stainless Steel by Plasma Nitriding	The tribological characteristics and corrosion resistance of Cr steel treated with plasma nitriding were enhanced after the treatment.	(Li et al. 2019)
Role of Martensite Structural Characteristics on Corrosion Features in Ni-Advanced Dual-Phase Low-Alloy Steels	Because of the considerable galvanic couple formation in dual phase weathering steel, as the quantity of martensite grew, so did the corrosion rate.	(Hai et al. 2021)
The Effect of Time and Temperature of Nitridation-Oxidation Process on Properties and Corrosion Resistance of AISI 316L Steel	The combination process of nitridation and oxidation increased both pitting corrosion resistance and wear resistance.	(Yazdkhasti et al. 2018)
The Effect of Post Weld Heat Treatment (PWHT) on the Microstructure, Microhardness and Sulphide Stress Corrosion Cracking (SSCC) of Ni-Base Superalloy IN625 Hot Wire TIG Cladding on AISI 4130 Steel	PWHT was done on AIS 4130 steel with an overlay of Inconel 625 using hot wire GTAW, however the PWHT performed at the maximum temperature revealed stress corrosion cracking in the base metal that spread into the weld overlay.	(Sedighi et al. 2021)
Pitting Corrosion Resistance and Bond Strength Of Stainless-Steel Overlay by Friction Surfacing on High Strength Low Alloy Steel	When SS 304 was applied to the surface of high-strength low-alloy steel by friction stir surfacing, the overlay demonstrated decreased pitting corrosion resistance. Induced martensite was also discovered at the microscopic level during the friction stir surfacing procedure.	(Singh et al. 2015)

(*Continued*)

TABLE 8.2 (*Continued*)
Review of the Literature on the Corrosion Resistance Overlay

Title	Remarkable Outcome	Reference
Structure–Property Relationships of Inconel 625 Cladding on AISI 316L Substrate Produced by Hot Wire (HW) TIG Metal Deposition Technique	Wear testing was performed (irrelevant).	(*Ac Ce Pte d M Us Pt*)
Comparison Study of Low-Heat-Input Wire Arc-Fabricated Nickel-Based Alloy by Cold Metal Transfer and Plasma Arc	Corrosion testing was not performed.	(Yin et al. 2020)
Microstructure and Selective Corrosion of Alloy 625 Obtained by Means of Laser Powder Bed Fusion	For intergranular corrosion tests, the LPBF and hot worked heat-treated grades of Alloy 625 showed the same corrosion rate, but the corrosion morphology was different.	(Cabrini et al. 2019)
The Electro-Slag Cladding Process: Application, Main Metallurgical and Testing Issues in the Fabrication of Reactors in Large Thickness 2¼Cr–1Mo–¼V	The practical aspects of installing the overlay on the LAS were investigated. The essay covered cracking phenomena, hydrogen-induced cracking, and other process-related topics.	(Taylor, Mandina, et al. 2013)
Effect of Welding Energy on the Corrosion and Tribological Properties of Duplex Stainless Steel Weld Overlay Deposited by GMAW/CMT Process	Four heat input values were examined for wear and corrosion resistance, as well as microstructural analysis. The abrasive wear resistance was excellent at the highest heat input value, 1.74 kJ/mm, while the corrosion resistance was the lowest.	(Luchtenberg et al. 2019)
Erosion–Corrosion Degradation Mechanisms of Fe–Cr–C and WC–Fe–Cr–C PTA Overlays in Concentrated Slurries	To resist erosion corrosion, 65% WC was added. Corrosion tests were performed at two different temperatures, and it was noted that at high temperatures, corrosion resistance decreased dramatically.	(Flores et al. 2009)
Influence of Heat Treatments on Microstructure, Mechanical Properties and Corrosion Resistance of Inconel 625 Overlay Cladded Using PTIG	After studying the effect of PWHT on corrosion qualities, it was concluded that lower PWHT temperatures gave improved corrosion resistance when compared to higher temperatures.	(Guo, Xiao, Wang, Wei, et al. 2020)
Effect of Post-Weld Heat Treatment (PWHT) on the Intergranular Corrosion of ENiCrFe-7 Weld Overlay Cladding Materials	After PWHT, only intergranular corrosion exhibited, other corrosion rate increased.	(Guo and Ling 2020)

TABLE 8.2 (*Continued*)
Review of the Literature on the Corrosion Resistance Overlay

Title	Remarkable Outcome	Reference
Investigation of Erosion-Corrosion Behavior of (WTi)C Based Weld Overlays	Erosion corrosion resistance is affected by particle size and dispersion. Small particles cause as much corrosion as larger particles in the matrix.	(Islam et al. 2017)
Effect of Heat Treatment on the Precipitation and Pitting Corrosion Behavior of 347 SS Weld Overlay	The sigma phase had a negative influence on corrosion resistance after PWHT on the overlay of SS 347.	(Lo and Tsai 2003)
Formation Quality Optimization and Corrosion Performance of Inconel 625 Weld Overlay Using Hot Wire Pulsed TIG	Overlay of 625 on the second layer was nearly about the wrought alloy 625.	(Longlong et al. 2016)
The Influence of Post Weld Heat Treatment Precipitation on Duplex Stainless Steels Weld Overlay Towards Pitting Corrosion	Solution annealing after the PWHT shown no weight loss.	(Sim et al. 2019)

Stainless steel 304L, 316L, SS 347 and duplex stainless steel (2205) are the most prevalent corrosion-resistant overlays used in industry. These are mostly austenitic and/or austenitic ferritic stainless steel grades that must be clad over ferritic pearlitic steels with greater carbon content than stainless steels (0.03% C with low carbon grades and 0.08% with standard grades). Because this clad/overlay technique is metallurgically incompatible, the first layer of SS 309L must be completed before any austenitic grade may be clad to achieve the final chemistry. The most significant constraint of austenitic steels is their use in high-temperature applications, where phenomena such as sensitization and sigma phase development may occur owing to the alloying elements present. Iron's alloying additives restrict its applicability to a large temperature range, while Ni superalloys have the greatest temperature range of any metal. This broad temperature range gives Ni alloys an edge over stainless steels in that they can endure both temperature and pressure. Given this, the researchers were particularly interested in the CRO of Ni alloys over carbon steel as shown in Table 8.2. As per Table 8.2, researchers worked on Ni–Cr–Mo alloys such as Inconel 625, Inconel 718, and so on using various techniques and under different environmental conditions. They studied the effect of welding parameters on the corrosion behaviour of the CRO (Guo, Xiao, Wang, He, et al. 2021; Guo, Zheng, et al. 2016; Ban et al. 2016; Alvarães et al. 2019; Liu, Wu, et al. 2022; Silva and Osorio 2022), and in some cases they also utilized the surface treatment for improvement in the corrosion resistance (Galedari et al. 2019; Maniee et al. 2020). Here is the detailed discussion on the research works which were performed to study the CRO using various methods.

In Table 8.2, nearly half of the CRO research works are from the Inconel group of alloys due to the combination of mechanical properties and corrosion behaviour of that group of alloys. Not only the combination of properties but also the challenges encountered during the welding overlay process of Inconel are very interesting because it consists of the multiple-phase microstructure. The microstructure is made up of a number of gamma phases, including laves phases and other carbides. This lack of consistency in the microstructure has a higher impact on corrosion resistance. Aside from Inconel alloys, certain works were also from the stainless steel overlay, in which SS 304L and SS 347 were overlayed with different welding conditions, such as PWHT, GTAW, ESSC, and so on. Figure 8.1 shows the number of items in which corrosion resistance was enhanced and corrosion resistance was lowered, regardless of the overlay material, because of welding parameters or PWHT. Figure 8.1 shows that about 56% of the items studied in Table 8.2 increased in corrosion resistance following CRO (Jung and Lee 2016; Ban et al. 2016; Zafari et al.; Ostovan et al. 2020; Avelar et al. 2020; Alvarães et al. 2019; Sathiya and Evangeline 2021; Maniee et al. 2020). The majority of the 44% articles with reduced corrosion resistance consist of PWHT, minimum two layers required to obtain desired corrosion resistance, and required chemistry, however due to carbide formation and other microstructural non-uniformities, i.e., presence of martensitse, or due to intergranular corrosion (Guo, Xiao, Wang, He, et al. 2021; Guo, Zheng, et al. 2016; Deacon et al. 2007; Liu, Wu, et al. 2022; Liu, Long, et al. 2022). The articles in which the corrosion resistance improved using electro-slag welding (ESW) (Mendes and Farneze 2020; Alvarães et al. 2019), Gas Metal Arc Welding (GMAW) (Jung and Lee 2016; Ban et al. 2016; Moshkbar et al. 2021; Silva and Osorio), Gas Tungsten Arc Welding (GTAW) (Zafari et al.; Ostovan et al. 2020; Arivazhagan et al. 2020; Guo, Xiao, Wang, Wei, et al.

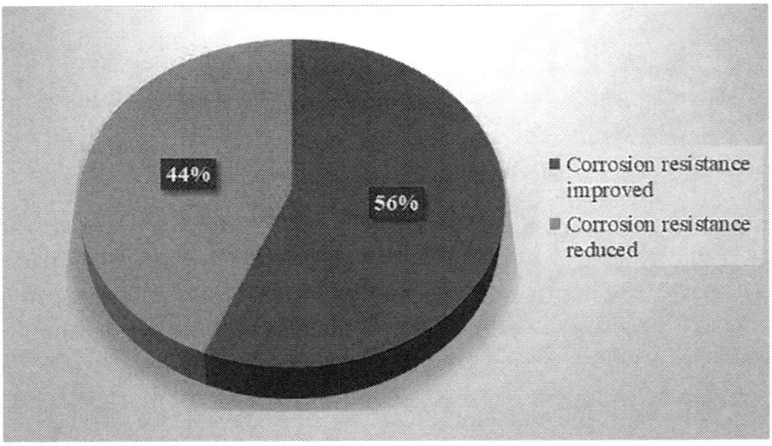

FIGURE 8.1 Comparison between the articles in which corrosion resistance improved and corrosion resistances reduced.

2020; Longlong et al. 2016), Plasma Nitriding (PN) (Naeem 2020; Maniee et al. 2020; Li et al. 2019; Yazdkhasti et al. 2018), Plasma Transferred Arc (PTA) (Flores et al. 2009; Islam et al.), Shielded Metal Arc Welding (SMAW) (Sim et al. 2019), some advanced process also included such as Cold Metal Transfer (CMT) (Sathiya and Evangeline 2021). In majority of the studies, 3.5% NaCl solution was used for the corrosion testing along with some other solutions.

8.4 ESW/ESSC PROCESS FOR THE CRO

When all of the procedures were compared, it was found that ESW was a lot superior process owing to the outcomes that ESW stated, such as the corrosion resistance being intact even at 1 mm height of the overlay, with the pitting resistance equivalent number (PREN), Fe content, and weight loss approach (Sandes et al. 2020; Alvarães et al. 2019). In each of the aforementioned studies, EQ NiCrMo-3 was utilized for the ESW process instead of the base metal SA 516 Gr 60/70 (carbon steel). Corrosion resistance is nearly entirely determined by the chemical composition, which is further determined by the dilution that occurs throughout the overlay process. Dilution occurs when the overlay material, which has a rich chemistry, transfers the elements to the base metal, which has a depleted chemistry. Taking this dilution into account in both investigations, the chemical composition acquired from 0.4 mm to 7 mm stays almost same. As a consequence, the corrosion resistance obtained from 0.4 mm and 7 mm has no change, as stated and claimed by the authors (Sandes et al. 2020; Alvarães et al. 2019). This is one of the properties of ESW, namely minimal dilution; as a result, most businesses today employ this method for large-scale equipment. The main advantage of the ESW process is the microstructure generated after the overlay. The microstructure obtained after the overlay was of dendritic and interdendritic regions with the least volume of second phases such as laves phases and niobium carbides, as mentioned in the earlier research works (Sandes et al. 2020; Alvarães et al. 2019). In fact, the amount of second phase was less than 2.5% (Sandes et al. 2020). Despite these benefits, it is not possible to conclude that this is the optimum procedure for executing the CRO on carbon steels and low-alloy steels since some difficulties are also linked with the ESW process. There was also evidence of hydrogen dis-bonding (Taylor, Mandina, et al. 2013; Gittos et al. 2008; Taylor, Casciaro, et al. 2015). This issue of hydrogen is observed due to the amount of hydrogen in the cladded portion (Taylor, Mandina, et al. 2013). The hydrogen-related issues are reported for the CRO of stainless steels overlay, similarly cracking due to hydrogen also reported in this process for stainless steels overlay (Taylor, Mandina, et al. 2013). Figure 8.2 illustrates the locations for the UCHC cracks that were observed for the cladding of stainless steel on the low-alloy steel material (2.25 Cr–1 Mo–0.25 V) under clad heating crack (UCHC). The primary cause of this sort of cracking was "fragilization" caused by unbalanced phases created by carbon and other alloying elements such as Cr, Mo, V, Nb, and Ti. At low temperatures, elements such as carbon and nitrogen undergo "fragilization" due to ageing. Secondary hardening and tempering may occur at higher temperatures, coupled with the complex carbide phases created by the carbide forming elements present in the steel, such as Cr, Mo, and V. These sorts of occurrences happened during the intermediate stress relief or

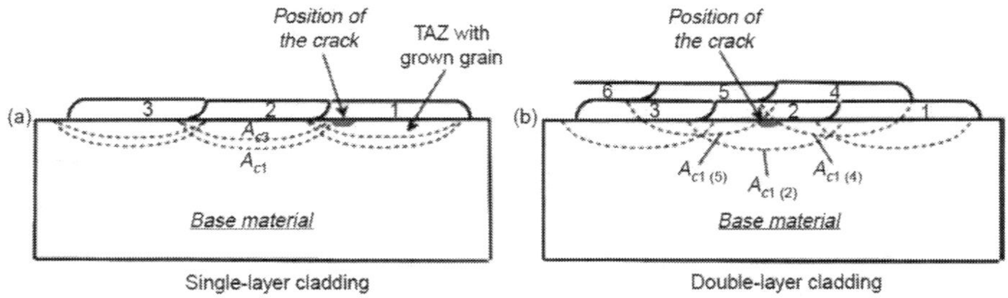

FIGURE 8.2 Schematic representation of the UCHC sites under electroslag strip cladding at thermally altered zones.

Source: (Taylor, Mandina, et al. 2013).

post-weld heat treatment, resulting in a complex interaction between the alloying components (Taylor, Mandina, et al. 2013). Because of these factors, the UCHC was seen at the locations that overlapped. Nonetheless, this may be prevented by managing the base metal chemical compositions. By limiting the components such as Pb, Bi, and Sb, the LAS becomes less vulnerable to the UCHC, and the possibilities of segregation of these elements are reduced. Another technique is to use low-heat input settings and attempt to limit the dilution such that the UCHC may be avoided with the smallest thickness of the CRO. To prevent this sort of cracking, use intermediate stress relief between the first and second passes (Taylor, Mandina, et al. 2013). Apart from this, cold cracking due to hydrogen and solidification cracking were also reported when considering the stainless steel overlay (Taylor, Mandina, et al. 2013). Hydrogen disbonding occurs when the overlayed components separate from the base material, typically owing to a local concentration of hydrogen that may enter by ambient conditions or faulty backing of the flux, etc. However, this may be avoided by using low cooling gradients or a long solidification period (Taylor, Mandina, et al. 2013). Magnetic stirring is presently employed by many heavy engineering businesses as a development or modification for the electroslag strip cladding, which assists in preserving the chemical composition and controlling the dilution. Evaluating the processes' advantages, this method is extremely useful for big-size equipment at the pressure portions, after which there are certain sections of the equipment such as manholes, nozzles, and so on. Where this process cannot be easily implemented due to certain practical parameters such as the set-up for this process is bulky because along with the strip, a flux feeder is attached, as is a magnetic stirrer, and the strips are available in widths of 30 mm, 60 mm, and 90 mm, making it very difficult to put CRO on each and every small portion of the large-size equipment. Nonetheless, over broad surface areas, this method may give the optimum option because of its high deposition rate, capacity to manage dilution, and ability to provide corrosion protection at a low CRO thickness.

As previously noted, additional techniques, such as GMAW, GTAW, PTA, CMT, and other surface treatment methods, such as plasma nitriding, are particularly beneficial for tiny-size portions of the equipment. Here is a breakdown of all of these techniques for CRO applications, one by one. A few sophisticated methods, such as hot

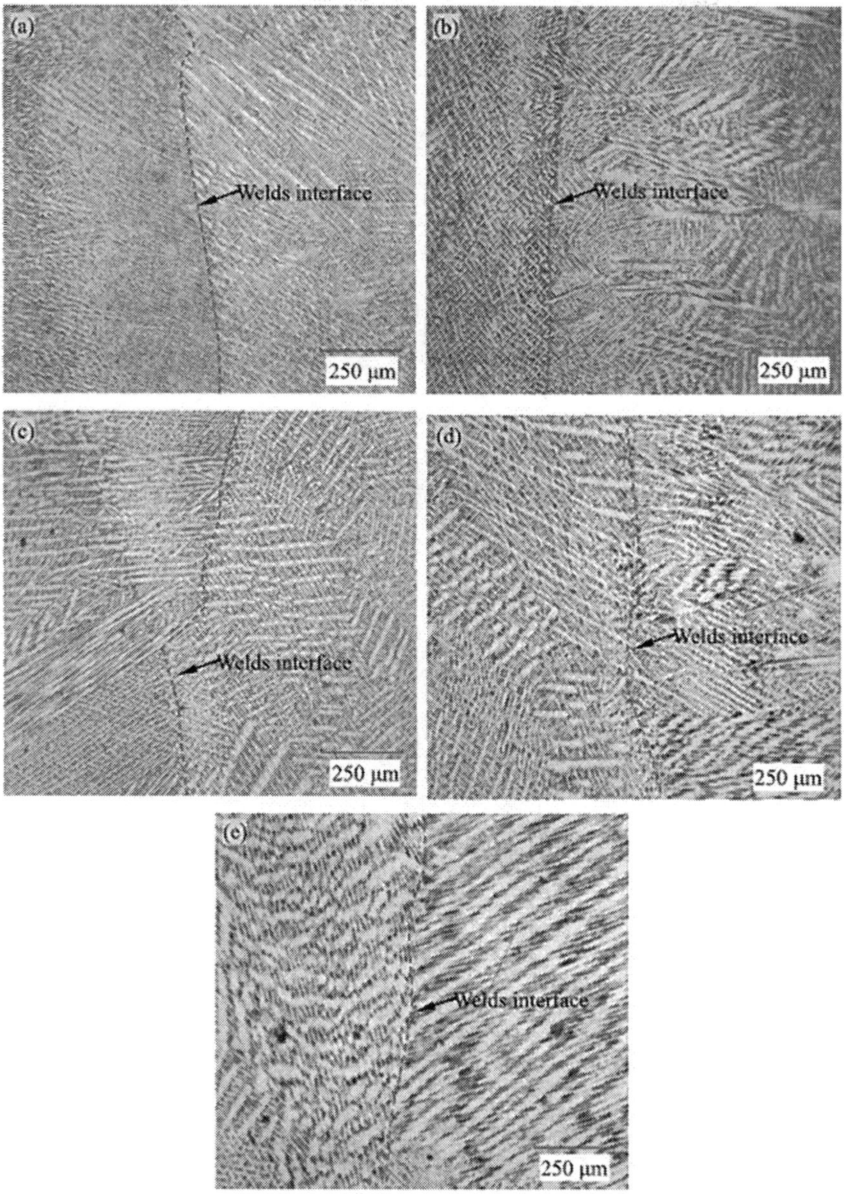

FIGURE 8.3A Microstructural comparison of Inconel 625 overlay using pulsed GTAW, (A) heat treated at 650°C and (B) at 750°C.

Source: (Guo, Xiao, Wang, Wei, et al. 2020).

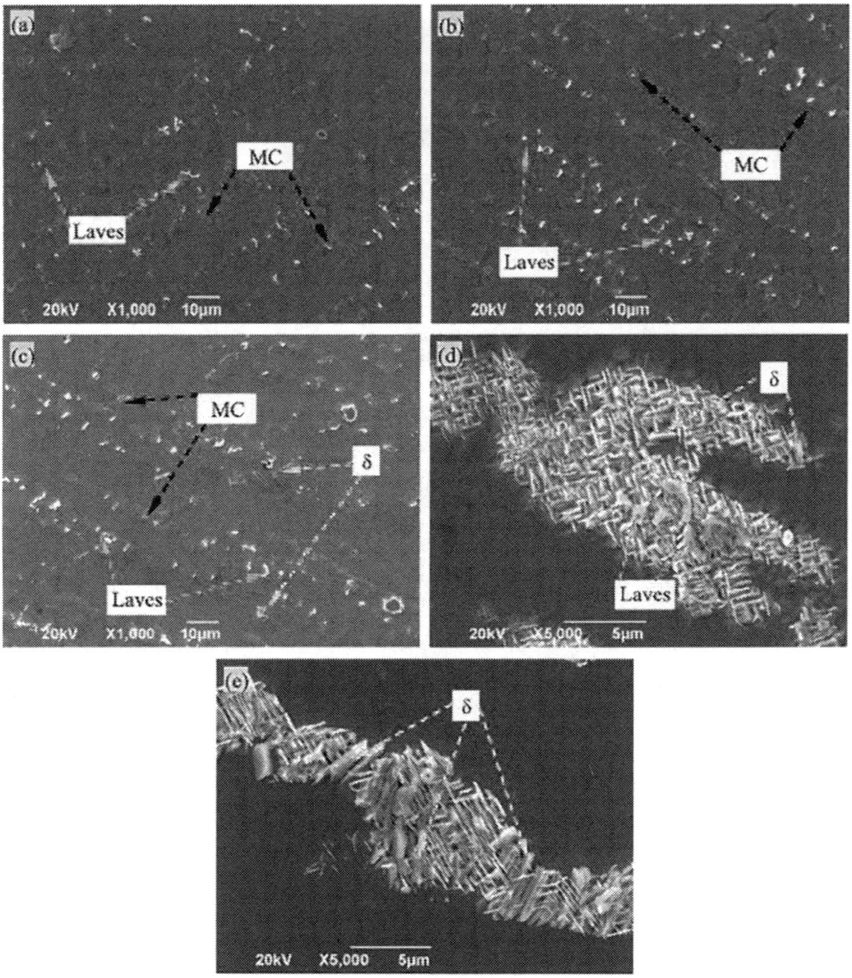

FIGURE 8.3B (Continued)

wire GTAW, were also investigated for CRO (Arivazhagan et al. 2020; Longlong et al. 2016). In this chapter, Inconel 625 was used under the GTAW for CRO on the various base materials such as P460N steel (Zafari et al.), A517 steel (Ostovan et al. 2020), and AISI 4130 steel (Guo, Xiao, Wang, Wei, et al. 2020). Taking the conventional GTAW process into account, the welding speed used was around 60 mm/min, and the welding current was ranging from 80 amp to 120 amp in both the studies (Zafari et al.; Ostovan et al. 2020). In both the studies, the overlay of Inconel 625 improved the corrosion resistance significantly, i.e., three times decrement in the corrosion rate mentioned by Zafari et al. (Zafari et al.), and negligible corrosion was reported by Ostovan et al. (2020). In both the studies, the weld microstructure was austenitic dendritic, with widmanstatten ferrite (Zafari et al.) and dendritic and presence of laves phases (Nb rich), along with metal carbides (Ostovan et al. 2020). The chemical composition of the overlay as a bulk was not reported. However, during the SEM and EDS analysis, it was reported that the oxide layer was present after the corrosion tests (Ostovan et al. 2020). Following the CRO with the pulse GTAW, PWHTs were performed on the samples to determine the effect of heat treatment parameters on corrosion resistance. The PWHTs were performed at 650°C and 750°C (two different temperatures), and the samples were tested under corrosion tests. Following the corrosion testing, it was discovered that the low PWHT temperature had no influence on corrosion resistance, however when comparing the PWHT temperatures, the higher PWHT temperature, i.e., 750°C, had the greatest effect. It was noted that the samples treated at 650°C showed no significant difference in microstructure from the as-welded samples. However, at 750°C temperature the laves phases dissolved, and needle-shaped precipitates which were of Ni_3Nb (Guo, Xiao, Wang, Wei, et al. 2020) were formed. Furthermore, when the PWHT temperature was raised to 850°C, the laves phase and metal carbides were totally dissolved, and the volume of the phase grew greatly, as shown in Figure 8.3. Because of these changes in the microstructure, the corrosion resistance of Inconel 625 overlay reduced significantly at high PWHT temperatures.

8.5 GTAW AND ADVANCED PROCESSES FOR CRO

While considering the GTAW, one advanced process called hot wire GTAW was also used for the CRO (Longlong et al. 2016; Arivazhagan et al. 2020). In one investigation, AISI 4130 (Longlong et al. 2016) and SS 316L (Arivazhagan et al. 2020) were utilized as base plates, with the same CRO, i.e., Inconel 625. Surprisingly, the hot wire GTAW produced disappointing results; the welding current was changed to 120 amp, 140 amp, and 160 amp, and the corresponding welding speeds were 175 mm/min, 150 mm/min, and 125 mm/min. It was critical to understand that the base material in this investigation was SS 316L, which already had considerable corrosion resistance. The corrosion resistance of the base metal was determined to be 1.065 mm/y, but the welding parameters were 2.364 mm/y, 1.915 mm/y, and 3.802 mm/y in all three trials. Comparing all these corrosion rates, it can be seen that the base metal has better corrosion resistance (Arivazhagan et al. 2020); this concludes that the hot wire GTAW (HW-GTAW) had not increased the corrosion resistance (Arivazhagan et al. 2020). A similar type of result was obtained by Guo et al. (Longlong et al. 2016) at the first layer of the overlay, where no improvement in the corrosion resistance was found. This was due to the high amount of dilution in the first layer of the overlay, especially the very high amount of Fe was reported in Longlong et al. (2016). This was the primary cause of the corrosion resistance decrease; otherwise, the microstructure and other characteristics were quite comparable and visible, as they should have been. Aside from that, the hot wire enhanced the current density of the preheated filler wire, allowing some interstitial components such as oxygen and other gases from the environment to enter the weld pool. However, when comparing GTAW with HW-GTAW, it can be determined that conventional GTAW is a far superior technique than HW-GTAW in terms of corrosion resistance at the moment. However, better results may be achieved by optimizing the parameters using HW-GTAW.

8.6 GMAW AND ADVANCED PROCESSES FOR CRO

When considering the GTAW process, the deposition rate remains extremely low, and when employing warmed filler wire, the corrosion resistance results are lowered, which is also a key restriction of the GTAW process. On the other hand, the GMAW generates the highest deposition rate, which lowers the time and speeds up the CRO process; this method is utilized only in limited areas of big vessels. Various factors, including shielding gas, PWHT, and others, were investigated in this GMAW process (Jung and Lee 2016; Ban et al. 2016; Moshkbar et al. 2021; Silva and Osorio). Under this process also, Inconel overlays were considered. In one of the studies (Jung and Lee 2016), the overall materials were the varied parameters, namely Inconel 600 and Inconel 625. It was obvious in this study that Inconel 625 provided better corrosion resistance than Inconel 600, this was due to the fundamental chemical composition of these two alloys, Inconel 625 has a richer chemistry than Inconel 600. Despite the differences in the findings, the quantity of dilution that happened during the overlay process was similarly quite minimal for both overlay alloys. The degree of sensitization was likewise confirmed to be within acceptable limits during the EPR test for both CROs (Jung and Lee 2016). In another study, the PWHT was performed on the CROed samples at 650°C, 750°C, 850°C, and 950°C after having 4 mm thickness of the overlay, and as previously discussed, at the higher PWHT temperature from 750°C onwards, the phase was observed in the microstructure; other than that the microstructures were quite normal or with no irregularities. The 850°C sample had the smallest difference between Ecorr and Epit, indicating that the passive film generated after 850°C was the least stable of the others. The most stable passive film was acquired at 650°C since it displayed the greatest difference between Ecorr and Epit, which was 786.2 (Jung and Lee 2016). Moura et al. (Silva and Osorio) conducted a research in which they changed the shielding gas for the Inconel 625 overlay of seven to eight passes utilizing the GMAW process. They employed two shielding gases, one that was 75% He + Ar and the other that was 25% CO_2 + Ar. The inert gas shielding caused little dilution on the top surface, but the active gas combination caused significant dilution, leaving the chemistry depleted on the top surface. Using inert gases for the bulk overlay, however, the quantity of dilution stays modest at any height. It was said that the presence of active gas promotes the formation of NbC precipitates prior to the laves phases, resulting in a decrease in corrosion characteristics (Silva and Osorio). In another study, it was observed that Inconel 625 increased the corrosion resistance of the base metal (Moshkbar et al. 2021), but curiously, just a single overlay was investigated for corrosion resistance and microstructure. Cold metal transfer (CMT by Fronius), an innovative method adapted from traditional GMAW, was also used to investigate the CRO of Inconel 625 over SS 316 (Sathiya and Evangeline 2021). Welding currents of 140 amp, 160 amp, and 180 amp were measured at the same voltage of 21 V, with various torch angles of 80, 60, and 70, and welding speeds of 175 mm/min, 150 mm/min, and 125 mm/min, indicating that there was no non-uniformity between the CMT parameters. Despite this, the corrosion resistance of the CRO was lower than that of the base metal. The authors found that corrosion resistance is determined by the dendritic arm spacing, with finer dendritic arms producing more corrosion resistance (Sathiya and Evangeline 2021).

8.7 SURFACE TREATMENT FOR CORROSION AND WEAR TESTING

The preceding description is for typical corrosion, which includes galvanic, pitting, crevice, and intergranular corrosions. Other than these, erosion corrosion is a subcategory of corrosion that frequently occurs in long pipelines or in areas where fluid is flowing through channels. Erosion corrosion occurs as a result of the combined effect of particles impinging on the structure's wall and the presence of corrosive media. To avoid this form of corrosion, surface qualities become very important, i.e., surface corrosion resistance and little erosion. To address this kind of corrosion, plasma nitriding was used (Naeem 2020; Maniee et al. 2020; Li et al. 2019; Yazdkhasti et al. 2018) to enhance surface hardness, which minimizes the probability of corrosive sites forming at minute areas and so improves corrosion resistance. Plasma nitriding was carried out at temperatures ranging from 400°C to 550°C in varied time durations. Nitride-producing elements must be present within the alloy or base metal so that nitrides may be created readily during processing, i.e., CrN were described in the studies (Naeem 2020). These carbides were far more resistant to wear, lowering the possibilities of erosion and corrosion. Because the base metal was stainless steel, the CrN formed in the homogenous matrix of austenite during the nitriding process boosted the wear resistance and corrosion resistance significantly (Naeem 2020). In other research works, the nitriding which was performed at 400–450°C, has shown better wear resistance and corrosion resistance (Maniee et al. 2020; Li et al. 2019).

8.8 DEMERITS OF THE VARIOUS WELDING PROCESSES IN THE CRO

According to the preceding discussion, it can be concluded that the overlay of Inconel 625 over any carbon steel and low-alloy steels produces the desired results; however, the overlay of Inconel 625 on stainless steel 316L does not produce the desired results; this is true not only for the aforementioned case, but there were certain cases identified in which no improvement, in fact a reduction in corrosion resistance, was found regardless of the process, base material, and CRO material (Guo, Xiao, Wang, He, et al. 2021; Guo, Zheng, et al. 2016; Liu, Wu, et al. 2022; Farzadi 2020; Sedighi et al. 2021; Cabrini et al. 2019; Lo and Tsai 2003; Luchtenberg et al. 2019). However, none of the study papers showed any aberrant outcomes, i.e., the microstructure was good, but the corrosion resistance was lowered. Many procedures utilized for CRO were significantly improved/changed in many of the aforementioned works, such as HW-GTAW (Sedighi et al. 2021), Laser Powder Based Fusion (Cabrini et al. 2019), and CMT (Luchtenberg et al. 2019). Concerning these advanced processes, it cannot be conclusively concluded that these processes, due to their inherent characteristics, reduce corrosion resistance because these processes were not thoroughly investigated for CRO but after optimizing the parameters and knowing the exact characteristics.

At the moment, these methods for CRO applications are completely incorrect. Aside from these sophisticated techniques, the inclusion of a few alloying elements in several of the works had a detrimental impact on corrosion resistance (Liu, Wu, et al. 2022; Liu, Long, et al. 2022). This was due to the formation TiC in the hypoeutectic matrix, which increases the susceptibility to corrosion (Liu, Wu, et al. 2022), as the Fe–Nb content increases, corrosion resistance reduces, and at 24 wt.% the worst corrosion resistance was observed (Liu, Long, et al. 2022). Guo et al. (2021) performed the Inconel 625 overlay using pulsed GTAW process; pulsing reduces the net heat input, which in turn results in the low dilution and better properties. Despite of that, it was concluded that the Fe dilution decreased with increasing the CRO layer thickness (after multiple pass). This conclusion might be the outcome of the higher dilution of Fe after the initial passes (Guo, Xiao, Wang, He, et al. 2021). These sorts of outcomes were also discussed before in the discussion of GTAW procedures. All of these findings indicate that the GTAW procedure is not very compatible with the CRO. In previous studies where PWHT was used, it was discovered that the corrosion resistance decreases following the PWHT due to microstructural changes (Guo, Zheng, et al. 2016; Sedighi et al. 2021; Farzadi 2020; Lo and Tsai 2003). These findings were achieved because of the microstructural changes that occurred during the PWHT. In the instance of Inconel 625, numerous alloying elements exist, and these multiple alloying components generate a multiphase alloy in which austenite may be the prime phase that is metastable. Phase changes are conceivable after heat treatment because to the metastable phase, additional carbides, and the presence of alloying metals. Because of these changes, micro galvanic corrosion sites emerge, resulting in galvanic corrosion at the micro level; when these sites multiply in number, the corrosion becomes severe (Guo, Zheng, et al. 2016; Sedighi et al. 2021; Lo and Tsai 2003).

8.9 SUMMARY OF THE DISCUSSION

Based on the discussions in the chapter, it is possible to infer that the CRO process is an excellent method for extending the life of large-scale equipment under any critical service circumstances. Furthermore, several procedures are accessible to execute CRO, such as ESSC, GMAW, GTAW, CMT, laser-based processes, and so on. The ESSC procedure, which was specifically conceived and built for the CRO, is the best of these processes. The primary benefits of this method are its capacity to perform CRO on vast areas with little dilution and to keep the chemical composition of the CROed metal/overlay extremely close to that of the wrought/as-cast chemical composition. The magnetic stirring was critical in regulating the dilution. Regardless of the planned PWHT, preserving the corrosion characteristics of the CRO because the microstructure may alter significantly throughout the PWHT. Aside from that, the GTAW procedure, which produces the highest-quality weld joint, is not ideal for CRO because of increased dilution. As a result of the significant quantity of dilution, the corrosion characteristics of the ESSC process remain poor. Similar findings were made for the GMAW process; however, increasing the CRO thickness leads to good corrosion resistance for both processes, whereas the addition of PWHT lowers the corrosion capabilities of the CROs. The laser-based approach is not a well-proven technology for the CRO, and the

HW-GTAW process is not a quote-worthy procedure for the CRO. Despite the negative results from the preceding two processes, by optimizing the parameters or modifying the process, these processes can produce some surprising results that will be very interesting to know while also opening the door to a new horizon of research. This chapter's goal was to examine the CRO's current processes and identify the constraints of new or improved procedures, as well as their benefits to the CRO.

8.10 ACKNOWLEDGEMENTS

The author of this chapter wishes to express his appreciation to the editors of this book, Dr. Vishvesh J. Badheka, Dr. Krishnan Sivaraman, and Dr. J. Krishnan, for allowing him to publish a chapter in this book that addressed the most practical elements of welding.

REFERENCES

Alvarães, Camila Pereira, et al. "Performance of the INCONEL 625 Alloy Weld Overlay Obtained by FCAW Process." *Materia, (Rio J.)*, vol. 24, no. 1, 2019. https://doi.org/10.1590/S1517-707620190001.0627.

Amani, Mahmood, et al. *A Comprehensive Review of Corrosion and Its Inhibition in the Oil and Gas*, SPE Kuwait Oil and Gas Show and Conference, Kuwait, 2015, pp. 1–14. https://doi.org/10.2118/175337-MS.

Arivazhagan, A., et al. "Laves Phase Formation and Segregation of Nb in Ni—Cr—Mo Superalloy Over 316L by Hot Wire (HW) TIG Cladding Process." *Arabian Journal for Science and Engineering*, vol. 45, no. 11, Springer Berlin Heidelberg, 2020, pp. 9685–98, https://doi.org/10.1007/s13369-020-04873-0.

Avelar, Alan Matias, et al. "Oxidation of AISI 304L and 348 Stainless Steels in Water at High Temperatures." *Materials Research*, vol. 23, no. 6, Materials Research, 2020, https://doi.org/10.1590/1980-5373-mr-2020-0373.

Ban, So-Dam, et al. "Corrosion Resistance of Inconel 625 Overlay Welded Inside Pipes as a Function of Heat Treatment Temperature." *International Journal of Electrochemical Science*, vol. 11, 2016, pp. 7764–74, https://doi.org/10.20964/2016.09.22.

Cabrini, Marina, et al. "Microstructure and Selective Corrosion of Alloy 625 Obtained by Means of Laser Powder Bed Fusion." *Materials*, vol. 12, no. 11, 2019, p. 1742. https://doi.org/10.3390/ma12111742.

Deacon, R. M., et al. "High Temperature Corrosion Resistance of Candidate Nickel-Based Weld Overlay Alloys in a Low NO x Environment." *Materials Science and Engineering: A*, vol. 461, no. x, 2007, pp. 392–402, https://doi.org/10.1016/j.msea.2007.01.150.

Farzadi, A. "Microstructural Development and Corrosion Behavior of 13Cr—4Ni-1Mo Martensitic Stainless Steel Clad." *Weld World*, 2020, pp. 1811–23, https://doi.org/10.1007/s40194-020-00961-9.

Feng, Z., et al. "Economic Impact of Corrosion in Oil Sectors and Prevention: An Overview." *Journal of Physics: Conference Series*, 2019, https://doi.org/10.1088/1742-6596/1378/2/022037.

Flores, J. F., et al. "Erosion—Corrosion Degradation Mechanisms of Fe—Cr—C and WC—Fe—Cr—C PTA Overlays in Concentrated Slurries." *Wear*, vol. 267, 2009, pp. 1811–20, https://doi.org/10.1016/j.wear.2009.02.005.

Galedari, Sahar A., et al. "A Comprehensive Review of Corrosion Resistance of Thermally-Sprayed and Thermally-Diffused Protective Coatings on Steel Structures." *Journal of Thermal Spray Technology*, 2019, https://doi.org/10.1007/s11666-019-00855-3.

Gittos, M. F., et al. "Disbonding of Austenitic Stainless Steel Cladding Following High Temperature Hydrogen Service." *Weld World*, vol. 52, 2008, pp. 54–67, https://doi.org/10.1007/BF03266632.

Guo, Longlong, Hualin Zheng, et al. "Effect of Heat Treatment Temperatures on Microstructure and Corrosion Properties of Inconel 625 Weld Overlay Deposited by." *International Journal of Electrochemical Science*, vol. 11, 2016, pp. 5507–19, https://doi.org/10.20964/2016.07.97.

Guo, Longlong, Fei Xiao, Fan Wang, Yutian He, et al. "Microstructure and Corrosion Resistance of Inconel 625 Overlay Welded by Pulsed TIG Process." *International Journal of Electrochemical Science*, vol. 16, 2021, pp. 1–12, https://doi.org/10.20964/2021.04.11.

Guo, Longlong, Fei Xiao, Fan Wang, Wenlan Wei, et al. *Influence of Heat Treatments on Microstructure, Mechanical Properties and Corrosion Resistance of Inconel 625 Overlay Cladded Using PTIG*, IOP Publishing, 2020.

Guo, P. L., and L. G. Ling. "Effect of Post-Weld Heat Treatment (PWHT) on the Intergranular Corrosion of ENiCrFe-7 Weld Overlay Cladding Materials." *Integrative Medicine Research*, vol. 9, no. 4, Korea Institute of Oriental Medicine, 2020, pp. 8636–45, https://doi.org/10.1016/j.jmrt.2020.05.101.

Hai, Chao, et al. "Role of Martensite Structural Characteristics on Corrosion Features in Ni - Advanced Dual - Phase Low - Alloy Steels." *Acta Metallurgica Sinica (English Letters)*, vol. 34, no. 6, The Chinese Society for Metals, 2021, pp. 802–12, https://doi.org/10.1007/s40195-020-01145-6.

Islam, Aminul, et al. "Investigation of Erosion-Corrosion Behavior of (WTi) C Based Weld Overlays." *Wear*, vol. 390–391, Elsevier B.V., July 2017, pp. 155–65, https://doi.org/10.1016/j.wear.2017.07.016.

Jung, Eui-Je, and Hae-Woo Lee. "Comparison of Corrosion Resistance and Corroded Surfaces of Welding Metal in Overlay-Welded Inconel 600 and Inconel 625 by Gas Metal Arc Welding." *International Journal of Electrochemical Science*, vol. 11, 2016, pp. 7125–38, https://doi.org/10.20964/2016.08.71.

Li, Yang, et al. "Effect of Electric Potentials on Microstructure, Corrosion and Wear Characteristic of the Nitrided Layer Prepared on 2Cr13 Stainless Steel by Plasma Nitriding." *Acta Metallurgica Sinica (English Letters)*, vol. 32, no. 6, The Chinese Society for Metals, 2019, pp. 733–45, https://doi.org/10.1007/s40195-018-0836-z.

Liu, Dashuang, Weimin Long, et al. "Influence of Nb Additive on Alkaline Corrosion Behavior of Slag-Free Self-Shielded Metal-Cord Welding Overlay." *Journal of Materials Engineering and Performance*, no. Ref 24, 2022, https://doi.org/10.1007/s11665-023-07862-0.

Liu, Dashuang, Yucheng Wu, et al. "Effect of Titanium Addition on the Neutral Corrosion Resistance of Welding Overlay." *Journal of Materials Engineering and Performance*, vol. 31, no. 7, Springer US, 2022, pp. 5212–20, https://doi.org/10.1007/s11665-022-06606-w.

Lo, I-Hsuang, and Wen-Ta Tsai. "Effect of Heat Treatment on the Precipitation and Pitting Corrosion Behavior of 347 SS Weld Overlay." *Materials Science and Engineering: A*, vol. 355, 2003, pp. 137–43, https://doi.org/10.1016/S0921-5093(03)00078-9.

Longlong, Guo, et al. "Formation Quality Optimization and Corrosion Performance of Inconel 625 Weld Overlay Using Hot Wire Pulsed TIG." *Rare Metal Materials and Engineering*, vol. 45, no. 9, Northwest Institute for Nonferrous Metal Research (China), 2016, pp. 2219–26, https://doi.org/10.1016/S1875-5372(17)30006-1.

Luchtenberg, Paola, et al. "Surface & Coatings Technology Effect of Welding Energy on the Corrosion and Tribological Properties of Duplex Stainless Steel Weld Overlay Deposited by GMAW/CMT Process." *Surface & Coatings Technology*, vol. 375, Elsevier, July 2019, pp. 688–93, https://doi.org/10.1016/j.surfcoat.2019.07.072.

Maniee, Arash, et al. "Improved Hardness, Wear and Corrosion Resistance of Inconel 718 Treated by Hot Wall Plasma Nitriding." *Metals and Materials International*, vol. 26, no. 11, The Korean Institute of Metals and Materials, 2020, pp. 1664–70, https://doi.org/10.1007/s12540-019-00476-z.

Mendes, Matheus C., and Humberto N. Farneze. "Microstructure and Corrosion Properties of the AISI 904L Weld Cladding Obtained by the Electro Slag Process." *Journal of Materials Research and Technology*, Elsevier B.V., 2020, https://doi.org/10.1016/j.jmrt.2020.11.048.

Moshkbar, M., et al. "Microstructure and Phase Analysis of Multilayer Ni—Cr—Mo Clad for Corrosion Protection." *Transactions of the Indian Institute of Metals*, vol. 74, no. 7, Springer India, 2021, pp. 1663–72, https://doi.org/10.1007/s12666-021-02256-z.

Naeem, M. "Improved Mechanical Properties, Wear and Corrosion Resistance of 316L Steel by Homogeneous Chromium Nitride Layer Synthesis Using Plasma Nitriding." *Journal of Materials Engineering and Performance*, vol. 29, February 2020, pp. 877–89, https://doi.org/10.1007/s11665-020-04653-9.

No Title, 2023, https://www.twi-global.com/technical-knowledge/faqs/what-is-corrosion.

No Title, March, 2020, https://www.financialexpress.com/economy/corrosion-eats-up-gdp-did-you-know-how-much-india-loses-every-year-to-wear-and-tear/1902183/.

Ostovan, Farhad, et al. "Microstructure, Hardness and Corrosion Behavior of Gas Tungsten Arc Welding Clad Inconel 625 Super Alloy over A517 Carbon Steel Using ERNiCrMo3 Filler Metal." *Journal of Materials Engineering and Performance*, vol. 29, no. 10, Springer US, 2020, pp. 6919–30, https://doi.org/10.1007/s11665-020-05178-x.

Peabody, A. W. *Control of Pipeline Corrosion*, second edition, Nace International, The Corrosion Society, 2001, ISBN 1-57590-092-0.

Sandes, Soraia S., et al. "Microstructural Characterization of Inconel 625 Nickel-Based Alloy Weld Cladding Obtained by Electroslag Welding Process." *Journal of Materials Engineering and Performance*, Vol. 29, May 2020, pp. 3004–15, https://doi.org/10.1007/s11665-020-04861-3.

Sathiya, A., and P. Evangeline. "Dissimilar Cladding of Ni—Cr—Mo Superalloy over 316L Austenitic Stainless Steel: Morphologies and Mechanical Properties." *Metals and Materials International*, vol. 27, no. 5, The Korean Institute of Metals and Materials, 2021, pp. 1155–72, https://doi.org/10.1007/s12540-019-00440-x.

Sedighi, M., et al. "The Effect of Post Weld Heat Treatment (PWHT) on the Microstructure, Microhardness and Sulphide Stress Corrosion Cracking (SSCC) of Ni-Base Superalloy IN625 Hot Wire TIG Cladding on AISI 4130 Steel." *Protection of Metals and Physical Chemistry of Surfaces*, vol. 57, no. 1, 2021, pp. 113–20, https://doi.org/10.1134/S2070205120060210.

Shekari, Elahe, et al. "Economic Risk Analysis of Pitting Corrosion in Processing Facility." *International Journal of Pressure Vessels and Piping*, Elsevier Ltd, 2017, https://doi.org/10.1016/j.ijpvp.2017.08.005.

Silman, H. "Corrosion and Corrosion Control: An Introduction to Corrosion Science and Engineering." *British Corrosion Journal*, vol. 7, no. 3, 1972, https://doi.org/10.1179/000705972798323134.

Silva, De Oliveira, and Alice Goncalves Osorio. "Effect of Inert and Active Shielding Gases in the Corrosion Resistance of IN625 Weld Overlays." *Journal of Materials Engineering and Performance*, vol. 31, July 2022, pp. 5886–97, https://doi.org/10.1007/s11665-022-06651-5.

Sim, Bernard-maxmillan, et al. "The Influence of Post Weld Heat Treatment Precipitation on Duplex Stainless Steels Weld Overlay Towards Pitting Corrosion." *Materials*, vol. 12, no. 20, 2019, p. 3285, https://doi.org/10.3390/ma12203285.

Singh, Amit Kumar, et al. "Pitting Corrosion Resistance and Bond Strength of Stainless Steel Overlay by Friction Surfacing on High Strength Low Alloy Steel." In *Defence Technology*, Elsevier Ltd, 2015, https://doi.org/10.1016/j.dt.2015.06.002.

Taylor, Publisher, C. Casciaro, et al. "Hydrogen Diffusion and Distribution in Weld - Clad Fabricated Products with Special Reference to Disbonding." *Welding International*, January 2015, pp. 37–41, https://doi.org/10.1080/09507119309548441.

Taylor, Publisher, M. Mandina, et al. "The Electro-Slag Cladding Process: Application, Main Metallurgical and Testing Issues in the Fabrication of Reactors in Large Thickness 2¼Cr—1Mo—¼V." *Journal of Modern Engineering Research*, June 2013, pp. 37–41, https://doi.org/10.1080/09507116.2012.753262.

Valdez, Benjamín, et al. "Corrosion Assessment of Infrastructure Assets in Coastal Seas." *Journal of Marine Engineering & Technology*, vol. 4177, 2016, https://doi.org/10.1080/20464177.2016.1247635.

Yazdkhasti, Milad, et al. "The Effect of Time and Temperature of Nitridation-Oxidation Process on Properties and Corrosion Resistance of AISI 316L Steel 1." *Surface Engineering and Applied Electrochemistry*, vol. 54, no. 5, 2018, pp. 508–17, https://doi.org/10.3103/S1068375518050125.

Yin, Xiaohui, et al. "Comparison Study of Low-Heat-Input Wire Arc-Fabricated Nickel-Based Alloy by Cold Metal Transfer and Plasma Arc." *Journal of Materials Engineering and Performance*, no. Ref 20, Springer US, 2020, https://doi.org/10.1007/s11665-020-04942-3.

Zafari, Masoud, et al. "Wear and Corrosion Behaviors of Inconel 625- Cladded Layer on P460N Steel in Different Environments." *Metallurgical and Materials Transactions B*, vol. 52, no. 4, Springer US, 2021, pp. 1972–84, https://doi.org/10.1007/s11663-021-02082-0.

9 Welding of Nickel Alloys

S. Bhattacharya

CONTENTS

DOI: 10.1201/9781003352419-9

9.1 INTRODUCTION

This chapter covers some of the basics of welding of nickel and nickel base alloys for those who are familiar with welding processes but not very conversant with welding of Ni base alloys. The basics of welding of other materials have already been covered in earlier chapters of this book.

Let us ponder over why welding of Ni base alloys are important from an industrial point of view [1]. It is because Ni base alloys can be used in a wide range of applications requiring properties to withstand extreme operating conditions. These alloys are used in a wide range of temperatures from cryogenic to high temperatures (approaching 1200°C/2190°F), for both aqueous and high-temperature corrosion resistance and many other extreme operating conditions. Also, it provides good ductility and toughness at low temperatures. One more important property of nickel is its ability to withstand dilution by many alloying elements without any serious adverse effect on its properties.

Given these extraordinary properties of nickel base alloys, it is the quality of the welding that then decides the overall performance of a welded product. The welding of nickel and its alloys is not difficult, though it is quite different from the welding of carbon and stainless steels. Therefore, all those who are familiar with welding of other base metals need to know the differences for a quality weld product. This requirement, thus, imparts importance to this chapter.

9.2 HISTORY OF NICKEL AND NICKEL BASE ALLOYS

Alloys of nickel have been known and used for thousands of years even before the identification of nickel as a metallic element. Nickel was first identified as an element by Cronstedt in 1751. Reliable evidence has been found to show that prehistoric iron–nickel alloys were of meteoric origin. The first man-made alloy containing nickel, believed to date back to 3500–3100 BC, is a bronze reamer.

In the late nineteenth century, commercially useful Ni base alloys were developed, and further higher level of sophistication was incorporated during the twentieth century. A large variety of nickel base alloys to meet the requirement of corrosion resistance and high- temperature applications was subsequently developed. Tremendous progress in many industries would not have been possible with other alloy systems. In case of nickel alloys the tonnage of commercial production does not justify the importance due to critical applications, such as magnetic alloys, controlled thermal expansion alloys, and alloys used in monitoring neutron dosage etc. of these alloys.

9.3 NI BASE ALLOY CLASSIFICATION

A nickel base alloy was earlier defined as any alloy containing more than 50% nickel as its constituent. The definition has changed with time, and now it is defined as any alloy which has nickel as the principal constituent. Addition of alloying elements change the properties of the main constituent, resulting in improved properties like corrosion resistance, increase temperature performance, pitting resistance, and so on.

Classification of nickel alloys is not so well defined as that of stainless steel and aluminium. However, the strengthening mechanism helps to loosely classify these alloys as given here.

1. Solid solution strengthened
2. Precipitation strengthened (hardened)
3. Super alloys—oxide dispersion strengthened speciality alloys

Commercial alloys though is not classified based on strengthening mechanism but are a group by themselves based on their specific properties and application and, therefore, has been covered in this section.

9.3.1 Solid Solution Alloys

Solid solution alloys are Ni–Cu and the simpler Fe–Ni–Cr alloys. These alloys can be welded by all fusion welding processes [2]. They are normally welded in the annealed condition. Since the heat affected zone (HAZ) does not harden during welding, heat treatment is not usually required after welding. Solid solution alloys have good corrosion resistance to both seawater and other general corrosion environments. Therefore, Monel is extensively used for such applications. The Ni–Cu alloys may be susceptible to porosity if proper shielding or well-deoxidised consumables are not used for welding. Other solid solution-strengthened Ni base alloys may contain only iron and are mostly used for their coefficient of expansion or electrical properties. The Ni–Fe alloys have reasonable weldability, but availability of good consumables is a constraint. Other solid solution alloys contain a variety of substitutional elements including chromium, molybdenum, and tungsten. Each element imparts particular characteristics and has the ability to alter the welding response. The maximum ultimate tensile strength values of the solid solution-strengthened Ni base alloys can be as high as 830 MPa (120 ksi) with yield strengths in the range of 345 MPa to 480 MPa (50–70 ksi). If higher strength levels are required, then precipitation-strengthened alloys are of choice.

9.3.2 Precipitation-Strengthened (Hardened) Alloys

This group of alloys contains alloying elements that form precipitates with nickel when subjected to an appropriate heat treatment. The most common of these elements are aluminium, titanium, and/or niobium. These precipitates are coherent with austenite matrix and thus strain the matrix, thereby providing strength and increasing it manifold. Gamma prime (γ') is the most common precipitate. The strength of such alloys can exceed 1,380 MPa (200 ksi) with yield strength (0.2% offset) of over 1035 MPa (150 ksi). Some of the common precipitation hardening alloys are Ni–Cu–Al–Ti, Ni–Cr–Al–Ti and Ni–Cr–Fe–Nb–Al–Ti. These alloys may be susceptible to post-weld heat treatment cracking. To overcome this deficiency, development took place which led to the nickel–chromium alloys that are strengthened by gamma double prime (γ'') precipitates. The gamma double prime precipitate forms more slowly than gamma prime which prevents post-weld heat treatment cracking and strain age cracking (SAC). The most popular of these alloys is Alloy 718. It is generally immune to SAC during post-weld heat treatment. One of the major applications of Alloy 718 is in the manufacturing of aerospace gas turbine shafting and pressure containment. This alloy has excellent fatigue life at service temperatures up to 760°C (1400°F).

9.3.3 Superalloys

While classifying a nickel alloy, the term superalloys is used to describe Ni base alloys with their superior strength properties provided by the gamma prime and gamma double prime phases as described earlier. These alloys generally have high strength and good corrosion-resistance properties at high temperatures. Initially, these alloys were difficult to weld, but after many years of technological developments, it is now possible with controlled welding process and keeping the residual stresses low. One such alloy is IN713C which is like Alloy X-750. Such alloys are extensively used for very high-strength, corrosion-resistant single-crystal turbine blades. The primary challenges to weld repair of these materials include the avoidance of stray grains in the melt pool and elimination of cracking.

There are other super alloys which have excellent specific properties such as very high creep strength. These are strengthened by oxides such as Yttria (Y_2O_3) dispersion. MA6000 and MA754 are some of such alloys. These alloys also exhibit excellent high-temperature oxidation but loses this characteristic when welded with conventional fusion welding techniques. In general, these alloys are difficult to weld. Another group of super alloys has been developed based on nickel-aluminides (NiAl or Ni_3Al compound). They have very high strength and corrosion resistance but are very difficult to weld because of their low ductility. Ni–Cr–B and Ni–Mo–Si alloys have also been developed for wear resistance in a variety of environments, but these alloys are also difficult to weld due to high hardness, low ductility, and the formation of low-melting range phases within their compositions.

9.3.4 Commercial Alloys

These alloys are not classified based on their strengthening mechanism but are based on their specific properties and are thereby application-dependent. These alloys are fully austenitic. They are better known by their trade name or alloy number.

Commercial alloys are classified as follows.

9.3.4.1 Commercially Pure and Low-Alloy Nickel

These alloys have excellent corrosion resistance to reducing atmosphere due to high nickel content (about 94% to 100%). They also have reasonable thermal transfer characteristics. Some of the examples of alloys of this group are Nickel 200, Nickel 201, Permanickel Alloy 300, and Duranickel Alloy 301.

9.3.4.2 Nickel–Copper Alloys

With the addition of copper to nickel, corrosion resistance in reducing chemical environment and seawater improves significantly. Its use is therefore extensive in nuclear submarines and ships. By altering the proportion, the properties can be further altered. Some of the nickel–copper alloys are Alloy 400 (66% Ni, 33% Cu), Alloy R-405, and Alloy K-500.

9.3.4.3 Nickel–Chromium and Nickel–Chromium–Iron

Addition of chromium improves the high-temperature oxidation resistance, and addition of iron improves strength. Addition of both these elements to nickel has led to the development of a large number of commercial alloys having a wide application range.

Addition of other alloying elements modified the properties to meet more stringent requirement of application. Such elements are Al, Si, Ti, Mo, and W in various proportions. Addition of molybdenum improved pitting resistance and together with niobium increased the strength of the alloy as in Alloy G3/G30 (Ni–22Cr–19Fe–7Mo–2Cu) and Alloy 625 by the addition of 9% Mo plus 3% Nb offers both high-temperature and wet corrosion resistance. Alloys of this group are extensively used in jet engines and in other severe corrosive environments.

9.3.4.4 Iron–Nickel–Chromium Alloys

This group is primarily iron base but nickel and chromium are the principal alloying elements and were developed for use in thigh temperature petrochemical environments where sulphur is the main corrosion agent. Alloy 800 (Fe–32Ni–21Cr) and its variants such as Alloy 800HT, 801 and 802 are in this category.

9.3.4.5 Nickel–Iron Low-Expansion Alloys

The applications of these alloys are extensive in lamp and electronics industries, where glass-to-metal sealing is important. Some of the alloys are Alloy 36 (Fe–36Ni), Alloy 42 (Fe–42Ni), Alloy 426, and Alloy 52 (Fe–51.5Ni).

9.3.4.6 Welding Alloys

Welding products for nickel alloys have similar compositions to that of base metals, although additions of aluminium, titanium, and other elements are made to the filler metals for specific purposes. Some of the important requirements for the addition of these element in the welding electrodes are to ensure proper deoxidation of the molten weld pool, to overcome any hot-short cracking, and malleability problems. More details of this group of alloys have been discussed in a later section of this chapter.

Controlled-expansion and soft magnetic alloys are some of the other alloys which have been developed for specific applications but are of less importance as a material of construction.

9.4 ALLOYING ELEMENTS IN NICKEL BASE ALLOYS AND THEIR IMPACT

9.4.1 Effects of Alloying Elements in Nickel Base Alloys

The alloying elements change the properties of the base metal and hence its intended application. Understanding the effects of alloying elements will help understand their impact on welding of nickel base alloys. The most common elements in nickel alloys are chromium, molybdenum, and iron. Other elements like tungsten, carbon, aluminium, titanium, copper, niobium, tungsten, and sulphur are also known to have a significant impact on the properties of nickel. Understanding the effect of alloying elements on the properties helps in selecting a particular alloy for a specific application.

Applications of nickel alloys as structural material are though not uncommon but are used only for critical applications due to cost considerations. Some of such applications are as follows:

- Aerospace and jet engines

  Nickel alloys are the natural choice within the aerospace industry due to

  - High temperature strength
  - Creep/stress rupture
  - Low thermal expansion
  - High-temperature oxidation

- Oil and gas extraction
  - Corrosion resistance to H_2S, chlorides, brines, elemental sulphur, hydrogen, seawater
  - Strength
- Pollution control and waste processing due to
  - Corrosion resistance

Process industry and refining

- High corrosion resistance (acids, chemicals, steam, and more)
- Automotive
- Marine industry
- Power generation
- Corrosion resistance to coal, steam, salt, oil, methane, and more
  - High-temperature strength
  - Creep/stress rupture

The unique property of nickel to dissolve a large number of alloying elements has led to the development of a wide spectrum of nickel base alloys giving diverse characteristics. Some of the significant impacts of important alloying elements in nickel alloys are given here.

9.4.1.1 Chromium (Cr)

Addition of chromium is generally in the range of 15% to 30%, though in certain specific alloys it could go as high as 50%. It enhances oxidation resistance by forming a tenacious oxide layer. Ni–Cr alloys are, therefore, extensively used in an oxidising atmosphere including nitric and chromic acids.

9.4.1.2 Molybdenum (Mo)

The addition of molybdenum significantly improves the resistance to pitting and crevice corrosion. Use of Mo-added alloys in non-oxidising acids such as hydrochloric and hydrofluoric acid is recommended. Such alloys are also used in other acidic applications like phosphoric and sulphuric acids (concentration below 60%). It has also been observed that addition of Mo improves high-temperature strength and creep resistance.

9.4.1.3 Iron (Fe)

Iron reduces the cost of the alloy besides improving high-temperature carburising resistance and controls thermal expansion.

9.4.1.4 Tungsten (W)

This element is similar to Mo in that it provides resistance to reducing acids and localised corrosion and imparts enhanced strength and creep resistance due to its low diffusivity in Ni and weldability. The critical pitting potential of Ni–W alloys was also found to become more positive (correspondingly a higher resistance to pitting) upon increasing the W content up to 3 wt.% but decreases with further additions.

9.4.1.5 Carbon (C)

Carbon can hinder corrosion resistance but adds strength at elevated temperatures.

9.4.1.6 Aluminium (Al)

Addition of Al can promote the formation of a tightly adherent alumina scale at high temperatures that resists attack by oxidation, carburisation, and chlorination. In combination with Ti, it also promotes age hardening in certain alloys.

9.4.1.7 Titanium (Ti)

As mentioned earlier, it promotes age hardening and can also combine with carbon to reduce susceptibility to intergranular corrosion because of the carbide that forms after heat treatment.

9.4.1.8 Copper (Cu)

It improves resistance to reducing acids and aqueous corrosion. Alloys that contain 30–40% Cu offer excellent resistance to all concentrations of non-aerated HF. If Cu is added to a Ni–Cr–Mo–Fe alloy, resistance to hydrochloric, phosphoric, and some concentrations of sulphuric acids is improved. It provides excellent resistance to corrosion by seawater.

9.4.1.9 Cobalt (Co)

Cobalt imparts unique strengthening characteristics to high-temperature alloys. Cobalt also increases the resistance in nickel alloys to carburisation, as Co increases the solubility of C in Ni base alloys and sulphidation because the melting point of cobalt sulphide is higher than nickel sulphides.

9.4.1.10 Niobium (Nb)

Nb has very high affinity to react with carbon which helps in stabilisation. It thus helps in resisting intergranular corrosion and localised corrosion. It also contributes to age hardenening and enhancement of strength. Elements like Cr and Mo sometimes contribute in solid-state precipitation reactions during long-time, high-temperature operations. Mo and W are also useful for increasing creep strength due to their low diffusivity in Ni. Although Ti and Al can be effective solid solution strengtheners, they typically improve strength by precipitation of $Ni_3(Ti,Al)$ phase in Ni base superalloys. Niobium is also an effective strengthener and is used both for solid solution strengthening and for its ability to strengthen by promoting precipitation of the Ni_3Nb phase.

Nickel and the effect of alloying elements could be summarised as follows.

- Ni—austenite stabiliser provides thermal stability and corrosion resistance to reducing acids and improves SCC.
- Cr—improves oxidation resistance and localised corrosion.
- Mo—improves corrosion resistance in reducing acids, localised corrosion, and strength.
- Nb—improves localised corrosion, stabilisation, age hardener, and strength.

- W—improves strength, weldability, and resistance to reducing acids.
- Cu—improves resistance to reducing acids and sulphide SCC
- Ti—improves IGA stabilisation and is an age hardener.
- Al—age hardener and improves oxidation resistance.
- N—austenite stabiliser, improves localised corrosion, strength.
- Fe—reduces cost, controls thermal expansion.

9.4.2 Phase Diagram

Phase diagrams are graphical representations of the phases present in a particular alloy being held at a particular temperature. Phase diagrams can be used to predict the phase changes that may occur in an alloy during a particular heating/cooling cycle including welding. In metallurgy, the term phase is used to refer to a physically homogeneous state of matter, where it has a certain chemical composition and a distinct atomic bonding and arrangement of elements. Within an alloy, two or more different phases can be present at the same time. The phases could be identified using metallography.

Phase diagrams are important for a metallurgist, as they provide the relationship between phases in a system as a function of temperature, pressure (generally considered 1 atmospheric pressure), and composition. Phase diagrams are useful for the selection of alloys with a specific composition and design and control of heat treatment procedures that will produce specific properties. They are also used to troubleshoot quality problems.

Each phase within an alloy has its own distinct physical, mechanical, electrical, and electrochemical properties. The property of an alloy is the composite property of all the phases present in the alloy. Thus, an alloy with more than one phase can be considered to be a composite material.

The phases present in an alloy depend not only on the alloy composition but also on the thermal treatment to which the alloy has been exposed. Thermal treatments are often used to alter the phase/s in an alloy to modify the properties.

Strengthening of nickel base alloys takes place either by solid solutioning or by precipitated phase hardening which enhances strength at peak temperatures. It also depends on the nature of solid solutions, pressures and environments, phase stability at different temperatures, and grain shape, size, and distribution. These are the important parameters for the study of nickel base superalloys. The performance of these alloys will largely depend on the composition. To impart all the required characteristics for a particular application, alloying elements should be added in appropriate proportions. When two or more alloying elements are used to develop an alloy, it creates a diagram giving the composition and consequently the microstructure and phase of the resultant alloy at a specific temperature since the solubility of one element in the other varies with temperature. It must be noted that each alloy system is different and behaves differently when welded.

The development of microstructures of an alloy of a particular composition at different temperatures is clearly depicted by a phase diagram. These diagrams also provide information on melting, casting, and recrystallisation of alloys. If more than two components are present in a system, the phase diagram becomes

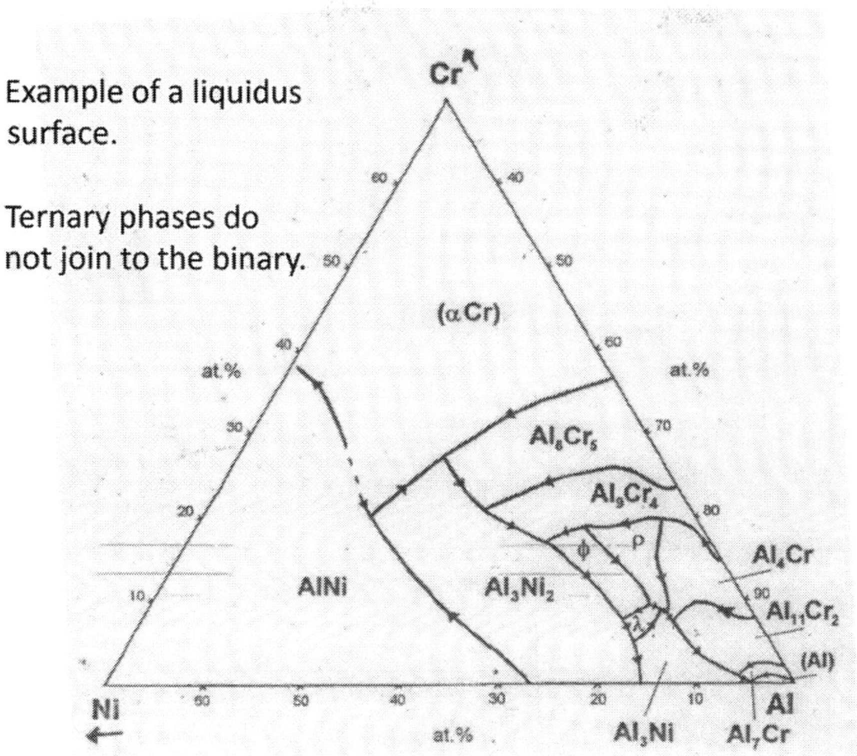

FIGURE 9.1 Three-component (Ni-Cr-Al) phase diagram projected in two dimention plane.

extremely complicated, and it is difficult to represent various phases as they are three-dimensional.

Early developers of Ni base superalloys relied predominately on available binary and ternary phase diagrams, since the multi-component thermodynamic routines were earlier not available.

9.4.2.1 Phase Diagram and Welding

During welding, the base metal and the weldment are subjected to a wide range of temperature cycles and consequently produces a range of microstructures during the course of solidification. The stability and properties of the welded structure depend largely on the microstructures. Hence, the importance of study of phase diagram of the alloy system being welded is underlined.

Figure 9.1 shows a typical nickel-alloy phase diagram.

9.5 WELDABILITY

The weldability, also known as joinability, of a material refers to the extent of its ability to get welded. Most metals can be welded, but some are easier to weld than others. A material's weldability is used as a guide to determine the welding process as well.

Weldability is hard to define quantitatively, so most standards define it qualitatively. The International Organization for Standardization (ISO) defines weldability in ISO standard 581–1980 [3] as:

> Metallic material is considered to be susceptible to welding to an established extent with given processes and for given purposes when welding provides metal integrity by a corresponding technological process for welded parts to meet technical requirements as to their own qualities as well as to their influence on a structure they form.

For all practical purposes, it is sufficient to remember the first definition rather than the technically all-encompassing aspects in the ISO definition. Weldability of nickel and its alloys is generally good. The composition, metallurgical structure, and its heat treatment and/or service history all affect its response to welding.

9.5.1 Effect of Grain Size

Wrought alloys generally have better weldability than castings. This is because wrought alloys have fine-grained components than cast items, as these (casting) generally have significant amounts of segregation. Coarse grains may lead to micro-fissuring in the HAZ; thus, high heat input should be avoided during welding. The castings often contain significant additions of silicon to provide fluidity. This addition can compromise the hot cracking resistance of some alloys and adversely affects their weldability.

Materials with a coarse grain structure (ASTM No. 5 or coarser) are generally found to be less weldable than those with finer grain structures. Coarse grain materials are particularly susceptible to base metal micro-fissuring. High-energy input processes, such as GMAW in spray mode transfer, are not recommended welding process for coarse grain materials. Table 9.1 provides a guideline to select the suitable welding process based on grain size for several common alloys. All the alloys

TABLE 9.1
Effect of Grain Size on Recommended Welding Processes

Common Alloy	Grain Size	GMAW	Elec. Beam	GTAW	SMAW
Inco 600	Fine	OK	OK	OK	OK
	Coarse	NR	NR	OK	OK
Inco 617	Fine	OK	OK	OK	OK
	Coarse	NR	NR	NR	OK
Inco 625	Fine	OK	OK	OK	OK
	Coarse	NR	NR	NR	OK
Inco 718	Fine	NR	OK	OK	OK
	Coarse	NR	NR	NR	OK
INCOLOY 800	Fine	OK	OK	OK	OK
	Coarse	NR	OK	OK	OK

are best welded in the annealed or solution-treated condition, and this applies particularly to the precipitation hardenable alloys.

Since rolling, drawing, or forming of the alloys generate high levels of residual stress, they are prone to cracking. Lower-temperature annealing (mill anneal) of alloys generally exhibit finer grain structures than those annealed at higher temperatures (solution anneal) and thus are found to offer better weldability. Some alloys must be solution annealed prior to final heat treatment (e.g., precipitation-hardenable alloys) or those used in high-temperature service conditions (e.g., INCO alloy 601 and 617 or INCOLOY alloy 800HT or 803). If possible, such alloys should be welded after a mill anneal and prior to solution anneal.

As mentioned in the earlier section, there are two families of nickel alloys, namely solid solution-strengthened alloys and precipitation-hardened alloys. They differ significantly in their metallurgical characteristics and, therefore, their weldability. The weldability characteristic of both these alloys are discussed here in detail.

9.5.2 Weldability of Solid Solution-Strengthened Alloys

The weldability of solution-strengthened (hardened) nickel alloys is best described on the basis of their susceptibility to cracking during welding. During solidification of the weldment, it results in a nickel-rich face-centred cubic (fcc) phase (austenite), leading to strong segregation of alloying and impurity elements. Strong segregation of alloying and impurity elements causes weld solidification cracking and embrittlement during PWHT. HAZ also becomes prone to liquation cracking due to such segregation at grain boundaries and residual segregation in the reheated weld metal during subsequent weld passes.

9.5.2.1 Fusion Zone Solidification Cracking

During the end stage of solidification, liquid films are distributed along the solidification grain boundaries and, in some cases, at interdendritic sites. These films are the last to solidify. Shrinkage stains develop at the partially solidified boundaries at this stage, and cracks form as the liquid film is unable to hold the grains together.

The metallurgical characteristic and the level of local strain present are naturally the two major factors influencing solidification cracking susceptibility. The solidification temperature range and the amount and distribution of interfacial terminal liquid are the key metallurgical characteristics. Solute distribution also influences the solidification temperature range and thus influences the cracking tendency. The solidification cracking can be controlled by increasing the terminal liquid during welding using backfilling technique. When the amount of terminal liquid is high, (greater than approximately 10 vol.%), it often flows into the cracks and provides the "crack healing" effect.

Some of the impurities have greater influence to make the nickel alloys more susceptible to solidification cracking, though most the impurities are generally harmful. The most critical amongst them are sulphur, phosphorus, and sometimes lead and silver. Boron and silicon can also increase cracking susceptibility. Elements like boron and sulphur are particularly harmful because they form low-melting point films. The addition of Mn helps in reducing the impact of sulphur, as it combines with sulphur to form MnS, which has a globular morphology and thus reduces the tendency to solidification cracking. If the impurity levels are kept low, all nickel alloys that have a single phase can be easily welded.

During welding, if proper care is not taken of issues such as high restraint, high heat input, or unfavourable weld shape, cracks are likely to take place in all engineering alloys.

9.5.3 Weldability of Precipitation-Hardened Alloys

As the name suggests, the strengthening mechanism of this group of alloys is by fine distribution of the precipitates of the compounds of the alloying elements such as Al or Ti produced by heat treatment. That means these alloys must be heat treated to develop its mechanical properties.

The weldability of precipitation-hardened alloys is not much different from solution-hardened alloys. However, since the metallurgical characteristics of this groups of alloys are significantly different from that of solid solution alloys, the precipitation-hardened alloys must be welded in the annealed or solution-treated condition.

While considering the weldability of precipitation-hardened alloys, a number of issues unique to precipitation-hardened alloys need to be taken into account. They include fusion zone solidification cracking, HAZ liquation cracking, strain-age cracking, and ductility dip cracking. Some of these issues are similar to solid solution alloys, but a few of the alloying elements present in the precipitation-hardened alloys may pose new challenges. The addition of niobium, for example, could cause HAZ liquation cracking due to the segregation of NbC at grain boundaries and constitutional liquation (i.e., the local eutectic melting of second-phase particles in a matrix at temperatures above the eutectic temperature and below the solidus of the alloy).

Strain-age cracking is unique to this class of alloys. It is a solid-state phenomenon. The addition of Al and Ti are particularly detrimental through the formation of γ' phase having the chemical formula T_3 BCx.

The common factors that affect solidification cracking susceptibility of nickel base alloys have been reviewed earlier. Several minor alloying elements such as P, S, B, C, and Zr influence the solidification cracking. As a rule, impurities such as P and S should be kept low as much as possible. The effects of minor alloying elements are discussed briefly here.

9.5.3.1 Effect of Minor Alloying Elements on Weldability

9.5.3.1.1 Boron and Zirconium

Boron, zirconium, and carbon are added as minor alloying elements to improve creep strength. They also affect the weldability. Boron affects the weldability very adversely. It also has a deleterious effect on solidification and liquation cracking. Zr and S combine to form ZrS in the grain boundary and interdendritic regions. Although Zr addition is generally known to have a deleterious effect on cracking resistance, Zr addition up to a limit of ~0.04 wt% is acceptable with no major adverse effect on cracking susceptibility when the B level is below 10–50 ppm (0.001–0.005 wt%).

9.5.3.1.2 Effect of Niobium Additions

Niobium is often intentionally added as an alloying element to improve localised corrosion resistance, stabilisation, strength, and age-hardening properties. The fusion zone solidification cracking susceptibility of Nb-bearing alloys strengthened by γ'' (gamma double prime) phase is primarily controlled by the formation temperature of the γ/Laves (Laves is an intermetallic compound that has a stoichiometry composition of AB_2) eutectic type constituent. The relative amount and morphology of the γ/NbC and γ/Laves constituents widen the solidification temperature range and affect the weldability among the commercial Nb-bearing alloys. Nb and Si have a negative effect on solidification cracking resistance also. However, increased C may blunt this effect. All these solidification cracking susceptibility studies have not taken into consideration backfilling, a common practice to prevent cracking.

9.5.3.2 Ductility-Dip Cracking (DDC)

Ductility-dip cracking (DDC) is a solid-state cracking phenomenon that occurs in the temperature range of 0.5–0.8 of the alloy solidus temperature that is observed in austenitic weld metals. In this temperature range, a sudden dip in ductility is observed causing embrittlement.

This is a constraint for many structural materials which are heat treated to strengthen through precipitation. It appears as intergranular cracks that form in the HAZ of the previous welding pass. This is particularly a difficult problem, where high chromium (~30 wt%) nickel base welding consumables are used.

DDC in nickel base alloys is essentially an elevated temperature, grain boundary creep phenomenon. A typical DDC in nickel base alloys occurs between 650°C and 800°C (1200°F and 1470°F). A ductility minimum occurs between 850°C and 1050°C (1560°F and 1920° F), and then there is a gradual increase in ductility above 1050°C.

Increases in sulphur content increased the tendency for a ductility dip between 950°C and 1150°C (1740°F and 2100°F), where sulphur segregated in to and embrittled the grain boundaries, which cracked under an applied stress. It is reported that the combined effects of grain boundary precipitation, grain boundary sliding, grain boundary migration, and grain boundary serration affect the DDC performance of the low-expansion alloys. It has been found that impurity (P, S, O, and H) segregation, grain boundary precipitation, and boundary "tortuosity" affect DDC susceptibility and that controlling the nature of grain boundaries is the key to avoiding DDC in nickel base weld metals.

DDC in nickel base weld metals can be a serious weldability issue, particularly in thick-section multi-pass welds where restraint levels are high.

The problem of DDC can be controlled by using high-chromium filler metals. These filler metals usually contain 4 wt.% Mo and additions of Nb up to 2.5 wt.%. Filler metals that contain sufficient Nb will form NbC precipitates at the end of solidification and are able to pin the migrated grain boundaries and impede their motion away from the parent solidification grain boundary. The net effect of this "pinning" is to produce a boundary that is not straight and referred to as "tortuous." The tortuous grain boundaries that result from boundary pinning provide a mechanical locking of the boundaries that effectively oppose grain boundary sliding or DDC. Thus, the addition of alloying elements, such as Nb, that promote the formation of carbides at the end of solidification appear to be an effective method of improving DDC resistance. It should be noted that while the addition of Nb appears promising with regard to the improvement in DDC resistance, the formation of low-melting Nb-rich eutectic constituents potentially increases the susceptibility to weld solidification cracking. A careful balance needs to be maintained so as not to solve one weldability problem while creating another.

The addition of Mo is also beneficial to reduce DDC. Mo may alter the nature of the $M_{23}C_6$ precipitation such that it forms in a temperature range that is more beneficial to the locking of grain boundaries or that Mo acts as a grain boundary strengthener to resist sliding and decohesion at elevated temperatures.

It has also been found that the use of hydrogen in argon shielding gas tends to increase the susceptibility to DDC. There is evidence that oxygen may contribute to DDC. The use of pure argon or helium shielding gases will generally improve resistance to DDC but may not be effective in totally eliminating it.

Where the use of filler metals which resist DDC is not permitted, minimising the weld restraint through a joint design will be helpful. If permissible, small stringer beads are suggested that minimise heat input.

9.6 WELDING METALLURGY

Nickel is a simple metal which does not undergo any phase changes during cooling from melting point to room temperature [4]. Nickel and its alloys do not, therefore, harden by quenching and so cooling rates are of less importance.

It is important that when a welding procedure is considered the metallurgical aspects of the operation must be considered. Often, to develop high strength and corrosion resistance, more complex alloys go through very intricate metallurgical phenomena. It therefore becomes helpful to a welding engineer to have adequate knowledge of metallurgy of the base metal as well of the welding product.

Metallurgical factors have a stronger influence on deciding the procedures for the welding of dissimilar materials and precipitation-hardenable alloys since their strength is developed by heat treatment, which makes metallurgical changes in their structure. It must be noted that any welded joint can undergo metallurgical changes when subjected to cold work and heat treatment. For high-temperature services, both metallurgical and design considerations are important for obtaining a longer life of the fabricated equipment.

During the process of welding, metallurgical changes take place in both the fusion zone and HAZ. The following sub-sections discuss these aspects of both the types of alloys.

9.6.1 Solid Solution-Strengthened Alloys

9.6.1.1 Microstructural Evolution in the Fusion Zone

These Ni base alloys solidify essentially as fully austenitic at the end of solidification. However, certain segregation during solidification of the alloying elements may result in local variation in composition at the sub-grain level, leading to the formation of some second phase(s) at the end of solidification. The influence of these factors are discussed here.

9.6.1.1.1 Interfaces in Single-Phase Austenitic Weld Metal

Various boundaries and interfaces exist when the weld metal solidifies, and many of the defects originate at these interfaces. It is therefore important to understand the characteristics of these interfaces.

At least three different types of boundaries of the weld metal may be present when observed metallographically.

1. Solidification Sub-Grain Boundaries (SSGBs)
2. Solidification Grain Boundaries (SGBs)
3. Migrated Grain Boundaries (MGBs)

9.6.1.1.2 Solidification Sub-Grain Boundaries (SSGBs)

Sub-grains form the smallest structure during solidification. Under normal conditions, the sub-grains are present as cells or dendrites. The boundaries that separate adjacent sub-grains are known as solidification sub-grain boundary (SSGB). They are distinguished from the bulk structure due to the difference in compositions. These boundaries can be easily identified by low misorientation.

9.6.1.1.3 Solidification Grain Boundaries (SGBs)

The solidification grain boundary (SGB) is the result of intersection of packets, or groups, of sub-grains. Because each of these packets of sub-grains has a different growth direction and orientation, high angular misorientation in a boundary is observed at their intersection. These are often called "high angle" grain boundaries. These boundaries occur along the trailing edge of the weld pool. It has been observed that often there is high concentration of solute and impurity at the SGBs. These concentrations result in the formation of low-melting liquid films along the SGBs, leading to solidification cracking.

9.6.1.1.4 Migrated Grain Boundaries (MGBs)

It has been observed that under certain conditions, the crystallographic components of SGB move away from the parent component, creating new boundaries. These new boundaries have a high degree of misorientation, and since they move, such boundaries are known as migratory grain boundaries. During reheating in multi-pass welding, these MGBs can migrate further. Sometimes, these grain boundaries segregate along the MGBs, known as sweeping. During this process of sweeping, the MGBs assimilate solutes and impurities. Amongst the impurities, the fast-diffusing ones are the most likely to be swept into the boundary. Austenitic weld metals are the most conducive to form MGBs. The grain boundary migration, a cause for DDC, is hindered, or "pinned," when a second phase forms during solidification. This pinning in effect stops THE migration, and thus DDC is significantly reduced. It may be worth noting that the actual migration distance of the MGB is only a few microns away

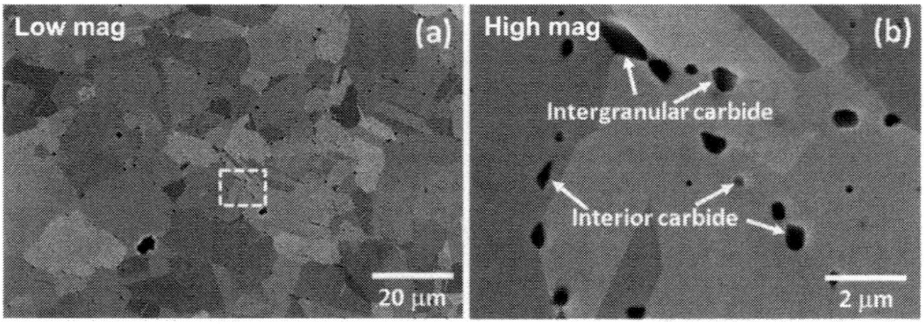

FIGURE 9.2 Grain boundary depicting the boundaries of grains.

from the SGB. DDC has already been described in greater details in Section 9.5.3.2. Figure 9.2 shows a diagram of the grain boundaries.

9.6.1.1.5 Elemental Segregation during Solidification

It is known that solidification behaviour and the resultant microstructure essentially control weldability, mechanical properties, and corrosion resistance of nickel base, which (microstructure) in turn are controlled by the solute distribution behaviour.

Studies have shown that a relationship exists between the solidification temperature range, cooling rate, and dendrite spacing. Knowing these data, it is possible to determine the solid-state diffusion parameter of the solutes. The results of extensive research have demonstrated that solid-state diffusion of substitutional alloying elements such as Fe, Cr, Nb, Mo, and Si in nickel alloys is insignificant during the solidification of fusion welds for elements. However, C diffusion is much higher, as it diffuses by interstitial mechanism as against the solid-state diffusion of other alloying elements. N also shows similar rates of diffusion. The importance of these studies is to show that some amount of migration of grain boundaries can always be expected.

These phases are located in the interdendritic region, pointing that they are associated with eutectic-like reactions that occur in the solute-enriched liquid that is the result of segregation at the end of solidification.

9.6.1.1.6 Dendrite Core Nb and Si

There is not much difference in segregation behaviour between multi-component alloys and the Ni-X binary systems, except for Nb and possibly Si and Ti. The segregation of Nb plays an important role in controlling microstructure in superalloys. The resultant double prime gamma phase (γ''-Ni_3Nb) so developed provides strengthening.

The solubility of Nb depends a lot of other alloying elements. The addition of Fe decreases the solubility of Nb. It has been found that elements of similar atomic radii (less than 1% difference) generally exhibit appreciable solubility, for example, elements like Fe, Cr, and Co. If there is a significant concentration gradient, the corrosion resistance behaviour tends to decrease. It has also been observed that preferential corrosion often takes place at the dendrite cores in welds of Mo-bearing alloys due to the micro-segregation of Mo.

9.6.1.1.7 Phase Transformation Sequence in Solid Solution-Strengthened Alloys

As the weldment solidifies, it undergoes phase changes. The weldment solidifies from the molten state, whereas adjacent to it, the HAZ undergoes phase transformation without going through the solidification process. Beyond that is the base metal which generally does not undergo any phase transformation. Thus, for phase transformation process, only the weldment and the HAZ need to be considered. The phases formed in HAZ are controlled by the alloy composition and the heat cycle it undergoes.

Let us first consider a two-component nickel base alloy such as MONEL 400, which is a Ni–Cu alloy. This alloy solidifies as austenite phase alone as can be seen from the microstructure (Figure 9.3). However, a simple Ni–Mo alloy such as B-2 generally solidifies as a single-phase austenite with small amounts of the M_6C carbide.

In some of the multi-component alloys too, similar phase transformation is observed in the fusion welds. In Fe–Ni–Cr alloys, where Ni is the predominant constituent, the solidification results primarily in the austenite phase. There are some more multi-component alloys which solidify as single-phase austenite. Ni–Cr–Mo alloys with Mo contents below ~15 wt% and very low Fe and W can also solidity as single-phase austenite without the formation of the brittle topologically closed packed (TCP) phases.

When carbon is added, carbides ($M_{23}C_6$) precipitate. This happens as the solubility decreases with decreasing temperature. On the other hand, if a strong carbide former is present as an alloying element such as Nb or Ti, an MC-type carbide is formed as in Alloy 800.

Microstructure evolution becomes more complicated in the presence of Mo due to the formation of intermetallic topologically closed packed (TCP) phases such as σ, μ, and P. The presence of W stabilises these phases which are undesirable from the

FIGURE 9.3 Microstructure of Monel 400 alloy, showing large phase boundaries of the two- component alloy.

weldability point of view. These phases extend the solidification range and thereby increase the susceptibility to solidification cracking. These phases make the weldment brittle and reduce the toughness and ductility.

The TCP phases contain a higher amount of Cr and Mo, and so they are less available in the austenitic matrix that results in reduced corrosion resistance. In the alloys containing higher Mo, the phases predominant are μ and P, while the σ phase is higher in Cr-rich alloys.

In case of Nb-bearing alloys, whether it is a solid solution-strengthening or precipitation-hardening alloy, the microstructural evolution in the fusion welds is very sensitive to Nb and C content and to a lesser degree to the Si content.

A small addition of Ti (~0.5 wt%) along with Si undergoes an interesting microstructure formation. The solidification starts with the formation of primary austenite dendrites where Si and Ti are discarded, and they move to the liquid. With gradual solidification, the liquid becomes richer in these elements till the solubility limits are reached. At that stage, the remaining liquid transforms to austenite, and the $(Ni,Co)_{16}(Ti,Cr)_6Si_7$ phase forms by a eutectic-type reaction. It has also been observed that the addition of Si and Ti has a significant effect on the weldability.

9.6.1.1.8 Elemental Segregation during Solidification

It has been found that the nominal alloy composition of most alloying elements in nickel base alloys does not significantly affect the segregation tendency but is controlled primarily by the pertinent values of equilibrium distribution coefficient and diffusivity in the solid.

Lower is the equilibrium distribution coefficient, steeper is the concentration gradients across the dendritic substructure of the weld. However, this can be eliminated if the solid-state diffusion of that element is high enough to promote back diffusion in the solid. Interstitial elements such as C and N can diffuse back almost infinitely fast under almost all fusion welding conditions.

Some of the alloying elements, namely Fe, Cr, and Co, exhibit appreciable solubility and that prevents a large concentration gradient. Carbon partitions strongly during solidification forming various carbides. The final carbon distribution in the solid weld is generally uniform due to the high solid-state diffusivity of C in Ni.

9.6.1.2 Microstructural Evolution in the HAZ

The second zone which experiences significant numerous metallurgical reactions is HAZ. This is more prominent in solid solution-strengthened alloys which affects the weldability of these materials. These reactions include recrystallisation, grain growth, particle dissolution, precipitate formation, grain boundary segregation, and grain boundary liquation.

The supply condition of the base metal is as important in the metallurgical reactions that take place during the fusion welding at HAZ. The recrystallisation reaction is not observed in annealed materials which is generally the condition of the supply of base material. In all other conditions of supply, recrystallisation reaction may take place. For supplies in solution-annealed conditions, some grain growth in the HAZ could be observed.

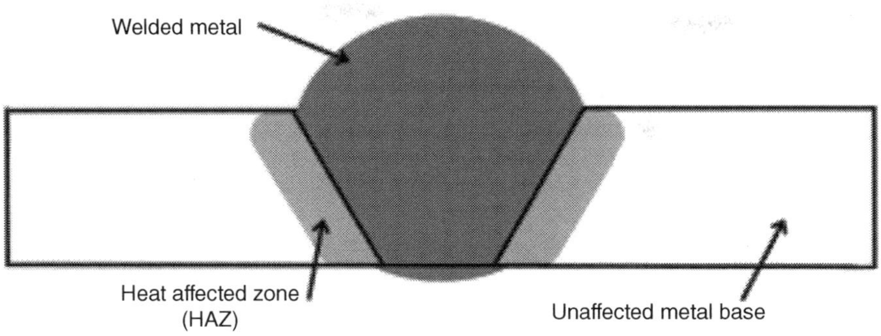

FIGURE 9.4 Schematic diagram of HAZ showing various zones formed during welding.

The extent of grain growth depends on the initial grain size, heat input, and the amount of initial plastic deformation. Substantial grain growth will take place for the smaller initial grain size as reverse is seen for the other two parameters. If the weld heat input is low and has a steep temperature gradient in the HAZ, not much of grain growth is expected in the HAZ. Grain size has a profound impact on deciding the welding process (Refer to Table 9.1). A typical HAZ presentation is shown in Figure 9.4.

Due to segregation of impurities and/or solute at the grain in the HAZ and fusion boundary region, liquation cracking may take place. S, P, Pb, and B are the most significant amongst the impurities/solute. B is sometimes added as an alloying element to improve creep properties. Course grains present in solution-annealed materials show a greater tendency towards liquation; therefore, high-input welding processes are not recommended for solution-annealed materials. The ultimate result of such liquation is cracking in the HAZ.

9.6.1.2.1 Phase Transformations—Carbide Formation

The γ and γ' (Ni_3Al) phases are the main constituents in alloys having low C (less than ~0.01 wt%). The Ni_3Ti phase can also form during solidification, but this phase adversely affects the mechanical properties and, therefore, is not wanted. During solidification, the first phase to form is primary γ. As the solidification proceeds, Al and Ti segregate in the liquid. Thus, the liquid composition moves away from being nickel-rich. Since γ' (and not Ni_3Ti) is the preferred phase, and Ti segregates to the liquid more aggressively than Al, this results in having alloys more of Al than Ti.

Near the end of solidification, alloys with high C content are known to form MC-type carbides. As mentioned earlier, the presence of other alloying elements in superalloys, most notably Al, will favour the formation of the γ' (Ni_3Al) phase over the Ni_3Ti phase. In addition, other carbide-forming elements such as Cr, Ta, Nb, and Mo can substitute for Ti in the TiC carbide. C will segregate strongly to the liquid phase during solidification along with the γ'-forming alloying elements such as Ti and Al. The Ni–Ti–C system exhibits a ternary eutectic point in which γ, MC, Ni_3Ti meet. This would leave the three solid phases intermixed in the final microstructure.

It should also be mentioned that the presence of small amounts of other elements such as B can lead to additional constituents like borides. These phases generally have a detrimental effect on the weldability because they form at lower temperatures and expand the solidification temperature range.

9.6.2 Precipitation-Hardened Alloys

9.6.2.1 Microstructural Evolution in Fusion Zone

As the solidification takes place, the composition changes as dictated by the solidus and liquidus lines in the phase diagram. The composition continues to change till the temperature comes down to a value at which eutectic reaction starts, and secondary phases starts to form. This is a simple progress for binary alloys, but in a multi-component alloy system, the process is more complex though the fundamentals remain the same. The composition of the liquid can thus be assessed using the phase diagram. The composition of the alloying elements such as Fe, Ni, and Cr in the solid as it forms will be proportional to the nominal content. With further progression of solidification, the solubility of carbide-forming elements and C reaches the maximum and carbides start forming. In Nb-containing alloys, at the maximum limit of solubility of Nb and C, the Laves phase transforms into γ + NbC at higher temperatures during solidification.

9.6.2.2 Microstructural Evolution in Heat Affected Zone

In precipitation-strengthened alloys, the base metal determines the microstructure of HAZ. The welding of these alloys can be done in both annealed and precipitation-hardened conditions with both having respective pros and cons. Solution-annealed condition is recommended when cracking in the fusion zone or partially melted zone (PMZ) of the HAZ is a concern. In this case, the entire welded structure must be post-weld heat treated to restore strength, which may not be feasible for bigger structures. Under such circumstances, the welding should be carried out in the precipitation-hardened condition. If the welding is carried out in the precipitation-hardened condition, then there is higher intrinsic restraint in the HAZ, which is not a desirable condition.

Some of the metallurgical reactions in the HAZ include grain growth, grain boundary segregation, grain boundary liquation, constitutional liquation, and, on cooling, precipitate formation. As mentioned earlier, the degree of grain growth is determined by the starting base metal grain size and the thermal history experienced by the HAZ. The heat input and the heat flow conditions during welding are the two important parameters which control the thermal history. The grain growth is not a desirable condition since they adversely impact the fatigue resistance.

To improve high-temperature creep resistance, B is added to some of the nickel alloys. This element if segregates at the HAZ boundaries may cause liquation and ultimately cracking. A similar situation may occur for other impurities like Al, Ti, Si, Nb, S, and P though the last two are usually very low to cause any significant problem.

Constitutional liquation (the constituent particle itself does not melt) is attributed to HAZ liquation. For constitutional liquation to occur, the particle must partially, but not completely, dissolve upon heating to the eutectic temperature. When the HAZ undergoes rapid heating, the microstructure is rapidly heated, and the particle starts to react with the surrounding matrix. As the heating of the microstructure continues, more of the liquid is formed. At still higher temperatures, additional constitutional liquation takes place around the particle until the alloy exceeds the solidus temperature, and bulk melting of the matrix begins. It (constitutional liquation) does not occur during either slow heating or highly rapid heating conditions. Interestingly, during arc welding this unique condition of heating rate exists where constitutional liquation can occur.

9.6.2.3 Post-Weld Heat Treatment

The precipitation-hardened alloys require heat treatment to have strength. The heat treatment cycles are dependent on the base metal and the filler metal used. Table 9.2 provides some typical recommendation of several precipitation-strengthened alloys.

These recommended heat treatments are meant to restore strength by precipitation and reduce residual stresses. Sometimes, multiple heat treatment is given to achieve the desired properties. It, therefore, is difficult to specify a common heat treatment schedule. For creep-limited service, a higher temperature anneal in the range from 1010°C to 1065°C (1850–1950°F) is chosen to produce coarse grains in the base metal, followed by a precipitation heat treatment. If impact strength is to be enhanced, a high-temperature anneal at 1065°C (1950°F) is chosen, and the precipitation treatment is conducted at 760°C (1400°F) for ten hours followed by a furnace cool and a hold at 650°C (1200°F) for 20 hours. These treatments are all variations to the PWHT recommendations listed in Table 9.2.

TABLE 9.2
Typical Recommended Post-Weld Heat Treatments for Several Precipitation-Strengthened Ni Base Alloys

Base Metal	UNS/AWS	Recommended Post-Weld Heat Treatment Procedures
K500	N05500	1100°F/16 hr, furnace cool (15°F/h) to 900°F, air cool.
750	N07750	1300°F/20 hr, air cool.
718	N07718	1325°F/8 hr, furnace cool to 1150°F, Held at 1150°F for 8 hours, air cool.
725	N07725	1900°F/1 hr plus 1350°F/8 hr, furnace cool to 1150°F, held at 1150°F for 8 hours, air cool.
80A	N07080	1700°F/1 hr, air cool to 1380°F, held at 1382°F for 4 hours, air cool.
90	N07090	1700°F/1 hr, air cool to 1380°F, held at 1382°F for 4 hours, air cool.
263	N07263	1472°F/8 hr, air cool.
909	N19909	1325°F/8 hr, furnace cool at 100°F/1 h to 1150°F, held at 1150°F for 8 hours, air cool.

9.7 WELDING PREPARATIONS

Weld preparation starts as early as from the design stage of the components to be welded. It should take into account such factors which ensure the welding operation is smooth and without any hinderance. Once the design considerations, which are outside the purview of this chapter, have been taken care of, the other aspects becomes important. The most of the important ones are discussed here [5, 6].

9.7.1 Cutting and Joint Preparation

Nickel can be cut to prepare the joint edges using the methods as those used for stainless steels. However, oxyacetylene cutting of nickel alloys (without iron-rich powder additions) results in the formation of refractory oxides, preventing accurate and smooth cuts. The choice of the cutting methods will largely depend on the thickness and shape of the parts being cut. Table 9.3 gives a summary of the cutting methods which could be used for the cutting of nickel and nickel alloys.

9.7.2 Surface Preparation

The surface preparation essentially consists of cleaning the surface of the parts being welded since the contaminants may lead to defects of several types. The contaminants on the surface generally comes from the previous operations or during the storage or transportation. They could be of different natures and may affect the welded part differently. Some of the most common contaminants and their methods of removal are discussed here.

TABLE 9.3
Nickel Alloy Cutting Methods

Method	Material Thickness	Comments
Shearing	Sheet/strip, thin plate	Prepare edge exposed to environment to remove tear crevices
Sawing and abrasive cutting	Wide range of thicknesses	Remove lubricant or cutting fluid before welding or heat treating
Machining	Wide range of shapes	Remove lubricant or cutting fluid before welding or heat treating
Plasma arc cutting (PAC)	Wide range of thicknesses	May be used for gouging backside of weld. Grind cut surfaces to clean metal
Powder metal cutting with iron-rich powder	Wide range of thicknesses	Cut less accurate than PAC, must remove all dross
Carbon arc cutting	Used for gouging backside of weld and cutting irregular shapes	Grind cut surfaces to clean metal

9.7.2.1 Oxides and Others on Surface Layers

Oxides on the surface of the parts to be welded are the heavy oxides generated either during the thermal cutting process while preparing the edge or joint or on the surface from earlier exposure of the components/base metal, which are generally thin. The surface oxides and dross generated during the thermal cutting must be removed by grinding or machining or abrasive blasting.

These surface oxides are usually of chromium, nickel, and particularly of titanium and aluminium in precipitation-hardening alloys. They have high melting points which do not fuse during welding. These oxides form either during heat treatment or in case of repair welding, during exposure of the parts to high-temperature applications. These oxide layers are highly tenacious and cannot be removed by mere wire brushing. Grinding, and in some cases machining, is required to remove these oxide layers. These oxides, if not removed, get trapped in the solidifying weld metal, leading to defects that are difficult to detect even by radiography. Surface oxides (or a rough surface) influence the depth of penetration and bead shape even in GTAW.

In some high-temperature service environments, a carburised or sulphurised surface layer may also develop. It is imperative that this surface layer must also be cleaned. Though wrought product forms or castings in the as-received condition are normally free of oxides, it is good practice to clean the surface of a 25 mm (1 in.)-wide band on both the top and bottom surfaces of the weld zone to bright metal with a medium grit flapper wheel or disc. Particular attention must be given to cleaning thin sheets and strips (> 0.05 mm). For cleaning of such thin materials, vapour blasting is a popular process.

While carrying out repair welding, the objective of cleaning is to remove the embedded contamination by grinding, abrasive blasting, or neutralising the surface. Acid-contaminated surfaces are neutralised with a mild basic solution, and alkaline-contaminated surfaces are neutralised with a mild acid solution. A thorough hot water rinse is a must following any neutralising operation. If caustic contamination is not removed before welding, often cracks develop at the weld and HAZ.

9.7.2.2 Contaminating Elements

There are a number of elements which penetrate into the grain boundaries and make the metal brittle. They may cause weld defects or reduce resistance to cracking. A summary of such elements and their deleterious impact and possible source are presented in Table 9.4. Once an alloy has become embrittled, it cannot be restored and has to be scrapped. Contaminants from oil or grease must be removed using solvents. Hand cleaning using a solvent-saturated cloth is used for large welded pieces, whereas small ones may be solvent dip cleaned.

Popular methods are immersion in, swabbing with, or spraying with alkaline emulsion, solvent, or detergent cleaners, or a combination of these. Vapour degreasing, steam, with or without a cleaner, or high-pressure water jetting can also be utilised. Guides as standard practice are available (ASTM A380, Standard Practice for Cleaning, Descaling, and Passivating Stainless Steel Parts, Equipment, and Systems). It is necessary to ensure at the end that complete cleaning has taken place, and the solvents have been removed. Even a residue on the drying cloth can indicate

TABLE 9.4
Embrittling Elements

Elements	Effect/Defect	Common Source/s
Sulphur, carbon	Reduces corrosion resistance	Hydrocarbons such as cutting fluids, grease, oil waxes, and primers
Sulphur, phosphorus	Cracking in welds and HAZ	Marking crayons, paints, and temperature-indicating markers
Lead, zinc, copper (low-melting point metals)	Cracking in welds and HAZ	Tools such as lead hammers, copper hold-down or backing bars, zinc-rich paint, galvanised steel
Shop dirt	Any of these	Any of these

incomplete cleaning. While selecting the solvent, it must be remembered that though a chlorinated solvent does not have any major effect on some of the nickel alloys, but other solvents may pose health hazards. All precautions must be taken to ensure that the solvents are not harmful or toxic to personnel and are not explosive.

All these cleaning requirements are applicable to components prior to any heat treatment as well.

9.7.3 Jigs, Fixtures, and Accessories

9.7.3.1 Fixtures and Positioners

Fixtures are used to hold the parts together throughout the welding and are job-specific, and when fixtures are attached to positioners, there is a further advantage in that welding can be done in the most convenient position. Some of the advantages of using fixtures are summarised as follows:

- Better joint match-up
- Less tacking and welding time
- Minimised distortion
- Accurate assembly

The mating pieces must be carefully aligned and matched for good quality welding. When one member is considerably thicker than the other, for example, a tank head that is thicker than the shell, the head-side should be machined to a taper of 3:1 or more to reduce stress concentrations. Joints with varying root openings require special adjustment by the welder to avoid burn-through or a lack of penetration. When the volume of identical parts is large, use of fixtures is more easily justified.

9.7.3.2 Backing Material

Copper, with its high thermal conductivity, is the material most often used as backing bars. Typical backing bars are designed for use with and without a backing gas. During welding, the copper bar chills the weld to solid metal without it being melted.

Argon when used as the backing gas provides excellent protection to the back side of gas tungsten arc welds. It helps control penetration and maintain a bright, clean undersurface. Nitrogen causes weld metal porosity in welds made with the GTAW process that have inadequate filler metal if used as a backing gas.

When copper backing bar or an inert gas backing purge is impractical, there are commercially available tapes, pastes, and ceramic backing products that can be utilised. These offer some protection from burn-through but little protection from oxidation, so final cleaning by an abrasive means or acid pickling is needed after welding when these backing materials are used.

9.7.4 Tools and Machines

Embedded iron is a serious concern for nickel alloys, which causes localised corrosion. It is, therefore, advisable to use tools that have not been used for other materials. Only stainless steel brushes should be used. Machines such as shears, punches, or rollers must be provided with some sort of cover (e.g., with felt, cardboard, films etc.) so that the workpiece surfaces are not damaged by such equipment due to pressed-in iron particles.

9.8 JOINT DESIGN

This section too does not describe the joint design in general but discusses the salient features which are to be considered for the welding of nickel alloys while designing weld joints [7]. The most economical joint is usually one that requires the minimum of preparation and the least amount of welding consumables and welding time while still resulting in the deposition of a satisfactory weldment. While designing a weld joint for nickel alloys, all these primary considerations are kept in mind.

In addition to these, the most important technical aspect of the design is to ensure that there is sufficient access for the welding torch and that full penetration may be achieved if it is required.

However, certain modifications of the designs may be required based on the welding process. This is accomplished by either increasing the root gap or increasing the included angle. For example, a spray arc GMAW has a deeper penetration weld bead than other arc welding processes so thicker root faces may be used.

9.8.1 Considerations

Molten nickel and its alloys are sluggish in character. It does not flow as easily as other ferrous metals. Therefore, the joint opening must be more than those of the ferrous metals so that the movement of the torch or filler metal is possible, and the operator is able to manipulate the weld puddle to reach the desired location of the joint.

Small included angles and narrow roots promote concave beads, an undesirable condition for nickel alloy welding. A concave bead often causes centerline cracking. Lower penetration is often observed while welding nickel alloys. To counter this, smaller land in the root of the joint is proposed.

9.8.2 Weld Joints

Butt welds should be such as to provide complete joint penetration welds to produce full strength and optimum performance in corrosion service. Beveling is not normally required for a material of 2.36 mm (3/32 in.) or less in thickness. A typical butt joint design is with Gap = 0.8 mm min., 2.3 mm max (0.03–0.093 in.).

Fillet weld generally does not need any joint preparation. However, single bevelled for T- joints between dissimilar thickness of material should have an angle of 45 degrees.

9.8.3 Groove Joints

Groove joints are recommended for materials thicker than 3/32 in. (2.4 mm). For full penetration, there are various types of groove designs such as V-, U-, or J- and should be welded from both sides. Erratic penetration often leads to crevices and voids causing corrosion.

For material over 9.5 mm (3/8 in.) thick, a double V- or double U-joint design is preferred. The increased cost of joint preparation is usually offset by savings in welding products and welding time. The double groove joint design also results in less residual stress than with a single-groove design.

A V-groove joint is normally bevelled to an 80 deg included angle, and U-groove joints to a 15-deg. side and a 4.8–7.9 mm (3/16 to 5/16 in.) bottom radius. The details are Gap = 0.8 mm min., 2.3 mm max (0.03–0.093 in.), Land = 1.6 mm to 2.4 mm (0.062–0.093 in.), Angle = 60° to 80° for a single U groove joint design.

Gap = 1.6 mm min., 3.2 mm max. (0.062–0.125 in.), Land = 1.6 mm to 2.4 mm (0.062–0.093 in.), Radius = 6.4 mm min. (0.25 in.) Half angle 15 Deg. are the details of double V groove joint design.

For a double U-groove joint design Gap = 0.8 mm min., 2.3 mm max (0.03–0.093 in.), Land = 1.6 mm to 2.4 mm (0.062–0.093 in.), Angle = 60° to 80° are recommended.

The details for a J-groove Gap = 1.6 mm min., 3.2 mm max. (0.062–0.125 in.), Land = 1.6 mm to 2.4 mm (0.062–0.093 in.), Radius = 6.4 mm min. (0.25 in.), Half angle 15 Deg. are as suggested. The bottom radius of a J-groove in a T-joint should be 9.5 mm (3/8 in.) minimum.

9.9 WELDING PROCESSES

This section does not describe in detail the welding processes in general but restricts within the scope of impact of welding processes and specific aspects that influence the welding of nickel alloys. It is intended to assist in working out the details of welding procedures for commonly used fusion welding processes used in the welding of nickel alloys [8]. The areas which are common to welding in general are not repeated here.

9.9.1 General Considerations

To develop a welding procedure, the engineer must first ascertain that the products of interest are indeed weldable and study the composition of the alloy, structure, and thermo-mechanical history and all those factors that affect its response to welding [6].

It is a very good practice to ensure that all weld beads have slightly convex contour, and flat or concave bead contours should be avoided while welding nickel alloys.

Preheating nickel alloys prior to welding is not normally required except when it is too cold (2°C or less). In that case, heating to an extent of about 300 mm of weld location to at least 10°C above the ambient temperature to prevent any condensation is recommended. These condensations often lead to the formation of porosity in the weld. Preheating of a steel component may be required when joining with a nickel alloy. Preheat is often beneficial when joining iron casting using a nickel consumable.

9.9.2 Tack Welding

Before welding starts, tack welding is required to maintain a uniform gap along the entire joints if not held in fixtures. The placing of tack should be such as to minimise the effect of shrinkage. While fitting two sheets, tack welds should be placed at each end and then in the middle section.

Carbon steel and nickel alloys generally have similar number and size of the tack welds. The length of tack welds may be as small as 3 mm (0. 125 in.) to over 25 mm (1 in.) long depending upon the thickness of the base metal. It is important to note that the shape of the tack does not cause a defect in the final weld. Heavy or high tack or abrupt starts and stops should be contour-ground before the final welding starts. Tack welds must be wire-brushed or ground to clean. They should be inspected for crater cracks, and if any crack is found, it should be ground out as well.

9.9.3 Welding Processes

Nickel alloys can be joined reliably by all of the welding processes or methods, with the exception of forge welding and oxyacetylene welding. The wrought nickel alloys can be easily welded under conditions similar to those used to weld austenitic stainless steels. However, cast nickel alloys, with a high silicon content, may present some difficulties in welding.

Nickel alloys are usually welded in solution-treated conditions. While welding, precipitation-hardenable alloys should be annealed if they were subjected to any operations that could introduce high residual stresses.

The most common welding processes for welding solid solution-strengthened wrought nickel alloys are shielded metal arc welding (SMAW), gas-tungsten arc welding (GTAW), gas-metal arc welding (GMAW), and flux-cored arc welding (FCAW). Submerged arc welding (SAW) and electro slag welding (ESW) have limited applicability, as does plasma arc welding (PAW). Although the GTAW process is preferred for welding of precipitation-hardenable alloys, both the GMAW and SMAW processes are also sometimes used.

9.9.4 Interpass Temperature

A maximum interpass temperature of 175°C (350°F) is widely used although one base metal producer is more conservative and recommends a maximum of 95°C (200°F).

9.9.5 Shielded Metal Arc Welding (SMAW)

SMAW is the most widely used manual welding process used for the welding of nickel alloys. The process details are well known and not described here. It is most suited for materials about 1.6 mm (1/16 in.) and above in thickness. Thinner materials, however, can be welded by the process if appropriate jigs and fixtures are used.

9.9.5.1 Electrode Types

The primary consideration in selecting an electrode for SMAW process is to ensure that the weld deposit matches the mechanical, chemical, and metallurgical compatibility and corrosion properties of the base metal. This is generally achieved through studying the chemical composition of the undiluted weld deposit by an electrode as given by the manufacturers and are generally governed by the classification of the electrode given in national and international standards. Overmatching, undermatching, or non-matching composition of the weld metal may be sometimes required for specific applications.

Classification of nickel alloy-covered electrodes is done as per their chemical composition of undiluted weld metal. The most popular classification system followed is as per American Welding Society (AWS) under ASME Part II Section C SFA 5.11. The details of recommended electrodes for some of the popular base metals are given in Table 9.5 [9]. The composition of the coatings on the electrodes is though an intellectual proprietary item but is classified based on its generic composition. Nickel alloy electrode coatings are best described as lime–titania type coatings; they cannot be classified as either lime or titania types because both ingredients are used.

TABLE 9.5
Matching Composition Filler Metals for Welding Nickel Alloys

Base Metal		Bare Electrode and Rod		Shielded Metal Arc Electrode		Flux-Cored Electrode	
Alloy	**UNS**	**AWS A5.14**	**UNS**	**AWS A5.11**	**UNS**	**AWS A5.34**	**UNS**
200	N02200	ERNi-1	N02061	ENi-1	W82141		
201	N02201	ERNi-1	N02061	ENi-1	W82141		
400	N04400	ERNiCu-7	N04060	ENiCu-7	W84190		
R-405	N04405	Note 1		ENiCu-7	W84190		
825	N08825	ERNiCrMo-3	N06625	ENiCrMo-3	W86112	ENiCrMo3Tx-y	W86625
G-3	N06985	ERNiCrMo-9	N06985	ENiCrMo-9	W86985		
G-30	N06030	ERNiCrMo-11	N06030	ENiCrMo-11	W86030		
33	R20033	AWS A5.9 ER33–31	R20033				
600	N06600	ERNiCr-3	N06082	ENiCrFe-3	W86182	ENiCr3Tx-y	W86082
690	N06690	ERNiCrFe-7A	N06054	ENiCrFe-7	W86152		
625	N06625	ERNiCrMo-3	N06625	ENiCrMo-3	W86112		
C-2/622	N06022	ERNiCrMo-10	N06022	ENiCrMo-10	W86022	ENiCrMo10Tx-y	W86022
C-4	N06455	ERNiCrMo-7	N06455	ENiCrMo-7	W86455		

TABLE 9.5 *(Continued)*
Matching Composition Filler Metals for Welding Nickel Alloys

Base Metal		Bare Electrode and Rod		Shielded Metal Arc Electrode		Flux-Cored Electrode	
C-276	N10276	ERNiCrMo-4	N10276	ENiCrMo-4	W80276	ENiCrMo4Tx-y	W80276
59	N06059	ERNiCrMo-13	N06059	ENiCrMo-13	W86059		
686	N06686	ERNiCrMo-14	N06686	ENiCrMo-14	W86686		
2000	N06200	ERNiCrMo-17	N06200	ENiCrMo-17	W86200		
B-2	N05500	ERNiMo-7	N10665	ENiMo-7	W80665		
B-3	N10675	ERNiMo-10	N10675	ENiMo-10	W80675		
B-4	N10629	ERNiMo-11	N10629				

Notes: 1. SMAW is the preferred welding process. 2. Group IV—Precipitation-hardening alloys. Contact the base metal producers for filler metal recommendations.

9.9.5.2 Electrode Storage

Flux-covered electrodes should remain sealed in their moisture-proof containers in a dry storage area before use. All partly used containers of electrodes should be stored in a cabinet with a desiccant or reheated to 6–80°C (10–150°F) above the highest expected ambient temperature. The flux coating is usually hydroscopic in nature and will absorb moisture. Therefore, it needs rebaking at about 315°C (600°F).

Proper storage of electrodes is integral to a good weld product. It is necessary that the welding electrodes are stored as prescribed by the manufacturer for its optimum use. Nickel alloy electrodes are generally supplied in hermetically sealed containers and are suitable for long storage period. Once the sealed container is opened for use and needs to keep the remaining unused electrodes, they should be stored in a heated container as recommended by the manufacturer and described earlier. Higher-temperature baking may be required if they get over-exposed to moisture. However, the general guidelines for storage and reuse are:

- Storage of opened electrodes: 110°C (225°F)
- Recondition bake: 260–315°C (500–600°F)

The low hydrogen type electrodes such as nickel molybdenum-coated ones need special attention to moisture pick-up. For such electrodes, the reconditioning temperature is 315–370°C (600–700° F) for two to three hours.

The moisture in the coating contributes to hydrogen gas generation during welding leading to weld porosity. This is more prominent in fillet welds than in butt welds.

9.9.5.3 Welding Current

Most nickel alloy electrodes are designed to operate on direct current electrode positive (DCEP), although some can operate on alternating current.

Each electrode diameter (core wire diameter) has an optimum range of operating current specified by the manufacturer. When operated within the specified current range, the electrodes have good arcing characteristics and burn with a minimum of spatter. Outside the specified range, the arc becomes unstable and the electrode tends to get overheated before it is fully consumed. Excessive current can also lead to porosity, compromised properties, and bend test failure because some of the alloying elements and deoxidisers are destroyed (oxidised) before they can be melted into the weldment.

The current density required for a given joint is influenced by many variables such as material thickness, welding position, type of backing, tightness of clamping, and joint design. For example, a slight reduction in current (5–15A) is necessary for overhead welding. Vertical welding requires 10–20% less current than welding in the flat position.

The operating current range is specific to the type and diameter of the electrode (diameter of the core wire) and is recommended by the manufacturer. Usually, they are printed on the package. Operating beyond the recommended current range overheats the electrodes and may cause reduced arc force and difficulty in controlling the arc near the end of the electrode.

9.9.5.4 Striking and Restriking the Arc

Striking the arc on the component surface should be avoided. Arc striking should be done in the seam area, e.g., on the seam flanks or on an outlet piece.

When the welder needs to break the arc, it should first be shortened slightly and the travel speed increased to reduce the puddle size. This practice reduces the possibility of crater cracking and oxidation.

Restrike also has significant influence on the soundness of the weld. A reverse or "T" restrike is recommended. This restrike method has several advantages. It establishes the correct arc length away from the unwelded joint so any porosity resulting from the strike is avoided.

Arc starting and stopping techniques are like those used for any low-hydrogen-type electrode. Some of the useful tips are given here:

- Strike the arc at some point in the joint so that the initial metal is remelted. An arc strike away from the weld may have cracks and, unless removed, may result in lower corrosion resistance in that area.
- Do not abruptly extinguish the arc, leaving a large weld crater. A depression will form as the metal solidifies, often with a slag-filled pipe or cracks in the centre of the crater depression. One acceptable technique is to hold the arc over the weld pool for a few moments and then move quickly back, lifting the arc from the completed weld. Another technique is to extinguish the arc against one of the joint side walls after filling the crater.
- Restrike the arc after a stop in the welding process at some point in the joint so that the initial metal is remelted.
- The preferred weld bead contour is one that is slightly convex to minimise the chance of centreline cracking. Centreline cracking sometimes occurs in fillet welds resulting from a concave bead contour.

9.9.5.5 Welding Procedure

As has been stated earlier, Ni and Ni alloy weld metals do not flow and spread as freely as steel weld metal. Therefore, the operator must direct the flow of the molten metal so that the weld metal wets the joint sidewalls, and the joint is filled appropriately. This is sometimes accomplished by slight weaving the electrode. The amount of weave will depend on such factors as joint design, welding position, and type of electrodes used. A stringer bead deposited without weaving may be used for single bead work or in close quarters on thick sections such as in the bottom of a deep groove. However, a weave bead is generally desirable. When a weaving is used, then the width of the weave should not exceed three times the electrode core diameter. Regardless of whether the welder uses weaving or the straight stinger technique, the contour of the bead should be convex.

Starting and restriking of an arc must follow the procedure mentioned earlier.

9.9.5.5.1 Weld Puddle Control

To eliminate slag entrapment and to get the desired bead contour, it is necessary to control the weld puddle. In downhand welding, the electrode is positioned ahead of the puddle (backhand welding). The electrode is generally inclined to the vertical by about 20 degrees for easily visibility of the puddle and its control.

9.9.5.5.2 Out-of-Position Welding

While welding out of position, smaller-diameter electrodes (3.2 mm/0.125 in. or less) should be used. The arc should be maintained short and the current on the lower side than while welding downhand welding. The electrode angle will have to be adjusted with $\pm$ 20° to vertical depending upon backhand or forehand welding.

The exact placing of the electrode is important since nickel alloy flowability is low and needs weaving for proper placement of the bead. The extent of weave will depend on many factors such as joint design, welding position, and the type of electrode to get the proper bead contour. As a general rule, the weave should not be wider three times the diameter of the electrode.

9.9.5.5.3 Weld Spatter

When the correct welding parameters are used not much of spatter is expected. A welding operator should be able to control the parameter such as excessive arc length, excessive amperage, incorrect polarity, or excessive moisture in the electrode coating which causes high spatter. Spatter due to magnetic blow can be controlled using similar techniques as that for ordinary steel.

9.9.5.5.4 Weld Slag Removal

The light cover of slag over the weld is inherent to the process as a desired requirement. Most often, it can be easily removed with hand tools or a wire brush. It could be self-peeling or if adherent may have to be removed using a chip hammer. It is essential to remove the slag layer before deposing the subsequent passes, or else it may result in slag entrapment.

9.9.6 Gas Metal Arc Welding (GMAW)

The general advantages of GMAW over GTAW and SMAW are faster welding speeds, no slag, meaning minimising post-weld clean-up, ease of automation, and good transfer of elements across the arc are also applicable to nickel alloy welding. Most of the nickel alloys can be welded using the GMAW process. Any of the three metal transfer modes, viz. spray, short circuiting, and pulse mode could be used, giving good results. The globular mode is not recommended for welding of nickel alloys, as this has an erratic arc that often gives inconsistent penetration and uneven bead contour.

To improve visibility and gas protection, smaller nozzles are recommended. An alternative is the use of gas lens which creates a laminar flow by using special screens inside the nozzle.

The criterion for the selection of the mode of transfer is similar to those for ferrous metals. To weld thinner sections [up to about 3.2 mm (1/8 in.) thick] short-circuiting arc is useful, as this generates a low heat input. This is also most suitable for single-pass welding. For welding of thicker sections and multi-pass welding, short-circuiting arc is not suitable, as it is prone to cause fusion defects. Spray transfer generates high heat input and is suitable for thicker sections and high deposition rates. However, this is generally limited to flat position welding. Table 9.6 gives guidelines for the setting of spray transfer. A gas flow of 35–60 ft^3, polarity DCEP is recommended for all these cases.

9.9.6.1 Shielding Gases

The selection of shielding gas is dependent on the metal to be welded and the type of metal transfer used. Most of the nickel alloys are welded using pure argon, helium, or a mixture of the two. Argon gas is preferred in manual welding which gives good penetration and less chance of melt-through at lower gas flow rates than helium. For automatic welding, where higher speed of welding is possible, helium is recommended since it produces higher heat input and deeper penetration. An improved bead contour and wettability are seen while using a mixture of argon and helium. The addition of hydrogen (up to 5%) produces a hotter arc, which produces a uniform bead surface in single-pass automated welds. While welding cast iron using nickel alloy filler metal, carbon dioxide or a mixture of argon and carbon dioxide could be used. Oxygen and nitrogen addition should be avoided. Oxygen addition creates porosity in alloys such as pure nickel and Monel alloy welds, whereas nitrogen

TABLE 9.6
Guidelines for Settings for Spray Transfer GMAW

Wire Diameter (in.)	Wire Feed (in./min)	Voltage (V)	Current (A)	Gas
0.030	550–700	26–32	175–240	Argon or in some cases Ar 25%-He
0.035	425–600	26–32	175–300	
0.045	250–350	26–32	225–325	
0.062	125–220	27–33	250–350	

addition has a deleterious effect on tungsten electrodes. Some of the most popular shielding gases used in GMAW are shown in Table 9.6.

9.9.6.2 Filler Metals

The recommended filler metals are given in Table 9.5. The selection of the wire diameter depends on the type of metal transfer selected and the thickness of the base material. The diameters used are usually 0.9 mm, 1.2 mm, and 1.6 mm (0.035 in., 0.045 in., and 0.062 in.) though other diameters are also in use for special occasions. In general, 1.2 mm (0.045 in.) diameter wire is used with spray transfer and 0.9 mm (0.035 in.) and 1.2 mm (0.045 in.) with short-circuiting transfer.

To achieve the desired properties, at least 50% of the weld metal should be of filler metal. A thorough mixing is therefore necessary to ensure no segregation and uniform property throughout the weld joint.

9.9.6.3 Current

For most welding processes, DCEP is used. This current gives deeper penetration than DCEN and therefore, is the most suited for overlay welding. Thinner sections too are welded using DCEN. In practice, DCEN is used for the welding of all nickel alloys irrespective of the mode of metal transfer.

The current to be used for spray transfer should be more than the transition point (the value at which transfer changes from the globular to the spray mode). The value of the transition point depends on the wire diameter, wire composition, and the power source characteristics. Separate slope and inductance control facilities in the welding equipment are required when short-circuiting transfer is used. In pulse mode transfer, the current usually is in square wave pattern. The power source for welding in the pulsing transfer mode is greatly different. At the peak current, it is spray transfer, whereas at the background current (lower current) the transfer is in globular range. This process can also be used in out of position unlike pure spray transfer. The best compromise between spray arc and short-circuiting arc modes is pulse arc mode and recommended for the general fabrication of nickel alloys.

9.9.6.4 GMAW Equipment

There is no special requirement for welding equipment for the welding of nickel alloys over those used for the welding of steel. For all GMAW, constant voltage power sources are recommended. The wire feed conduit should be plastic lined to reduce drag. The synergic pulse power source is simpler to operate where there are very few dials to adjust.

Power sources for welding nickel alloys in the short-circuiting mode of transfer should have good slope control. The current generation of power sources for GMAW in the pulsing transfer offers excellent solid-state controls and very pleasing welding characteristics. Their arc wave control makes them particularly useful for out-of-position welding.

9.9.6.5 Torch

To weld nickel alloys, the torches should be correctly rated. They are generally air-cooled. Often, wire feed problem is experienced due to overheating of the torch.

When inert gas is used for shielding, the torch rating is reduced to approximately half of the standard duty cycle rating with carbon dioxide since carbon dioxide itself acts as a cooling agent.

9.9.7 Gas Tungsten Arc Welding (GTAW)

GTAW is the most widely used process for welding thin sections of nickel alloys. Materials even as thin as 0.05 mm (0.002 in.) can be welded using the GTAW process. GTAW process is the process of choice used for welding precipitation-hardened alloys. GTAW is performed with DCEN.

The inert shielding gas is usually argon. Filler metal in the form of bare wire is fed into the arc as needed, either manually or automatically. The process details are not covered in this chapter, which remain same for all other metals.

The following summarises some of the advantages of using GTAW process for welding nickel alloys:

- Practically no post-weld clean-up to remove slag
- All position welding capability; particularly useful for pipe welding
- No weld spatters
- No alloy loss during welding
- Good as-welded surface; minimal finishing required

Some of the aspects of importance in nickel alloy welding are discussed here.

9.9.7.1 Shielding Gases

The choice of shielding gas is similar to that of GMAW. Nickel alloys are not very reactive, but it is always advisable to amply shield both the weld puddle and the backside. The filler metal should preferably also remain within the gas cover during welding.

A number of gases and their mixers are used as shielding gases in this process. Some of them are helium, argon, or a mixture of the two. Small quantities (up to 5%) of hydrogen can be added to argon, which produces a hotter arc and more uniform bead surfaces for single-pass welding. The use of hydrogen is normally limited to automatic welding. It enables higher travel speeds up to 40% over those achieved with only argon.

While using only helium, a higher travel speed is possible because of the higher heat input generated due to longer arc length. However, at lower welding currents (below 60 A), the arc is more difficult to start and maintain while using only helium gas. When joining small parts or thinner materials where lower currents are required, the problem could be overcome either by using argon gas shielding or by using a high-frequency current arc-starting system.

Nitrogen additions are not recommended because of rapid deterioration of the tungsten electrode and also because they introduce the possibility of weld metal porosity in the non-chromium-bearing nickel alloys.

For a proper shielding of the weldment, gas flow rate is critical. Low gas flow rates will not protect the weld, while high rates can cause turbulence and aspirate air, thus, making the gas shield ineffective. A 0.28–0.57 cu.m./hr (10–20 cu.ft/hr) is typical for manual welding using argon gas as shield. Automatic welding may require considerably higher rates of gas flow. Helium should flow at 1½–3 times the rates for argon to compensate for helium's higher buoyancy. To ensure better shielding and visibility, the largest gas cup practical for the job at the minimum practical distance from the work should be used.

The shielding gas should be pure, or else it may cause porosity. Such contaminations are usually from air, which is often sucked into the shielding envelope or may be due to draft, wind etc. Loose fittings of the electrode extension cap of the gas cup may also cause air being sucked in. Use of a gas lens will help in avoiding such contamination.

9.9.7.2 Electrodes

Tungsten electrodes or those alloyed with 2% thorium may be used. The alloyed electrodes have a longer life with greater current-carrying capacity that make them more economical in the long term. It is important to avoid overheating of the electrodes at excessive current levels.

The depth of penetration and the width of the bead, especially with welding current over 100 A, are significantly influenced by the electrode tip shape. A tapered tip produces the best arc stability and penetration control. For most work, the vertex angle should be between 30 degrees and 120 degrees with an area of about 0.38 mm (0.015 in.) diameter at the tip. Larger angles (blunter tips) can be used to produce narrower beads and deeper penetration.

9.9.7.3 GTAW Equipment

A GTAW equipment is in no way different from the standard ones. It is though preferred to have it fitted with high-frequency arc start system. It eliminates the need to contact the work with the electrode to start the arc. Contact starting can damage the electrode tip and may result in tungsten inclusions in the weld metal. Another advantage of a high-frequency circuit is that the starting point can be chosen before the welding current starts, which eliminates the possibility of arc marks on the job. Units with stepless control are preferred over those using step reduction.

9.9.7.4 Current

DCEN is recommended for both manual and automated welding of nickel alloys. A high-frequency circuit assists in starting the arc, and a current-decay unit for slowly stopping the arc should also be used with GTAW nickel base alloys. Contact starts and "pull away" arc stops techniques should not be used. An abrupt arc break may cause rough, porous, or fissured craters. The use of a current decay unit helps in the gradual reduction of current and thereby eliminating these defects.

To weld thinner material or with poor fit-up, pulsing current mode is most suited. Pulsed-current is also useful in making the root pass of pipe joints as well.

9.9.7.5 Current Controls

Besides the power source control, additional current control devices are often provided as foot pedal or torch-mounted hand current control devices. These devices help in on-job current controls such as poor fit-ups. A further advantage to this type of control is that the welder can slowly reduce the current (and the weld pool) at arc stops to eliminate crater cracks.

9.9.7.6 Filler Metals

The recommended filler metals for GTAW welding of nickel alloys are shown in Table 9.5 as per AWS A5.14. However, it is always advisable to verify from the manufacturer of the filler metal for its suitability for the purpose.

Straight lengths are normally used for manual welding, and spool or coil wire is for automatic welding. Bare wire for GTAW should be wiped clean before using and stored in a covered area. It should be noted that the core wire of coated electrodes used in SMAW should not be used as GTAW filler metal.

Nickel alloy filler metals are not exactly the same as that of the base metal. It often contains elements that are not in the base metal to control porosity or to improve resistance to cracking. At least 50% of the weld metal (preferably 75%) should be from the filler metal to get the required composition. Proper mixing of the filler metal is very important to avoid segregation.

9.9.7.7 Arc Initiation

As described earlier, high-frequency arc start system is the best. In its absence, the technique of striking a pilot arc or lift arc techniques may be used. A scratch start arc technique is though not desirable, but if it had to be used, then the use of staring tabs adjacent to the weld joint is useful to eliminate damage to the base metal.

9.9.7.8 Arc Stopping

Arc extinguishing is an important step to check crater crack. Good arc-stopping practice is particularly important in the root pass of welds that are welded from only one side. If cracks occur in this situation, they may extend completely through the root. In the absence of any system for the current decay, decrease the arc pool by increasing the travel speed before lifting the electrode from the joint. After the arc is broken, hold the torch over the crater for several seconds to allow the weld to cool under the protection of the argon atmosphere.

9.9.7.9 Welding Procedure

A slight inclination to 90 degrees to the work in the forehand position is necessary for good visibility during manual welding. However, it should not be too acute an angle, as it can cause aspiration of air into the shielding gas.

The electrode extension beyond the gas cup should be as short as possible. It varies with the joint design. For example, an extension of 4.8 mm (3/16 in.) maximum is used for butt joints in thin material, whereas 9.5–13 mm (3/8–1/2 in.) may be required for fillet welds. Minimum arc length as possible is recommended. When no filler metal is added, it should be 1.27 mm (0.05 in.). A longer arc during autogenous GTAW can cause porosity. It should be of minimum length for the thicker material. The filler metal is added at the leading edge of the puddle, and care must be taken not to contact the tungsten electrode. Any agitation of the puddle must not take place to prevent possible burning of the oxidising elements.

In autogenous welding (i.e., where no filler metal is added), an optimum welding speed must be maintained, or else there is high possibility of porosity in the weld. Grooved back-up bars or inert gas backing is required where full penetration welding is made. In their absence, the weld underside will likely discolour (oxidise), and porosity will appear.

9.9.7.9.1 Arc Shielding

Nickel alloys are not as reactive as titanium or zirconium. Even then it is a good practice to shield the weld puddle, backside, and filler metal.

9.9.7.9.2 Filler Metal Addition

Inadequate filler metal addition is one of the constraints of this process. Sometimes, it results in a concave bead, an undesirable contour, and centreline cracking. In non-chromium nickel alloys, it may even cause porosity.

9.9.8 Flux-Cored Arc Welding (FCAW)

Though the FCAW process has many advantages, the use of this process is still not very widespread due to constraints in the availability of welding consumables. FCAW process is becoming more widely accepted for the welding of nickel alloys.

The FCAW process is very similar to the GMAW process. The feature that best distinguishes this process from other semi-automatic arc welding processes is the flux in the core of the filler metal. The electrode in this process uses flux-filled tubular electrode that provides shielding of the weld metal. Sometimes, additional shielding is provided by using inert gas. This depends on the composition of the flux. The type of flux also determines whether or not the process can be used in the downhand position or out-of-position welding. Most of the nickel alloy electrodes use a shielding gas of CO_2 or Ar–CO_2 mixture. This process has higher productivity than GMAW and SMAW. Many of the electrodes can be used in all positions.

9.9.8.1 Electrode/Wire

Currently, the specifications includes only chromium-bearing solid solution alloys, but other nickel alloy products such as a nickel–copper electrode are commercially available. The electrode manufacturer should be consulted for shielding gas and amperage range recommendations. Before use, the electrode should be properly stored to prevent moisture absorption. A wire which is not being used should be stored with a desiccant or heated to 6–8°C (10–15°F) above the highest-expected ambient temperature. Flux-cored wires which have absorbed moisture may be re-baked at 204°C (400°F) for about six hours.

9.9.8.2 Welding Equipment

There is no difference in the power source used in FCAW and GMAW or SAW. Direct current power sources with constant potential are generally used for welding of nickel alloys. However, there is a need for knurled and U- or V-grooved wire feed rolls to avoid flattening of the wire, as it could happen in smooth-roll wire feeders.

The use of four-roll wire feeders is more reliable. Contact tips should be of the same diameter as that of the wire used.

9.9.8.3 Shielding Gas

Additional gas coverage for shielding is mostly used with FCAW wires. Argon, carbon dioxide, or a combination of the two are the most common shielding gases used. A gas flow rate of 25–50 ft^3/hr (0.7–1.4m^3/hr) is recommended.

The self-shielding type wires contain such ingredients which provide the protective shielding to the puddle. This type of wires is more useful in field welding, where providing gas cover is difficult. The flux also assists in producing a slightly convex bead shape or profile.

9.9.8.4 Flux

The slag from the flux in the wire protects the deposit from reacting with air, shapes the bead contour, stabilises the arc, and deoxidises the molten weld puddle. Also, the flux assists in producing a slightly convex bead shape or contour. The flux ingredients are so formulated as to make the cored wires operable for downhand or out-of-position welding. All position wire slags freeze more quickly and generates a smaller weld puddle. The slag should be removed from the deposit before the next pass reaches, as described in GMAW section.

9.9.8.5 Welding Parameters

Welding parameters for FCAW are similar to those of SMAW. The weld finish and deposition rate are highly influenced by the current setting. DCEP is used while welding using this process. An increase in current increases penetration and dilution. Whereas, a decrease may improve out-of-position welding and lowers dilution. Excessive current is unwanted, as it causes a lack of slag cover, increased spatter, and increased crack susceptibility.

High voltage is also not desirable, as it increases the arc length and thereby increases the chance of porosity. Weld spatter and bead width also increase. Nickel alloys are welded using short arc length with an average welding voltage of about 28 volts.

TABLE 9.7
Recommended Welding Parameters for Flux-Cored Wire

Wire Diameter In./mm	Wire Feed In/ min (m/min)	Welding Current - DCRP (A)	Welding Voltage (V)	Electrode Extension (stick out) In./mm
0.093 (2.4)	100–200 (2.5–5)	250–350	28–33	0.75–1 (19–25)
0.078/2.0	150–250 (3.75–6.3)	225–325	28–32	5/8–7/8 (16–22)
0.062 (1.6)	200–300 (5–7.5)	200–300	27–32	½–3/4 (12.5–19)
0.045 (1.1)	250–350 (6.3–9)	130–180	26–30	3/8-5/8 (9.5–16)

For welding of nickel alloys using self-shielding electrodes, a longer electrode extension is used, whereas shielding types use a shorter extension. A shorter extension may be useful for better penetration and reduced spatter. A longer extension causes overheating of the wire, leading to excessive spatter.

The impacts of travel speed are similar to SMAW process. It is adjusted depending upon the requirements of penetration and dilution since the travel speed affects the heat input. The torch angle should preferably be perpendicular to the work; however, a slight drag or push may be used based on the joint design. Table 9.7 gives the welding parameter recommendation for this process.

9.9.8.6 Welding Procedure

Welding using the FCAW process can yield high-quality welds if proper procedures are followed. The welding of nickel alloys requires a clean base metal before welding even though the flux in the wire does assist in this action. The surface of all nickel base alloy metals should be free from oxides, especially scale formation from heat treatment. A light sanding or grinding may be used to clean the surface. The slag from the flux may get entrapped and through proper joint design may reduce the risk.

9.9.9 Submerged Arc Welding (SAW)

To weld thicker sections (over 6.5 mm or 0.25 in.) SAW is the most suitable process, as it has very high deposition rate. This process is used in the downhand position also extensively for the overlays. This process generates less amounts of fumes and so is more operator-friendly.

9.9.9.1 Flux

For most nickel alloys, welding flux is available in the market, though AWS specification does not exist. Like welding of any other alloy using SAW process, the flux melts and protects the weld metal from interaction with atmosphere, provides stable arc, and may add alloying elements to the weldment. In most nickel alloys welding fluxes are of mechanically mixed type.

Submerged arc fluxes are somewhat hygroscopic and susceptible to moisture absorption so it must be stored in a dry area. In certain cases, re-baking may be required. Reclaimed flux is generally re-baked at 315°C to 480°C (600° to 900°F) for two hours. Unfused flux is usually reclaimed by recycling. Although continual recycling of the flux can alter the balance of fine/coarse particles ratio.

Flux burden is an important parameter to get a good weld. Excessive amount of flux may cause craters and embedded flux. The flux should be added at the front of the torch and not into the arc, as it may cause excessive flux burden and cause the defects as described earlier.

9.9.9.2 Filler Metal

The filler metal specification as per AWS is same as those used for GMAW, i.e., A5.14 and shown in Table 9.5. Though the wire diameters generally used in SAW process are bigger than the wires used in GMAW. The range varies from 1.1 mm to

2.4 mm (0.045–0.93 in.). The most common diameter used is 1.6 mm. To achieved the desired weld property, it is the wire–flux combination which is important.

9.9.9.3 Current

Nickel alloy welding may be carried out using both the polarities, i.e., DCEN and DCEP. DCEP is preferred for welding butt joints, whereas DCEN is used for overlaying purpose with oscillating technique which gives a higher deposition rate and lower penetration. DCEN requires a deeper flux burden which results in increased flux consumption. Further modification of SAW is in use for overlays of large surface with thin strips. Electro slag welding (ESW) is another variation of SAW having still higher deposition rates.

Depositing stringer beads with DCEN is not recommended for nickel alloy welding.

9.9.9.4 Welding Procedure

To weld nickel alloys by the SAW process, manual welding is not recommended, as the viscous nickel alloy puddle requires precise electrode positioning. Almost always SAW is carried out using automatic equipment. Si pick-up during welding is a major concern, and so a flux giving low Si should be used while welding nickel alloys. As already mentioned, the wire–flux combination is critical. The manufacturers must be consulted for a suitable combination.

Slag entrapment is a regular problem but can be reduced by appropriate joint design and bead placement. In a multi-pass welding, beads should be placed so as to provide an open or reasonably wide root for the next bead. Bead contour can be controlled by travel speed and voltage to get a desired convex shape.

SAW is one of the most suitable processes to weld heavy sections of nickel alloys which has about 50% higher deposition rate than the GMAW process. The SAW process is operator-friendly and less hazardous to them, as they are not exposed to arc. Besides, this process generates fewer fumes.

9.9.10 Other Welding Processes

There are other welding processes such as plasma arc, electron beam, and laser welding which are used for welding nickel alloys but not very widely. The resistance welding processes like spot, seam, projection, and flush welding are though suitable for most nickel alloys, but they find limited use in welding for corrosion-resistant services.

9.9.11 Welding Nickel Alloy Pipes

Nickel alloy pipes are of vital importance in industry. Most of them are of small inner diameter. Welding techniques for pipe are significantly different from that of large-diameter equipment such as a pressure vessel. The major difference is that in most pipes it is very difficult to have access to the inner side. There is not much difference in the welding of nickel alloy pipes from and that of other steels.

The use of GTAW orbital pipe-welding units for welding nickel alloy is reasonably common. Internal purging is used to prevent the formation of oxide at the underpass. To carry out circumferential welding of ductile cast iron pipes, NiFe welding consumables are used.

The thinner wall chromium-bearing solid solution alloys can readily be autogenously welded and filler metal added on heavier wall tubes to complete the weld after the root pass. Nickel and nickel–copper solid solution alloys are more sensitive to porosity from gases such as oxygen, nitrogen, and CO_2 in making autogenous welds. Therefore, the addition of filler metal or the use of consumable inserts may be required for nickel and nickel–copper solid solution alloys.

Manual welding is more common for welding of nickel alloy piping. There are three options in the manual welding of nickel alloys:

- The use of backing rings
- Consumable inserts
- Open root joints with hand-fed filler metal

Backing rings are not much recommended due to increased possibility of slag entrapment between the backing ring and the pipe. The rest of the two procedures give high-quality root welds. The standard nickel consumable insert shapes are produced as per ANSI/AWS A5.30. Class 1, 3, and 5 (inverted T, I, and T shape respectively). Classes 3 and 5 are often easier to fuse than the larger volume of Class 1.

9.9.12 Strip Welding Overlays

Electroslag strip welding (ESS) and SAW surfacing (SAS) are two of the good overlaying welding processes. The ESS and SAS processes utilise identical equipment and always use DCEP. Both processes offer high-deposition rates 30–50 lbs/hr and are limited to use in flat 1G position. The suggested parameters for ESS are presented in Table 9.8.

9.9.13 Dissimilar Welding

One of the most common uses of nickel alloys is the welding of dissimilar metals. The combinations of dissimilar metals are too many to recommend individual

TABLE 9.8
Typical Parameters for ESS

Wire Size In,/mm	Current (A)	Voltage (V)	Travel speed In,/min (mm/min)	Strip speed In./min (mm/min)	Strip Stick out In./mm	Flux Burden In./mm
1.2/30	600–800	24–25	5–8 (127–203)	60–70 (1524–1778)	1–2 (25–50)	0.5–11 (3–25)
2.4/60	100–1300	24–25	5–8 (127–203)	60–70 (1524–1778)	1–2 (25–50)	0.5–1 (13–25)

combination. The welding is also very complex and difficult to predict the result, particularly dilution.

9.9.13.1 General Guidelines for Welding of Dissimilar Metals

Dissimilar welding involves complex metallurgical considerations. The final composition of the weld deposit depends not only on the composition of the electrode or the filler metal but also onthe amount of dilution of elements from the two base metals joined, the process, and the technique used. All of them determine the selection of welding material that will produce the properties required by the application. It is possible that more than one welding product meets the requirement. Then the choice is on the user to pick the product.

Sometimes, the suitable welding product may have to be selected based on its ability to accept dilution from the base metal without causing any harmful effect. Very often, high nickel welding products are used while welding nickel base alloy with another material. Dilution of nickel by copper, chromium, and iron are generally of concern. If the other metal is steel, then a buffer layer of nickel (min. 93% Ni) should be applied over the steel to restrict dilution of the nickel base metal by iron. The dilution by iron should not be more than 5% in Cu–Ni weld deposit, whereas in SMAW deposits of Ni–Cr, the acceptance limit is 50%. While joining cast material which usually contains higher silicon, the dilution by it should not exceed 0.75%.

Non-matching or over matching nickel consumables are often used to weld dissimilar metals to meet specific mechanical or environment requirements. The use of high molybdenum or Inconel welding consumable for welding of super austenitic steel or Ni–Cr for 9% steel for cryogenic applications is to offset the elemental segregation of elements during solidification. The examples are numerous to show the versatility of nickel alloy weld metals in dissimilar welding applications.

9.9.14 Practical Welding Hints

In many respects, the welding methods of stainless steels and nickel alloys are similar with a few differences which have been highlighted in the preceding sections. They can be summarised as follows:

- Tack welds must be done with short intervals.
- Nickel has a coefficient of thermal expansion less than that of stainless steel; so it is recommended that the distortion and distortion control measures to be applied should be similar to those of carbon steel.
- The nickel weld pool is sluggish and does not flow as freely as carbon or stainless steel. This may result in a lumpy and very convex weld bead and a poor toe blend unless the welder manipulates the weld pool to avoid such defects.
- Although stringer beads are recommended, a slight weave to assist the weld metal to wet the side walls of the preparation is beneficial. If weaving, the oscillation should be limited to 2.5 times the core diameter of the electrode. A further characteristic of nickel alloys is that the amount of penetration

is lesser than for carbon or stainless steel. This means that the penetration cannot be increased by increasing the welding current. The root face thickness should be reduced to lower than stainless steel to a value of less than 1.5 mm (0.059 in.)

- Removable backing strips are very useful to control the root bead shape. These can be made from copper, stainless steel, or a nickel alloy. Carbon or low-alloy steel backing strips must be avoided.

9.9.15 Post-Fabrication Cleaning

This step is as important as any of the welding steps to achieve the final requirements of the fabricated product. The most common cleaning operations involve the cleaning of surface contaminants and the removal of embedded iron.

Some guidelines on post-fabrication clean-up are discussed here.

9.9.15.1 Surface Contaminants

These contaminants are similar to those of pre-welding surfaces such as grease, oil, crayon markers, paint, and so on. Typically, these contaminants can be removed by spraying or scrubbing with a detergent or solvent [10].

9.9.15.2 Embedded Iron

Iron particles get embedded during the process of welding and pose a very serious problem during operation, as they are more prone to corrosion. When fine grit flapper wheels or disks are used, then care must be taken to assure that the abrasive is free from iron or other foreign material that could contaminate the surface.

ASTM A380 and A967 recommend some tests to detect embedded iron particles. Some of the useful tests to determine the presence of iron are:

- Spray the surface with clean water and inspect for rust streaks after 24 hours.
- Immerse the surface in a 1% sodium chloride solution or use as a spray in enclosed spaces such as tanks. Inspect for rust spots after 12–24 hours.
- For small areas, use a ferroxyl test. The formation of blue spots indicates the presence of free iron.

9.9.16 Post-Weld Heat Treatment

The requirement of post-weld heat treatment is dependent on the type of alloy. Solid solution alloys generally do not require any post-weld heat treatment, whereas precipitation-hardening alloys require heat treatment after welding to develop full strength. If for any specific application, heat treatment such as annealing or stress relieving is needed to avoid age hardening, it is advisable to contact the manufacturer of the base metal. If welding induces moderate-to-high residual stresses, then the precipitation-hardned alloys would require a stress-relief anneal after welding and before ageing. As already mentioned, before any heat treatment, all alloy surfaces should be thoroughly cleaned of oil, grease, paint, or markings, and similar contaminants to avoid catastrophic corrosion during heat treatment. The method of heating and cooling and the amount of sulphur in the furnace atmosphere must be controlled, or the alloys can get damaged.

The solid solution alloys such as Alloy 200 or Alloy 625 do not require post-weld heat treatment to maintain corrosion resistance but may be subject to PWHT either to reduce the risk of stress corrosion cracking if the alloy is to be used in caustic soda service or in contact with fluoro-silicates or to provide dimensional stability. A typical stress relief treatment would be 700°C for half an hour for Alloy 200, whereas 790°C for four hours for the higher-chromium content alloys such as Alloy 600 or Alloy 625.

9.10 TESTING AND INSPECTION

The testing and inspection procedures are similar to those for carbon and stainless steels. These techniques include visual examination, radiography, mechanical testing, e.g., tensile and impact, and metallography.

Visual examination is the most widely used method of weld inspection. It is simple visual examination. Etching of the specimen enhances better visibility of details of the weld structure. The defect becomes clearer after etching. Nickel alloys are normally non-magnetic, and hence magnetic particle inspection is of no use. Pure nickel like Nickel 200 and Nickel 201 are weakly magnetic, so this technique could be used. To determine local small-surface defects, liquid penetrant inspection is extensively used.

Mechanical testing such as bend testing is often used to evaluate the soundness and ductility of the welded joints. Both free and guided bend tests are used. Specimens may be tested with the weld oriented either transverse or longitudinal. Transverse specimens are commonly used for weld qualification. Tensile and impact strengths are conducted of undiluted weld metal whenever it is required. These are destructive tests, and so these are conducted on specifically prepared samples.

9.11 SAFETY

The welding of nickel alloys does not require any special safety measures other than those used for all other metals. The general safety precautions required for the welding of other metals are adequate for these purposes as well. In the areas such as electrical equipment, gas equipment, eye and face protection, fire protection, labelling hazardous materials, and similar items extra precautions need to be taken. The biggest safety concern is from gases and fumes. Besides the shiedling gases, many welding and cutting processes produce fumes and gases which are harmful to the operators. Fumes are mixtures of fine solid particles and gases which originate primarily from welding consumables with a minor amount from the base metal.

The operators must be aware of the adverse effects of these gases. The constituents of these fumes essentially originate from the welding consumable and partly from the base metal and other factors such as welding process, current level, and arc length. Compounds of chromium are the most harmful. Immediate effects of overexposure to welding fumes containing chromium and nickel are like effects produced by fumes from other metals. Nausea, headaches, and dizziness are common symptoms of fumes inhalation. Skin rashes due to chromium and nickel are also not uncommon.

It is advisable not to breath fumes and gases containing chromium and nickel. Ventilation or exhaust at the arc, or both, to keep fumes and gases away from the breathing zone and general area is a good practice. In some cases, natural air movement will provide enough ventilation. Smoke extraction equipment or ventilation systems improve operator comfort and are essential for operator safety. Where ventilation is a process constraint due to draft generated and not practicable, use air sampling to determine if corrective measures should be applied. The SAW process is operator-friendly and less hazardous to them as they are not exposed to arc; besides, this process generates lesser amounts of fumes.

The American Welding Society publishes reference guide on welding safety, titled "Safety in Welding and Cutting" (American National Standard Institute/Accredited Standards Committee, ANSI/ASC Z49.1) [11]. For additional safety information, refer to the current edition of AWS G2.1M/G2.1 [12]. While selecting the solvent, it must be remembered that though chlorinated solvent does not have any major effect on some of the nickel alloys, other solvents may pose health hazards. All precautions must be taken to ensure that the solvents are not harmful or toxic to personnel and are not explosive.

REFERENCES

1. Nickel Institute Publication, *Guidelines for the Welded Fabrication of Nickel Alloys for Corrosion-Resistant Service: A Practical Guide for Welders, Material Engineers and Design Engineers Knowledge for a Brighter Future*, second edition, Nickel Institute Publication, 2018.
2. Nickel-Alloys.net. (2025). Nickel and nickel alloys: Everything you wanted to know about nickel and nickel alloys.
3. ISO. (1980). *Heat-treated aluminium and aluminium alloy castings – Specification (ISO 581:1980).* Geneva: International Organization for Standardization.
4. John N. DuPot, John C. Lippold, Samuel D. Kiser, *Welding Metallurgy and Weldability of Nickel-Base Alloys*, John Wiley & Sons, Inc., Hoboken, NJ, 2009.
5. Andersen, L. (n.d.). *Repair Welding of Seawater Pipes* (Technical Update). TE Andersen Consulting.
6. Special Metals Welding Products Company. (n.d.). *Joining*. New Hartford, NY: Special Metals Welding Products Company.
7. TWI Global. (n.d.). *Weldability of materials—Nickel and nickel alloys.* Cambridge, UK: The Welding Institute.
8. NeoNickel. (n.d.). Welding of nickel alloys. NeoNickel Technical Resources.
9. ASME. (2023). *ASME Boiler and Pressure Vessel Code: Section II, Part C—Specifications for welding rods, electrodes, and filler metals* (including SFA-5.11, SFA-5.14, SFA-5.34). New York: American Society of Mechanical Engineers.
10. ASTM International. (2017). *ASTM A380/A380M-17: Standard practice for cleaning, descaling, and passivation of stainless steel parts, equipment, and systems.* West Conshohocken, PA: ASTM International.
11. NSI/ASC Z49 Committee. (2012). *ANSI/ASC Z49.1: Safety in welding, cutting, and allied processes.* Miami, FL: American Welding Society (secretariat for ANSI).
12. American Welding Society (AWS). (2014). *AWS G2.1M/G2.1: Guide for the evaluation of hot cracking in welds.* Miami, FL: American Welding Society.

Index

Note: Page numbers in *italics* indicate a figure, and page numbers in **bold** indicate a table on the corresponding page.

D

E

F

G

M

N

O

P

U

V

W

Z

Made in the USA
Las Vegas, NV
19 September 2026

74a9bdcf-4be3-4485-8279-1c71150baf17R01